UNINTELLIGENT DESIGN

UNINTELLIGENT DESIGN

MARK PERAKH

59 John Glenn Drive
Amherst, New York 14228-2197

Published 2004 by Prometheus Books

Inquiries should be addressed to
Prometheus Books
59 John Glenn Drive
Amherst, New York 14228–2197
VOICE: 716–691–0133, ext. 207
FAX: 716–564–2711
WWW.PROMETHEUSBOOKS.COM

08 07 06 05 04 5 4 3 2 1

Library of Congress Cataloging-in-Publication Data

Perakh, Mark, 1924–
Unintelligent design / by Mark Perakh.
p. cm.
Includes bibliographical references and index.
ISBN 1–59102–084–0 (cloth : alk. paper)
1. Religion and science. 2. Bible and science. 3. Science—Philosophy. 4. Intelligent design (Teleology) I. Title.

BL240.3.P47 2003
291.1'75—dc21

2003008111

Printed in the United States on acid-free paper

CONTENTS

PREFACE

A couple of years ago I received a phone call from somebody whose name was not familiar to me. He introduced himself as a geneticist affiliated with Cornell University, and told me that he had come across my Web site and was intrigued by a section titled "Texts." This section contains several articles in which a new method of statistical analysis of text, labeled Letter Serial Correlation (LSC), and its application to texts in various languages are described. The man from Cornell asked me whether or not the method (which was developed as a joint effort by a prominent Australian mathematician and myself) could also be applied to a DNA strand, if the latter is represented by a string of alphabetic symbols. Although the LSC method had so far been applied only to texts in natural languages and pieces of gibberish composed by various means, I saw no reason why it would not work for the DNA strand. I briefly discussed with the man from Cornell the prospects of cooperation in such a study. Several months later, the man from Cornell called me again and said that he happened to be visiting the state I live in, and was at that time not far from my home. I happily invited him to come over. In an hour he showed up, accompanied by his son, a youngster about sixteen or eighteen years old.

We spent some time discussing the structure of DNA and the features of the LSC method. In the course of that quite friendly discussion my guest mentioned that the DNA strand contains, in addition to identified genes, multiple sequences which do not seem to perform any function. The LSC method supposedly could distinguish between gene sequences and "meaningless" parts of the strand. In passing, I mentioned that such a structure seems to testify against the idea of DNA having been designed, pointing rather to its origin in a random process. Suddenly, the expression on my guest's face changed. He mumbled something, hastily concluded our discussion, and both visitors left.

About twenty minutes later the man from Cornell and his son reappeared at the entrance to my home and requested to talk to me. Of course I let them in, puzzled by their strange behavior. When we took our seats, the geneticist from Cornell said that he was struck by my words in regard to DNA not being a result of design. Quite emotionally, he and his son gave something which could best be characterized as a sermon, using various arguments to prove to me that their Christian faith was based on irrefutable evidence. He expressed as a question one of the arguments his son seemed to consider very convincing: "How can you explain that the body of Jesus disappeared from the cave where it was placed after the crucifixion?" The geneticist from Cornell did not seem to view his son's argument as feeble. Having politely thanked my guests for their proselytizing effort, I said that I did not find their arguments convincing. In a scientific study, much stronger proof would be required to seriously consider evidence as reliable. Before leaving, the man from Cornell asked whether or not our disagreement on a religious issue would prevent our cooperation in studying the DNA strand by means of the LSC statistics. "Not at all," I answered. "I'll be happy to cooperate and I am looking forward to hearing from you soon. Your religious views have no bearing whatsoever on the prospects of our possible fruitful scientific study of that fascinating subject."

My guests left. I never heard from them again.

Here is one more story. Several months ago, I happened to read a book by a professor of philosophy from a college in the United States. In this book, its author's goal was to lay a foundation for the "intelligent design theory" (which will be discussed in detail in the following chapters). Although I did not agree with the main thesis of this writer, I was impressed by his logic and overall objectivity, which, in my view, favorably distinguished his philosophical treatment of the subject from the multitude of other authors propagating the intelligent design theory. I placed a brief, very positive review of his book on the Internet. While giving that book high marks, I mentioned, without elaboration, that I was inclined to dispute certain elements of the writer's discourse.

Having become aware of my review, the author of the book in question emailed me a message in which he thanked me for the positive review and also requested I share with him my doubts in regard to some points of his discourse. In response, I sent him three pages of comments, wherein I explained the points of disagreement with his ideas and also mentioned that I had written two essays in which some of those points were discussed in detail. A few days later, he mailed me a brief reply to my comments and requested that I send him the two essays I mentioned. I sent a rejoinder to his counterarguments as well as the two essays he wished to read, and requested, as a reciprocal favor, comments on my essays.

I never heard from him again. He did not even bother to acknowledge the receipt of my message. Several weeks later I sent him a short reminder, asking for at least a confirmation that he received my essays. He remained silent.

I did not initiate the discussion in either of the above cases. I did not argue against the beliefs of these three people. Our discussions were strictly limited to the nonreligious subjects which were pertinent to the areas of mutual interest. I never overstepped the boundaries of a polite exchange of views, and made no disparaging remarks in regard to their views or ideas. My only guilt in both cases seemed to be in having briefly mentioned that I did not share their religious faith, although I did not say or write even a single word which would indicate disrespect for their beliefs.

While the lack of elementary politeness on the part of both the geneticist from Cornell and the philosopher from another U.S. college is more their problem than mine, their behavior apparently reflected their strong aversion to dealing with somebody who does not share their religious beliefs, even if the discussion is not about their faith. What makes these believers so strongly refuse to cooperate or even communicate with anybody who has or seems to have a different view? The man from Cornell preferred to give up the prospective study of a fascinating subject which obviously interested him very much rather than cooperate with a supposed unbeliever. The philosopher from the other U.S. college chose disregard for an elementary politeness rather than engaging in a dispute with a supposed unbeliever about the subject in which this philosopher is obviously deeply interested.

These two events seem to exemplify the abject intolerance often displayed by proponents of intelligent design, irreducible complexity, and other similar ideas, which, despite their various disguises, are essentially aimed at imposing a religious outlook on society.

Why are these design theorists and religious writers so strongly confident in their opinions that they disdainfully dismiss any opposing views? Are the arguments in favor of their views indeed so convincing?

In this book, I will try to answer that question.

The literary production by design theorists and those who advocate the compatibility of science with the Bible comprises hundreds of books and papers. Every month, scores of books in this vein are printed by multiple publishers who are confident of the assured profitability of this genre. It is impossible to review all those hundreds of books and articles. I had to choose a few, those which have gained a considerable popularity and are more or less representative of the entire literature in question.

All chapters of this book are connected by the same general idea—testing the validity of the arguments by religious writers in regard to the relation between science and religious faith. While there are many cross-references between the chapters, each chapter can also be read separately, as each covers a specific subject which can be discussed independently of the rest of the book.

The book consists of three parts. Part 1 (chapters 1 through 3) is devoted to a discussion of the recent incarnation of creationism under the label of intelligent design theory. In the three chapters of this part, I analyze the literary output of the three most influential promoters of intelligent design theory, William Dembski, Michael Behe, and Phillip Johnson. My conclusion is that the intelligent design theory is an example of pseudoscience.

In part 2 (chapters 4 through 11) I review a number of books and articles devoted to the attempts to reconcile the biblical story with scientific data. My conclusion is that these attempts, both by writers who approach it from a Christian perspective and by those who do so from the standpoint of Judaism, have failed, since the arguments in these books and papers are often arbitrary. Some of these books contain elementary errors and often substitute the desired for the actual.

In part 3 (chapters 12 through 14) I offer a general discussion of the ideas which underlie the particular topics of the preceding chapters. To this end, chapter 12 contains a discussion of what constitutes good science from the viewpoint of a scientist rather than that of a philosopher of science. In chapter 13, I discuss the concept of probability, which is frequently misused in many of the books reviewed in parts 1 and 2. In chapter 14, I review a rather vivid example of what started as an attempt to do good science, but almost at once degenerated into bad science and finally into pseudoscience—the rise and fall of the so-called Bible code.

Finally, in the afterword I briefly summarize the conclusions which in my view stem from the preceding discourse.

In this book I try to explain the gist of the dispute in terms as simple as possible, to make the discourse comprehensible to nonexperts, while also trying to avoid oversimplification which reduces the discussion to platitudes. Let the readers judge whether or not I have succeeded.

ACKNOWLEDGMENTS

While writing this book, I was lucky to enjoy support from many friends and colleagues who generously gave their time to review many parts of my discourse, offered numerous comments which helped eliminate errors, small and large, and suggested various ideas which I happily incorporated into my text.

Some of them read every chapter of this book, offering many insightful comments. Others commented only on selected parts of the book, those which were closer to their own fields of interest. I owe a debt of deep gratitude to each of them, since they sacrified their time to help me save my own and improve my book without expecting anything in return. Brendan McKay, Gert Korthof, Alec Gindis, Eliezer Reinstein, Althea Katz, Alexander Eterman, Yigal Bloch, Matt Young, Richard Wein, and Sergei from Russia whose last name I don't know, I would like to tell you that I highly value your input.

I would like also to thank the anonymous reviewer who pointed out weaknesses in the first version of the manuscript and made helpful suggestions.

I would also like to thank my wife of over forty years for her limitless patience and the encouragement she invariably offered during the many months I spent at my computer.

PART 1
UNINTELLIGENT DESIGN

1

A CONSISTENT INCONSISTENCY

How William Dembski Infers Intelligent Design

Christ is never an addendum to a scientific theory but always a completion.

—W. Dembski, *Intelligent Design*, p. 207

As Christians we know naturalism is false.

—W. Dembski, "Mere Creation," p. 14

. . . (S)cientific creationism has prior religious commitments whereas intelligent design has not.

—W. Dembski, *Intelligent Design*, p. 247

The above quotations from the writings of William A. Dembski illustrate two things: first, his actual agenda as a Christian preacher who presents his thesis in seemingly scientific terms, and second, his inconsistency, as the third quotation negates the two preceding ones.

Dembski is a prolific writer whose literary production—while covering an extensive span of subjects, from history of philosophy to probability theory, theology to information theory—seems to be all devoted to one idea: to prove that the universe in general and life in particular are the results of a design by an unnamed intelligent mind.

In this chapter I shall discuss three of Dembski's books, *The Design Inference: Eliminating Chance through Small Probabilities*, *Intelligent Design: The Bridge between Science and Theology*, and *No Free Lunch: Why Specified Complexity Cannot Be Purchased without Intelligence*, as well as a number of his papers.[1]

It seems that Dembski is one of the most prominent participants in the intelligent design movement. The articles and books of his colleagues and supporters are full of praise for Dembski's discourse. Here is just one example, written by Rob Koons, professor of philosophy at the University of Texas (quoted from the blurb on *Intelligent Design*): "William Dembski is the Isaac Newton of information theory, and since this is the Age of Information, that makes Dembski one of the most important thinkers of our time. His 'law of conservation of information' represents a revolutionary breakthrough."

Here is one more quotation. Professor of biochemistry Michael J. Behe (see chapter 2), also often referred to as a pioneer in the modern revival of the intelligent design, in his foreword to Dembski's *Intelligent Design* wrote, "I expect that in the decades ahead we will see the contingent aspects of nature steadily shrink. And through all of this work we will make our judgment about design and contingency on the theoretical foundation of Bill Dembski's work."[2]

I could easily quote many more examples of high acclaim bestowed on Dembski's work by his colleagues. It has to be said, though, that all of Dembski's admirers limit their acclaim of his work to general statements rather than offering arguments in support of his thesis.

While Dembski's colleagues so highly admire his contribution to intelligent design theory, there have been critical voices; unlike his supporters, his critics often suggest detailed arguments aimed at showing the weaknesses and errors in his output. For example, in *Tower of Babel*, professor of philosophy Robert T. Pennock offered some critical discussion of certain parts of Dembski's work.[3] Pennock did not, though, discuss Dembski's work in a systematic way. Moreover, Pennock's criticism of Dembski is buried among many other topics discussed in the mentioned book. Another book, *Nature, Design, and Science*, in which we find a more detailed and systematic criticism of Dembski's work, was published by professor of philosophy Del Ratzsch.[4] In an appendix to the mentioned book, Ratzsch subjects some parts of Dembski's work to a strong critique.

In a review of *The Design Inference*, probability theorist and professor of philosophy Ellery Eells concludes that Dembski's theory is not "on the mark."[5] A detailed critical analysis of Dembski's theory from a philosophical and Bayesian standpoint was offered in a paper by professors of philosophy Branden Fitelson, Christopher Stephens, and Elliott Sober.[6] Highly critical reviews of Dembski's work were published by professor of ecology Massimo Pigliucci.[7] One more critical discussion of this work was offered by professor of philosophy Philip Kitcher.[8]

Other critical reviews of Dembski's work have appeared on the Internet, posted by Eli Chiprout, Taner Edis, Wesley R. Elsberry, Gert Korthof, Thomas D. Schneider, Jeffrey Shallit, Howard J. Van Till, Richard Wein, and Matt Young.[9] While there are certain points critical of Dembski that are common to more than one of these authors in their reviews, one also finds among them a variety of approaches and viewpoints, all of which nevertheless agree that Dembski's work contains many weaknesses and inconsistencies.

While I largely agree with the critical comments by the above writers (except for some minor points, some of which will be discussed later), I intend to offer in this chapter my own more or less systematic critical analysis of Dembski's theory, including his theoretical treatment of probability, complexity, information, and design. I will try to make my critical analysis of Dembski's work as simple as it is reasonably possible, thus making it comprehensible for nonexperts. In some instances, such an approach requires substantial simplifications without which a person having no extensive educational background in certain fields will not be able to comprehend the gist of the dispute. Whenever it will be impossible to avoid using some concepts or terms with which unprepared readers may be unfamiliar, I will try to explain these concepts or terms in plain words.

Dembski seems to be a well-educated man of many talents, who, in my view, was led astray by his desire to promptly develop a neat theory of design that would support his preconceived views and beliefs. Instead of following the logic of an objective analysis, he attempted to squeeze the enormous variability of real situations into the Procrustean couch of a one-dimensional theory. The real world, however, rarely fits a neat scheme.

DESIGN WITHOUT A DESIGNER

Almost at the very beginning of *The Design Inference* we discover a peculiar feature of Dembski's discourse. Its succinct expression is given in the following statement: "Design therefore constitutes a logical rather than causal category."[10]

What is the meaning of that statement? If design is disconnected from any causal history, it seems to mean that Dembski's concept is that of a *design without a designer*. Indeed, the quoted assertion is preceded by the following statement: "Although a design inference is often the occasion for inferring an intelligent agent . . . as a pattern of inference the design inference is not tied to any doctrine of intelligent agency."[11] It is hard to read that quotation as anything other than an assertion that at least in some cases design does not imply a designer.

For centuries, the battle cry of intelligent design proponents was, "If there is design, there must be a designer." The proponents of intelligent design viewed that slogan as logically unassailable. Now Dembski, the new champion of intelligent design, announces that the hypothesis of a designer is not necessary.

My interpretation of Dembski's assertion finds confirmation in his other statements. On the same page as his first assertion, he writes: "Thus, even though a design inference is frequently the first step toward identifying an intelligent agent, design as inferred from design inference does not entail an intelligent agent."[12]

I submit that the design inference, whether according to Dembski or any other author, is aimed at distinguishing events that are designed by an intelligent agent from events that occurred without such an agent. Design inference is really interesting only if it is *inference to a designer*, either human, alien, or supernatural. (In order to stay within the framework of Dembski's concepts, I am not mentioning here the very interesting questions about design stemming either from artificial intelligence or from natural processes, since the latter has been discussed by Massimo Pigliucci and Wesley R. Elsberry.[13])

The reason for Dembski's approach may be his desire to avoid accusations that design theory is just a disguised religion. However, to claim that design has meaning without a designer can hardly sound credible either to proponents or to opponents of the intelligent design hypothesis.

Just two pages after Dembski's quoted claim that design does not necessarily imply an intelligent agent, he seems to have forgotten this claim. He discusses an example of an election fraud committed by one Nicholas Caputo. As we will discuss later in detail, Dembski's method hinges on a triad of explanatory options which are, according to Dembski, regularity, chance, and *design*. However, when discussing the Caputo case, Dembski presents this triad as regularity, chance, and *agency,* replacing *design* with *agency.* The meaning of the term *agency* is unequivocally explained by Dembski in the next paragraph as an action "of a fully conscious intelligent agent."[14] Hence, in Caputo's example, Dembski uses *design* and *agency* as synonyms, where *agency* means *actions of an intelligent agent.*

This is just one example of inconsistencies found in many parts of Dembski's work.

Dembski's Explanatory Filter

Description of the Explanatory Filter

Dembski suggests that his explanatory filter is a versatile tool for identifying design. He also maintains that the procedure encapsulated in his filter has been used routinely in many fields of human endeavors, without realizing it.

So far, Dembski has published his description of the explanatory filter at least six times.[15] The schematic presentations of his filter are slightly different in these six publications, but essentially they all are just variations of the same scheme.

There are several points underlying Dembski's scheme. One is that every event can be attributed to one of only three possible sources. The first such source Dembski calls *necessity* (in four of the published schemes of his filter) or *regularity* (in one of the published schemes) or *law* (in one more of the published schemes). The second possible source of events is *chance*, and the third is *design* (sometimes also referred to as *agency*). According to Dembski, these three possible sources of events cover all possibilities and are clearly distinguishable from each other. If, according to Dembski, an event can be attributed to law (regularity, necessity) then its causal connection to chance or design is unequivocally excluded. Likewise, if an event can be attributed to chance, a possibility of its causal connection to law and/or design is eliminated. Finally, if an event can be attributed to design, this automatically excludes its possible causal connection to chance and/or law. Indeed, here is a quotation from Dembski's *The Design Inference*: "To attribute an event to design is to say that it cannot reasonably be referred to either regularity or chance. Defining design as the set-theoretic complement of the disjunction regularity-or-chance guarantees that the three modes of explanation are mutually exclusive and exhaustive."[16]

The second fundamental point of Dembski's scheme is the dominant role of probability of an event in the process of the filter's application. The event to be analyzed is subjected to three tests, aimed at determining whether it can be attributed to regularity (law, necessity), chance, or design. Correspondingly, the filter comprises three so-called nodes, i.e., three steps of testing. At the first node the choice is made between attributing the event in question either to law (regularity, necessity) or to absence of law. If law (regularity, necessity) is determined as the source of the event, the procedure stops at that step, while chance and design are eliminated as possible causal antecedents of that event. If, though, the law (regularity, necessity) is excluded as a causal antecedent, the event passes to the second node.

At the second node the choice is made between either attributing the event unequivocally to chance, or, without eliminating the possibility of chance, also allowing for its possible attribution to design. If chance has been determined unequivocally as the causal antecedent, while the possibility of design is eliminated, the test stops at that step. If, though, neither chance nor design can be eliminated as possible causal antecedents, the event passes to the third, ultimate node. At this step, the final choice is made between attributing the event either to chance or to design, the two alternatives being, according to Dembski, mutually exclusive.

According to Dembski, the criteria determining the choice between the two alternatives at each node of the filter are different for the first and the second nodes, on the one hand, and for the third node, on the other hand.

At the first and the second nodes there is, according to Dembski, one and only one criterion, which is the value of the event's probability. At the first node, law (regularity, necessity) is determined as the causal antecedent of the event if and only if the probability of that event is large. Dembski omits the question of what should be the lower bound on the probability in question in order for the event to qualify for being attributed to law (regularity, necessity).

At the second node, the only criterion for either unequivocally choosing chance as the causal antecedent of the event, or passing it to the third node, is again solely the value of the event's probability. If this probability is determined as being, in Dembski's terms, intermediate, the event is kicked out from the filter, being thus attributed to chance. Again, Dembski avoids indicating what is quantitatively the lower bound for the probability to be viewed as "intermediate." If, though, the probability of the event in question turns out to be low (whatever this term means quantitatively), the decision about the event's causal connection is postponed and the event passes to the third node.

At the third node, the crucial choice is made between attributing the event to chance or to design. Unlike at the two preceding nodes, where the sole criterion in use was the value of the event's probability, the criterion at the third node is twofold. To qualify for being attributed to design, the event in question must (a) have a low probability and (b) be "specified." Each of these two conditions is necessary, but neither of them alone is sufficient to attribute the event's origin to design. Only the two listed conditions together are both necessary and sufficient. If at least one of the two conditions is not met, the event is attributed to chance. If both conditions are met, the event is attributed to design.

Specification According to Dembski

As indicated in the preceding section, Dembski's criterion of design entails two necessary elements, one being the low probability of the event in question, and the other the event's specification.

Dembski first explains that specification of an event means that it displays a pattern. Hence, Dembski's criterion of design is the combination of a very low probability with an identifiable (recognizable, specified) pattern.

Dembski spends a considerable effort to elaborate his requirement of a recognizable pattern (specification). In order to serve as a specification, the pattern, according to Dembski, must meet an additional condition of "detachability." While Dembski offers a rather convoluted analysis of detachability, he also provides an example clarifying that concept. He writes:

> [S]uppose I walk down a dirt road and find some stones lying around. The configuration of stones says nothing to me. Given my background knowledge I can discover no pattern in the configuration that I could have formulated on my own without actually seeing the stones lying about as they do. I cannot detach the pattern of stones from the configuration they assume. I therefore have no reason to attribute the configuration to anything other than chance. But suppose next an astronomer travels this same road and looks at the same stones only to find that the configuration precisely matches some highly complex constellation. Given the astronomer's background knowledge, this pattern now becomes detachable. (p. 17)

From this example, it is evident that by detachability Dembski actually means a *subjective* recognizability of the pattern in question. In order to decide that the pattern discerned in a low probability event is detachable, and hence serves as specification (i.e., points to design) we must be able to recognize that pattern as matching some already familiar image. For that to happen, we must have certain background knowledge.

In order for an event to be detachable, Dembski teaches us, it must meet several conditions. The first condition is *conditional independence* of the background knowledge. This condition means that the background knowledge which we utilize to recognize the pattern must not affect the probability of the event in question estimated on the assumption of it being produced by chance. For Dembski, the probability of an event and its specification are two independent categories that do not affect each other.

The second condition is *tractability*. This term means, in Dembski's words, that "by using *I* it should be possible to reconstruct *D*," where *I* is the background "information" and *D* is the pattern in question (p. 149).

While conditional independence and tractability are, according to Dembski, the constituent parts of detachability, to qualify for specification the pattern must meet one more condition, referred to by Dembski as *delimitation*. That concept is explained by Dembski as follows: "[T]o say that *D* delimits *E* (or equivalently that *E* conforms to *D*) means that *E* entails *D** (i.e. that the occurrence of *E* guarantees the occurrence of *D**)"(p. 152). In that definition, *E* means an event, *D* means the pattern, and *D** means "the event described by *D*" (p. 151).

Dembski's main idea has been succinctly expressed under the label of "Law of Small Probability": "Specified events of low probability do not occur by chance" (p. 48).

"Mathematism" as a Tool of Embellishment

Before discussing in detail the inconsistencies in Dembski's explanatory filter theory, I wish to first comment on one striking feature of Dembski's writing that is especially pronounced in his highly technical monograph *The Design Inference* as well as his latest book, *No Free Lunch*. If the quality of a mathematical treatise were evaluated by the number of mathematical symbols, Dembski's book *The Design Inference* would qualify as a great achievement in mathematics. This may be one of the reasons why many of Dembski's colleagues in the so-called intelligent design movement so admire his opus. They commonly praise the supposed great rigor of Dembski's mathematical analysis. However, reviewing all these extensive collections of mathematical expressions in Dembski's book reveals that only a few of them are anything more than a simple illustration of whatever Dembski states in plain words. Except for a few cases (of which some are not quite relevant to Dembski's thesis), his mathematical exercise neither proves any new mathematical theorem nor derives any new formula. Actually, the removal of most of those formulas would hardly make much difference except for depriving Dembski's book of its mathematical appearance.

When a new mathematical theorem is proven, it advances the mathematics itself, thus possibly opening new vistas for additional applications. If a mathematical formula is derived in physics, some technical science, or engineering, it compresses into easily comprehensible form certain essential relations between various data, which otherwise would be much harder to review and manipulate. This immensely facilitates some useful procedure. If, though, mathematical symbolism is used for the sake of symbolism itself, it does not advance the understanding of a subject; at best it simply saves some space and time in the discussion of a subject, and at worst it makes the matter more obscure because the esoteric symbolism requires lengthy deciphering.

Actually, *The Design Inference* contains little of genuine mathematics, but is full of *mathematism*, that is, the use of mathematical symbolism as embellishment, often only to create an impression of a scientific rigor of the discourse.

To illustrate my point, consider the following example. In *The Design Inference*, Dembski offers the following argument:

Premise 1:	E has occurred.
Premise 2:	E is specified.
Premise 3:	If E is due to chance, then E has small probability.
Premise 4:	Specified events of small probability do not occur by chance.
Premise 5:	E is not due to regularity.
Premise 6:	E is due either to a regularity, chance, or design.
Conclusion:	E is due to design. (p. 48)

I am not yet discussing either merits or drawbacks of the above argument, since my goal at this point is simply to illustrate the mathematism employed by Dembski throughout his book.

Next Dembski writes:

> The validity of the preceding argument becomes clear once we recast it in symbolic form (note that E is a fixed event and that in Premise 4, X is a bound variable ranging over events):
>
> Premise 1: $\text{oc}(E)$
> Premise 2: $\text{sp}(E)$
> Premise 3: $\text{ch}(E) \rightarrow \text{SP}(E)$
> Premise 4: $\forall X[\text{oc}(X)\ \&\ \text{sp}(X)\ \&\ \text{SP}(X) \rightarrow {\sim}\text{ch}(X)]$ (A)
> Premise 5: ${\sim}\text{reg}(E)$
> Premise 6: $\text{reg}(E) \vee \text{ch}(E) \vee \text{des}(E)$
> Conclusion: $\text{des}(E)$. (p. 49)

The above argument, now rendered in a mathematically symbolic form, exactly reiterates the preceding plain-word rendition of the same argument. A question is: In what way does representing the same argument in a symbolic form make its validity clear? I submit that reiterating the above argument in a symbolic form adds nothing to its interpretation, neither supporting nor negating its validity. Moreover, this rendition in itself does not even save space or time since the symbols used in it require explanation in plain words. In order to make the symbolic rendition understandable, its author had to explain to readers that

> oc(E) = E has occurred, sp(E) = E is specified, SP(E) = E is an event of small probability, reg(E) = E is due to regularity, ch(E) = E is due to chance, and des(E) = E is due to design. The sentential connections ~, &, $\vee$, and $\rightarrow$ denote respectively *not*, *and*, *or*, and *if-then*. $\forall X$ is the universal quantifier (read "for all X" or "for every X"). (p. 49)

As can be seen, the symbolic rendition not only does not add anything of substance, it actually has no advantages over the preceding plain-word rendition even from the viewpoint of brevity. It seems to me that its only purpose was to impart on the discourse a rigorously mathematical appearance.

Moreover, still not satisfied with the above symbolic rendition of his design inference, Dembski offers several modifications of that rendition, gradually making its appearance more and more complex.

I can envision a possible suspicion that my criticism of Dembski's extensive use of mathematical symbolism stems from my own discomfort with mathematics. I don't think this is the case. While I am a physicist rather than a mathematician, I enjoy mathematical treatment of various problems. I have derived hundreds of formulas which have been published in several hundreds of articles and monographs. They cover a rather wide range of topics.[17] I have no objections to Dembski's extensive use of mathematical symbolism, which is his right and often looks quite attractive, but I don't think this extensive *mathematism* justifies viewing his discourse as "mathematically rigorous." Many parts of that mathematical symbolism seem to serve no useful purpose.

Can Probability Be Separated from the Event's Causal Antecedents?

I will discuss now a point, which, in my view, entails a rather general fault of the approach embodied in Dembski's explanatory filter.

Suggesting his explanatory filter as a versatile tool for discriminating between law, chance, and design, Dembski bases the process of such discrimination on the evaluation of probabilities of events. One moves from one node of the filter to the next according to the estimated value of the event's probability. Dembski's entire chain of arguments presumes that probability is an independent category which may be estimated by itself without accounting for the possible cause of the event in question. For example, in *The Design Inference* we read, "Thus, if *E* happens to be an HP event, we stop and attribute *E* to a regularity." In this sentence *E* stands for "event" and HP for "high probability" (p. 38).

Actually we can't assert that "*E* happens to be an HP event" if we have not first assumed that it is due to law (regularity, necessity). In fact, probability does not exist by itself as an abstract concept, and can only be estimated by accounting for information about the event in question. Dembski seems to realize that fact when he discusses probability in its own chapter, but seems to forget about it when he turns to his explanatory filter.

According to Dembski, at the first node of his filter we attribute events to law (regularity, necessity) *because* their probability is high. Actually the procedure is opposite to his scheme: we conclude that the probability of an event is high, *because* it is due to law (regularity, necessity).

Possibly Dembski's reversal of the normal order of inference in this case stems from his confusion of two different procedures—one of postulating a certain law (via scientific induction) and the other of attributing a particular event to some law. Obviously, the procedure at the first node of Dembski's explanatory filter is of the second type. (Procedures of scientific induction which are common in scientific research are discussed in detail in chapter 12.) Despite the superficial similarity between the procedure of scientific induction and Dembski's alleged attribution of an event to law because its probability is high, these two procedures are principally different. At the first node of Dembski's filter, we have to decide whether or not a particular event has to be attributed to regularity, while in the procedure of a scientific induction we postulate a definite regularity after having observed multiple repetitions of occurrences of certain events. In the latter case the tentative conclusion of a researcher is, "Under these particular conditions the probability of a certain event is very high." On the other hand, at the first node of Dembski's filter the conclusion, according to his scheme, has to be, "The probability of that particular event is high, *therefore* it must be attributed to regularity."

However, we can't conclude that the probability of a particular event is high unless we know it is due to regularity. Assume that we observed a particular event—a piece of the metal gallium in a vessel melted when the temperature reached about 302.5 K. Observing that event does not provide a clue regarding its probability. Unless we already know the law—the transition from solid to liquid in the case of pure gallium, at atmospheric pressure, always occurs at about 302.5 K—we cannot assert that the observed event has a high probability and therefore has to be attributed to law. On the other hand, if we know the law—pure gallium under atmospheric pressure melts at about 302.5 K—then we can confidently attribute the observed event to a law, and hence estimate its probability as being high.

Even if an event has been observed many times, this in itself is not sufficient to assume that its probability is high. As discussed in chapter 12, there is a necessary intermediate step—postulating that the observed repetition of the event was a manifestation of a law. It is not an uncommon situation in scientific research when a repetition of a certain event is observed but nevertheless no assumption is made that a new law is at work.

In order to assign to an event a high probability, first a law has to be accepted.

Likewise, at the second node, according to Dembski, we attribute an event to chance *because* its probability is intermediate. Again, the common procedure is just the opposite: we estimate the probability of a particular event *assuming first* that it is due to chance (see the example with a raffle described below). Note that at the third node of the filter, Dembski himself suggests that we estimate the probability of an event by first assuming that it is due to chance, which is contrary to the procedure he suggests for the first and the second nodes.

As can be seen from Dembski's own definition of probability (which will be discussed in detail later in the chapter), he defines probability as being conditioned "with respect to the background information" (p. 123). I believe that if Dembski has adopted a certain definition, he is supposed to stick to it throughout his discourse. However, when Dembski turns to his explanatory filter he seems to forget his own concept of probability.

Imagine that we estimate the probability of John Doe's winning in a raffle. Let us assume that there are one million tickets distributed in that raffle, each with the same chance of winning. What is our estimate of John Doe's probability of winning? Can we say unconditionally that the probability in question is 1/1,000,000?* If we adopt Dembski's definition of

* For the sake of uniformity and ease of mathematical calculation, probabilities will be rendered in the form *x/y* and can be read "*x* in *y*."

probability, we can't say that. Based on his definition, we must say instead: "John Doe's probability of winning is 1/1,000,000 upon the assumption that the drawing is random." In other words, the estimation of probability incorporates an assumption regarding the nature of the event in question, namely its being the result of chance.

Imagine, though, that we have information about John Doe being in cahoots with the organizers of the raffle who have a record of earlier frauds. This background information must be incorporated in our estimate of probability. Upon the assumption that the new information *obtains,* the new estimate of probability of John Doe's winning is immensely higher than before. Based on the new information, we assume that John Doe's win is due to design (in this case, fraud); that new assumption leads to a drastically increased estimate of the probability of his win.

The situation is different for the third node of Dembski's filter where the probability is first estimated upon the assumption of chance as the cause of the event, and then the situation is reconsidered accounting for the *side information*. The latter, though, is assumed not to affect the probability. I will discuss this assumption later in this chapter.

It does not matter for the estimation of probability whether background information is actually available or is assumed for the sake of estimation. We estimate probability on the basis of certain background information, either actually available, or assumed for the sake of estimation. Consciously or subconsciously, the assumption about the cause of the event is incorporated into the estimate of probability.

In particular, to conclude that an event is due to law, according to Dembski, we have to first find that its probability is high. However, if we do not assume a priori that the event is due to law, so that we estimate its probability upon the assumption that it is due to chance, we will often arrive at a small probability which, according to Dembski, would point to either chance or design rather than to law. Here seems to be a vicious circle and to break out of it, there seems to be only one way—to get out of the confines of Dembski's scheme.

Law versus either Chance or Design

Another weakness of Dembski's scheme seems to be that, while attributing each event to either law, chance, or design, he fails to account for the taxonomy of events according to any other criteria. It seems rather obvious that there are classes of events for which it may be impossible to identify their causal antecedents as belonging to only one of the three distinctive categories.

Consider Dembski's example of an archery competition. If an archer shot an arrow and hit a target, it is, according to Dembski, a specified event which definitely must be attributed to design. In Dembski's scheme, design excludes both chance and law. Can we really exclude law as a causal antecedent of the event in question? I submit that the archer's success was the result not of design alone, but of a combination of design and law. Indeed, the archer's skill manifests itself only in ensuring a certain velocity of the arrow at the moment it leaves the bow. This value of velocity is due to design. However, as soon as the arrow has separated from the bow, its further flight is governed by laws of mechanics. The specified event—the perfect hit—was due to both design and law. The arrow would not hit the target if any one of these two causal antecedents were absent. In this case *design operates through law* and would be impossible without law. Therefore Dembski's scheme, which artificially divorces law from design, viewing them as two completely independent explanatory categories, does not seem to jibe with reality. (Besides law, chance may also contribute to the occurrence of a hit; for example, an accidental gust of wind may affect the flight of the arrow.)

Likewise, there is a class of events for which it is impossible to separate law from chance as causal antecedents. Here is an example. There is a machine used for training tennis players. It randomly hurls tennis balls toward a player. There may be a large number of balls flying every minute, and it is impossible to predict the exact direction of each next flying ball. Choose an area anywhere within the court, say, of one square meter. Assume a particular ball landed within that area. Is that event due to chance or law? Let's say that over the course of a certain period of time the total number of flying balls was 1,000, and that only 50 of those balls landed within the selected one square meter. I believe that in such a situation most of the observers will attribute the event in question to chance. In fact, though, chance only determines the initial velocity of each ball. Upon leaving the machine, the flight of the ball and hence the location of its landing are determined by laws of mechanics. In this case, *chance operates through law*, so the location of the ball's landing is determined by both chance and law. The event most reasonably has to be attributed to a combination of law and chance.

Furthermore, as statistical science shows, random events follow certain laws; therefore, even if an event is viewed as random, it cannot be completely divorced from a (statistical) law which is instrumental in causing the event in question. For example, recall the so-called Galton board, a device that demonstrates the normal (Gaussian) distribution of chance events. In this device, hundreds of small balls are placed in a hopper which has an opening in its bottom. Pulled by gravitation, the balls fall down one by one. On their way down, the balls encounter a grid of hexagonal baffles. At each baffle, each ball has the same probability of 1/2 to pass the baffle either on the left or right side of the baffle. After passing several rows of baffles, the balls fall into a row of bins. Which ball happens to get into which bin is determined by chance. However, regardless of the absolute sizes of the device or of its parts, the overall result is always the same: when a sufficiently large number of balls fill the bins, their distribution between the bins meets the normal (Gaussian) distribution. In this case, the situation is in a sense opposite to the case of the tennis balls: while for the tennis balls *chance operated through law*, now the *law (Gaussian distribution) operates through chance*.

Moreover, if we review again the example with tennis balls, it is easy to see that, since the machine that hurls the balls has been designed by a human intelligent agent (an engineer), the event in question may be viewed in a certain sense as a causal consequent of all three sources—design, chance, and law, whose contributions to the occurrence of the event cannot be separated from each other since each of them is necessary for the event to occur.

There is an enormous number of situations wherein regularity, design, and chance are intertwined in various combinations, each contributing to varying degrees to the occurrence of events. Moreover, more than half a century after the formulation of the principles of cybernetics, Dembski's scheme seems to be too simplistic in that it views the causal history of events as a one-directional, straightforward process, thus ignoring feedbacks, conditional causes, superimposition of multiple causes of events, etc.

Therefore, in my view, Dembski's scheme—based on the uncompromising demarcation between law, chance, and design which are viewed as clearly separate causal categories, being always completely independent from each other—seems to be rather off the mark.

Unequivocal Chance versus either Chance or Design

Now let us review what happens if an event passes to the second node of Dembski's filter. At this step, the probability of the event, which was found to be "not large" at the preceding step, is reevaluated in order to determine whether it is "intermediate" or "small." We know already that Dembski does not offer a definite quantitative criterion for classifying probability as either intermediate or small. Of course, without such a criterion the procedure becomes uncertain, since what seems small to John may seem very large to Mary.

The more important objection to Dembski's scheme at this point, though, is that according to the analysis outlined above, attributing an event to law or chance is normally not based on a prior estimate of probability, as Dembski suggests, but rather quite the opposite—probability can be estimated only *after* either law or chance have been determined as the event's causal antecedents. Therefore I submit that the first and the second nodes of this filter offer an unrealistic scenario and hence play no useful role for the design inference. Thus if any meaningful design inference takes place, it can only occur within the framework of the third node of the filter. Of course, if that is the case, the filter loses its impressive trinodal appearance, which so neatly matches the three supposedly independent causes of events.

Assume, though, that we follow Dembski's scheme and that, having arrived at the second node of the filter, we have somehow determined that the probability of the event in question is not intermediate but small, in which case we proceed to the third node.

Design versus Chance

The criteria of design according to Dembski

At the third node of the filter, according to Dembski's scheme, the choice is made between design and chance. Before analyzing the details of Dembski's procedure for discriminating between design and chance, let us briefly discuss a few general points.

One such point is the nature of design, and another is what can be called the degree of design.

Regarding the nature of design, it seems reasonable to distinguish between various types of design. Even if we omit the host of vexing questions related to the possible design by artificial intelligence, we still can imagine at least three different kinds of design, namely a human design, an extraterrestrial's design, and a supernatural design. This question has been very thoroughly analyzed by Ratzsch.[18] (I am omitting the discussion of the design by either artificial intelligence or by natural processes because these types of design are absent in Dembski's theory.)

Dembski does not seem to acknowledge the differences between these three versions of design. On the contrary, he seems to stress the features common to all types of design. Remember Dembski's statement that design is a logical rather than causal category and that design does not necessarily entail a designer?

When we are dealing with a human design, usually we recognize design quite easily. Neither a design theorist such as Dembski nor the opponents of that theory will argue about the source of a poem or a novel, both readily attributing it to design and rejecting chance as a possible source of the text in question.

In the case of a hypothetical extraterrestrial design, the situation is more complex. Since we have no experience with this type of design, we may be at a loss when encountering certain objects which may look to us as having emerged through some chain of chance events, whereas they may be products of a mind whose mental processes can be immensely different from ours. Dembski's filter hardly seems to be of help in such a situation.

If we turn to supernatural design, the problem is both similar and different as compared with extraterrestrial design. In the case of aliens we can at least reasonably assume that their designing activity is constrained by the same laws of physics we are familiar with. If we assume, as it is commonly done, that the supernatural designer is omnipotent, i.e., is not constrained by natural laws and is capable of creating new laws at will or breaking the existing laws in any particular case, then the distinction between law and design, as applied to a supernatural design, becomes meaningless, since the natural laws themselves are assumed to have been created by the supernatural designer. Again, Dembski's filter does not seem to be of help in this situation either.

Because of Dembski's generalization of the supposed indications of design, without accounting for differences between human, alien, and supernatural design, his filter is useless for the most interesting discrimination—between the three listed types of design.

In relation to Dembski's concept of specification, let us take a look at an example which has been discussed by Dembski several times. In this example, two strings of Scrabble letters, both of equal length, are compared, one a meaningless combination of letters and the other a phrase from *Hamlet.* According to Dembski's explanation, both strings have equally low probability of emergence by chance. We recognize design in the meaningful phrase because, according to Dembski's scheme, it is *specified,* i.e., it conforms to a recognizable pattern, while the line of gibberish is not specified and therefore is attributed to chance.

I submit that the explanation by Dembski's explanation is not quite adequate. I believe it is more reasonable to conclude that if we see a string of Scrabble letters on a table, we attribute its occurrence to *agency* regardless of its being a quotation from Shakespeare or a piece of gibberish. Remember that in *The Design Inference* Dembski used the term *agency* as a synonym for *design* (p. 11), although elsewhere he distinguishes between these two concepts.

Since Dembski's approach entails separation of design inference from an inference to a designer, obviously the question of a designer's *purpose* becomes moot. This discussion is about Dembski's theory; therefore I will assume that the only question we are really concerned with is whether an event has occurred by chance or its causal antecedent can be traced to an intelligent agent. Whatever purpose such an agent might or might not have, while may be of interest, will be a separate issue. Hence I will use the term *design* simply to mean that the event in question occurred because of an action by an intelligent agent, leaving out the question of purpose.

Moreover, I believe that the common concept of purpose entails the concept of a *conscious* action. If an event resulted from a subconscious action, it can hardly be attributed to a purpose even if the action was by an intelligent agent.

It is easy to imagine situations when a meaningful phrase resulted from a purposeless action, while a gibberish phrase has been created for a purpose. There are many examples of the former. Anyone who has taken part in lengthy and boring meetings knows that very commonly the participants, while listening to the discussion, absentmindedly doodle and scribble on pieces of paper. The products of these subconscious actions are most often meaningless figures and nets of curves, but not too rarely they form some meaningful words and even phrases, created without consciously realizing it—phrases which their creators would not be able to remember a minute after the meeting is over, not to mention explain.

Now turn to an example of a gibberish phrase created for a purpose. Look at the following line: "Epsel mopsel raisobes." This line is a quotation from a poem by a Russian poet, A. Zakharenkov, printed in a collection titled *Strofy Veka*.[19] This sequence is gibberish—it has no meaning either in Russian or in any other language. Its author deliberately wrote this line as gibberish to create a certain comic effect. It was designed for a purpose.

Back to the example with the two strings of Scrabble letters, we do not think even for a minute that the letters in the gibberish string have lined up on the table by themselves, due to some chance process. Somebody had to make these letters, bring them to the room, place them on the table, and arrange them in a straight line. We are confident that all of this was done by a human; i.e., the occurrence of that piece of gibberish was due to design (in the above defined sense) no less than the occurrence of the phrase from *Hamlet*.

Let us review the question of whether or not a string of letters must necessarily have an identifiable semantic meaning in order to be viewed as "specified."

Here is an example. Since 1912 many scholars all over the world have been investing a considerable effort trying to decipher the so-called Voynich manuscript (VMs). A slightly magnified reproduction of a segment of that manuscript is shown in figure 1.1, and a computer-generated rendition of another segment of that manuscript in figure 1.2.

Neither the language nor the alphabet of that manuscript is known. All attempts to decode it have so far been unsuccessful. Therefore, some scholars have suggested that it has no meaningful contents but is instead a hoax, with just over two hundred pages of gibberish. I am of the opinion, based on a statistical analysis of the VMs's text and shared by the majority of those who have tried deciphering the VMs, that it is a meaningful text. On the other hand, some scholars, including Brendan McKay, my colleague in the effort to apply the Letter Serial Correlation test to the VMs, are inclined to think that it is gibberish. However, regardless of the choice between the two mentioned views, nobody has ever doubted that the VMs was written by some medieval author, i.e., that it is a product of design.

A glance at the text in figures 1.1 and 1.2 makes it immediately obvious that we deal with an artifact, designed by a human mind, even though it is unknown whether or not the text is meaningful. Contrary to Dembski's scheme, the design is identified in this case without having available any "detachable" pattern, which, according to Dembski, is a necessary condition for recognizing design.

Let us note that Dembski's view of the difference between the two strings of Scrabble letters seems to indicate that he considers *meaningfulness* of the string as an indication of design and the *absence of meaning* as an indication of chance. We will remember that when discussing Dembski's treatment of information.

An important point seems to be also that all of the above discussion is relevant only to human design. In the case of alien design, and even more of supernatural design, not to mention design by artificial intelligence, we may not know what the signs of design really are.

Let us now discuss specification from another angle. According to

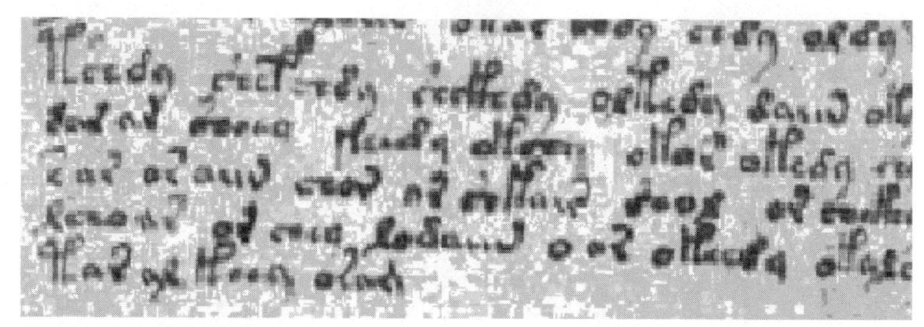

Figure 1.1. A segment of the Voynich manuscript (the original is in color). (Courtesy of Beinecke Rare Book and Manuscript Library, Yale University.)

Dembski, to qualify as specification, the event must be detachable and meet the condition of delimitation. In its turn, to be detachable, the event must meet the conditions of epistemic independence of the side information and of tractability. While this multistep scheme looks rather complicated, especially when Dembski renders it in a heavily symbolic mathematical form, when we review examples provided by Dembski himself or by his colleague Michael J. Behe, we see that actually the idea underlying the discrimination procedure is not very complicated. In one example an astronomer *recognized* the configuration of a constellation in a pile of stones. In another example, we *recognize* a quotation from *Hamlet* in a string of Scrabble letters.

Actually, all those convoluted notions of detachability, tractability, and delimitation seem to be superfluous and the criterion of specification seems to boil down to the simple requirement that can be expressed as: *an event is specified if it displays a recognizable pattern*. Of course, if Dembski limited his discourse to such a brief and easily comprehensible assertion, he would not be able to write a whole book with its seemingly sophisticated mathematical apparatus.

What does recognizability entail? To recognize a pattern we must have in mind some image, independent of the pattern actually observed, to which we compare the observed pattern. That is actually the idea of "detachability," stripped of its sophisticated embellishments.

In view of the above, we can discuss Dembski's criterion of design without delving into the intricacies of his convoluted mathematical discourse.

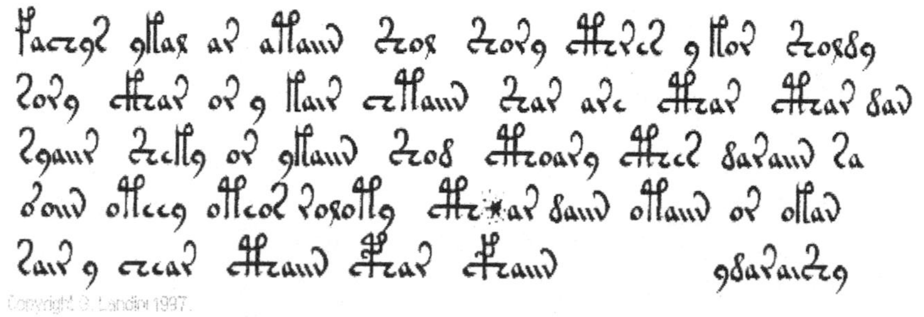

Figure 1.2. A computer-generated rendition of a segment of the Voynich manuscript. Copyright 1997 by Gabriel Landini (reproduced with permission).

False positives

Dembski admits that intelligent agents can, in his words, "mimic" chance and that in such cases his filter produces *false negatives*.

However, Dembski insists, his filter never produces *false positives*. In other words, if at the third node of the filter the conclusion is that the event is due to design, this conclusion is reliable.

To support his assertion, Dembski suggests two lines of proof. The first proof of the filter's reliability, according to Dembski, is a "straightforward inductive argument: in every instance where the explanatory filter attributes design and where the underlying causal history is known, it turns out design is present; therefore design actually is present whenever the explanatory filter attributes design."[20]

While Dembski devotes several pages to the elaboration of this assertion, he does not substantiate it by providing any *record* which would indeed show his filter's impeccable reliability. How can he prove that, indeed, his filter *correctly* indicates design *in every instance*? At best, he may assert that in those examples he has investigated, his scheme indeed correctly identified design, but how can he be sure that it is true for every instance? Indeed, he has reviewed in his publications only a very few examples, thus hardly providing a basis for sweeping generalization (not to mention that we don't know whether or not his examples were deliberately selected to meet his requirements).

Generally speaking, anecdotal evidence is not proof. However, when a categorical statement like that by Dembski is offered, anecdotal examples can legitimately serve as a rebuttal. In a few paragraphs, I will describe instances of false positives, which, in my view, exemplify the lack of substantiation in Dembski's categorical assertion.

Here is the second argument offered by Dembski to prove the immunity of his filter to false positives: "The explanatory filter is a reliable criterion for detecting design because it coincides with how we recognize intelligent causation generally."[21]

Remember that Dembski precisely suggested that his filter is a better and more reliable tool for recognizing intelligent causation. Now he justifies its alleged perfection by comparing it to how we do it without his filter. If we can do it without his filter, and we all know how we do it, what then is his filter for? If his filter, though, is indeed a hitherto unknown perfect tool for recognizing intelligent causation, which is superior to "how we do it generally," then how can the comparison to an inferior way vouch for the filter's reliability? Both arguments in favor of his filter's reliability only express Dembski's own personal view, but hardly have evidentiary significance.

Let us see if indeed false positives are never produced by the explanatory filter.

I submit that Dembski's explanatory filter produces false positives in many common situations. One example of a false positive produced by Dembski's filter was suggested by Ratzsch.[22]

He was once driving on a deserted road. Along the roadside there was a long fence with just one tiny hole at a certain location. A gust of wind happened to drive a tumbleweed across the road in front of Ratzsch's car. The tumbleweed rolled through the fence precisely where the sole small hole was located. This event obviously met the requirements for a design inference according to Dembski's explanatory filter. The event had a very small probability because the area of the hole was only a very small fraction of the fence's area. The event was specified in the same sense as a target painted on a wall is specified. Indeed, like the painted target, the hole existed before the gust of wind had occurred. In fact, though, the event was due to chance alone, so the explanatory filter produced in this case an obvious false positive. Ratzsch, who himself is a prominent colleague of Dembski in design theory, has thus correctly demonstrated the lack of substantiation in Dembski's assertion that the explanatory filter never produced false positives.

Here is another example. This is a real story. It is true and I will tell it, trying to recall it as accurately as possible. My late cousin Joseph (nicknamed Kot), who lived in a different city, used to visit us from time to time, and the last such visit took place in 1939. When the German army invaded Russia in 1941, I lost all trace of him. In 1949 I lived in the city of Odessa, Ukraine. In April of 1949 I went to Moscow for a few days to give a talk in one of the research institutions. On my second day in Moscow, around noon, I was walking in the Okhotny Ryad street, which was crowded by people. At some moment the crowd around me slightly receded for a few seconds and I saw a man walking toward me. To my amazement, I recognized Kot. I learned that he lived for the last couple of years in the city of Balkhash, several thousand miles from Moscow. Balkhash is a small town situated in the central area of Kazakhstan. My cousin wound up in that city due to some peculiar circumstances which are irrelevant to this story.

He was traveling with his wife and two daughters from Balkhash to his native city of Kharkov in the Ukraine for a vacation. On the morning of that day, on his way from Balkhash to Kharkov, he arrived in Moscow, where he was to stay for only a few hours, and then depart for Kharkov in the afternoon. One of his daughters caught a cold. He left his family at the railway station and took the subway to Okhotny Ryad to find a pharmacy.

It is obvious that the probability of a chance encounter with Kot in the described circumstances was minuscule. It is easy to verify that the described event also met Dembski's conditions of detachability (which comprises conditional independence and tractability) and delimitation. Instead of delving into Dembski's detailed definitions of these concepts, let us recall his own example of an astronomer who recognized the configuration of a constellation in a pile of stones. The astronomer in that example recognized the pattern because he had the proper background knowledge—he had in his mind the image of that constellation. This image did not affect the probability that the configuration of stones happened by chance (i.e., conditional independence). He could easily create in his mind the image of the constellation in question (i.e., tractability). The configuration actually observed was among those he had in his mind (i.e., delimitation). Likewise, I recognized my cousin because I already had in my mind the image of him, which knowledge did not affect the probability of our chance encounter (i.e., conditional independence). I could easily create in my mind the image of Kot (i.e., tractability). The image of Kot was among all those images I had in my mind (i.e., delimitation). If, according to Dembski, the astronomer's recognition of the configuration of a constellation was a specified event of low probability, then so, too, was my encounter with Kot.

In Dembski's view, recognizing the configuration of a constellation in a pile of stone leads inevitably to inferring design, i.e., to the conclusion that somebody intentionally arranged the stones in the observed configuration. Hence, if we accept Dembski's scheme, we have to conclude that my encounter with Kot was designed. It was not. This is a clear case of a false positive.

The story, however, had a continuation. In 1969 I lived in the city of Tver, some 120 miles northwest of Moscow. In April of 1969, exactly twenty years after my encounter with my cousin, one morning I took a train to Moscow where I planned to stay for just a few hours and return to Tver the same evening. Close to noon I was walking on *the same block of the same Okhotny Ryad street* where twenty years earlier I had met Kot. Suddenly somebody hugged me from behind. I turned and, to my amazement, recognized my old friend Karl F. (Karl is alive and well and now lives in Brisbane, Australia.) Karl and I climbed many mountains together in Pamir and Tien-Shan. The last time I met him before that encounter in April 1969 was in Siberia in 1959. After that, I had not known his whereabouts. But in April of 1969, I learned from Karl that at that time he lived, of all places, in the city of Balkhash! He wound up in that forlorn city due to peculiar circumstances which are irrelevant to this story. He had come to Moscow for only a few days.

I leave it to the readers to estimate the probability of our encounter at exactly that minute at exactly that particular spot, exactly twenty years after a similar encounter at the same location with my cousin, a man whose name also began with the letter K, each time ten years after my previous meeting with each of them, both having come from the same remote town. The precise calculation of probability for the described events is difficult because many details of the situation have to be assumed without a verifiable information (for example, I can't confidently assert what exactly were the frequency and durations of Kot's and Karl's visits to Moscow, how many streets there are in Moscow, etc.). Therefore I will not provide specific numbers for that probability, which obviously was exceedingly small.

Let us now apply Dembski's design inference scheme to the described event. To this end, let us use the design inference argument from Dembski's *The Design Inference* in its symbolic form (p. 49) and add to it the description of the event in plain words. In the following scheme, Dembski's argument in its mathematically symbolic form, which was shown earlier in this chapter and designated as set (A) of formulas, is on the left, while my addition of the particulars of the event in question is on the right in the double square brackets.

Premise 1: oc(E): [[Event E occurred—I met Kot and Karl]].
Premise 2: sp(E): [[Event E is specified—I recognized Kot and Karl from among all the pedestrians]].
Premise 3: ch(E) $\rightarrow$ SP(E): [[If our encounters were due to chance,

the probability of that event would be vanishingly small]].

Premise 4: $\forall X[\text{oc}(X)$ & $\text{sp}(X)$ & $\text{SP}(X) \rightarrow \sim\text{ch}(X)]$: [[Specified events of small probability do not occur by chance]].

Premise 5: $\sim\text{reg}(E)$: [[My encounters with Kot and Karl were not due to a regularity]].

Premise 6: $\text{reg}(E) \vee \text{ch}(E) \vee \text{des}(E)$: [[My encounters with Kot and Karl could occur only due to either law, chance, or design]].

Conclusion: $\text{des}(E)$: [[My encounters with Kot and Karl were due to design]].

Indeed? Isn't this a false positive at its extreme? Contrary to Dembski's confidence in the reliability of his scheme, which allegedly never produces false positives, such false positives can be expected in many situations.

If my extremely improbable encounters with Kot and Karl can be viewed as a rare exception, a rather more common example that immediately comes to mind is a raffle. Imagine a raffle in which ten million tickets have been sold, each bearing a seven-digit number, from, say, 0000000 to 9999999. Sweepstakes in which up to thirty million and even more tickets are distributed, for example by magazine peddlers, are common in the United States. The winning number is usually determined in advance; say, in our example it is 9765328. The probability of winning for each individual player is of course the same—1/10,000,000, provided fraud is excluded. While this probability is not as exceedingly small as other probabilities sometimes calculated—for example, for the spontaneous emergence of a protein molecule—it is small enough to exclude law as a cause of winning. If John Doe won in an honestly conducted raffle, it was due to chance. However the event—John Doe's winning—is clearly specified. Both the player and the winning number are specified. In particular, the winning number constitutes a recognizable pattern. The combination of small probability and recognizable pattern, according to Dembski's filter, determines design (in this case fraud) as the cause of that event. It seems to be a false positive.

Of course, Dembski might say that the probability in this case is not small enough to warrant the conclusion of design. Indeed, this is his argument when he discusses the case of the Shoemaker-Levy comet (wherein he estimated the probability as being 1/100,000,000 in *The Design Inference*, insisting that his filter does not yield design easily (p. 228). The probability of an event, according to Dembski, must be very low indeed to infer design. However, when Dembski discusses what constitutes a sufficiently small probability, he characterizes the probability of 1/100,000, which is one hundred times larger than in our example, as sufficiently "small to eliminate chance in case the conditional independence and tractability conditions are also satisfied" (p. 189). The listed conditions are satisfied in our example. Anyway, the numbers in themselves are not crucial; it is the principle which is under discussion. Indeed, it can easily be shown that the absolute value of probability is of secondary significance, and that 1/10,000,000—and sometimes much larger numbers—can in many instances be justifiably viewed as a sufficiently small probability for the purpose of design inference.

To this end, consider a small raffle in which only one hundred tickets are sold, which, however, is played more than once. The probability of John Doe's winning once in that raffle, if it is conducted honestly, is, of course, 1/100. Assume now that John Doe won that raffle three times in a row. The probability of such a triple win is $(1/100)^3$, which is 1/1,000,000. That is ten times larger than in the above-mentioned sweepstakes where it was 1/10,000,000. However, despite the larger probability of the triple win in the latter case, we now suspect fraud, i.e., design, and justifiably so. In the latter case the probability of one in a million is obviously small enough to suspect fraud, while the smaller probability in the sweepstakes with ten million players is justifiably used for assuming chance. This shows that the meaning of a certain value of probability is not absolute but has to be viewed in relation to the specific circumstances of the event, which does not seem to be accounted for by Dembski's filter. (A detailed discussion of the reasons for the different intuitive interpretation of the results in a large raffle played just once and small raffle played several times in a row will be given in chapter 13.)

It should be stressed that the particular event to be analyzed is a *specified* player winning in the raffle/sweepstakes, not *someone* winning. If in the case of the large sweepstakes we decide that 1/10,000,000 is not small enough to warrant the design inference, then it should hold even more for the small raffle played three times, where the probability of the event in question is ten times larger. On the other hand, if we decide that the probability of 1/1,000,000 is small enough to infer design in the case of a triple win, it should hold even more for the case of a large sweepstakes where the probability of the event is ten times smaller.

I believe the above examples show some of the deficiencies of

Dembski's filter, which can easily produce both false negatives and false positives.

Illusory patterns

I believe that there are classes of situations in which the explanatory filter is fully expected to produce false positives. One such situation is that of an *illusory pattern.*

Consider the following example. The Caucasus mountain range extends for several hundred miles. It comprises thousands of peaks, passes, dales and valleys, gorges and chasms, glaciers and moraines, etc. All these relief elements have different shapes, most of them quite irregular, although here and there parts of a mountain may form some more or less regular geometric patterns. The particular shape of this or that relief's element depends on an enormously large number of accidental factors, so when we observe a particular mountain we realize that its particular shape has an extremely small probability. If we apply Dembski's filter to find whether or not that particular mountain's shape is due to chance or design, it would pass the first two nodes of the filter and lead right to the third, the crucial test in which we will look for a recognizable pattern. For the overwhelming majority of the mountains no pattern will be recognized, so the emergence of the particular shape of any particular mountain will be justifiably attributed to chance.

(Of course, proponents of intelligent design can insist that everything is the result of design by a supernatural mind, so the irregular or sometimes seemingly regular shapes of the hills and gorges were designed that way. There is no rational way to reject such a statement. If, though, such a claim were made, it would make Dembski's filter unnecessary, since that filter allows for chance events to occur, while the above claim denies their existence altogether. Our discussion is within the framework of an approach which allows for chance and deals with the question of how to rationally distinguish between chance and design, in particular using Dembski's criterion.)

There is, though, in the Dombai region of the Caucasus range a mountain named Sulakhat. This word is a woman's name in the local language. Anybody looking up at Sulakhat from the valley circling that mountain immediately understands the reason for that name. From the valley the mountain looks like the perfect profile of a woman on her back, with the clearly delineated features of a young pretty face, neatly combed hair, taut breasts, arms crossed over her stomach, and slim legs slightly bent at the knees. The contours of the woman's body display all the features of a fine work by an accomplished sculptor. Many first-time visitors to Dombai refuse to believe that all they are seeing is an accidental combination of rocks and ice fields. Indeed, if we apply Dembski's filter, we clearly see a combination of improbability (complexity) with a recognizable pattern, which, according to Dembski, indicates design.

If, having climbed the mountain and having accurately measured the "body" of Sulakhat, we discovered that it is indeed a figure carved from a giant slab of stone, we could reasonably decide that the filter provided good reason to conclude that we saw the product of design.

Actually, this is an example of an illusory pattern. Indeed, as mountain climbers walk up the slopes of Sulakhat, they gradually discover that the sculpturelike shape is an illusion. The figure which from the valley looks like a human body turns out to be a combination of various rocks, scattered over a wide plateau and separated from each other sometimes by hundreds of feet. At a closer look, the alleged sculpture breaks down into an incoherent conglomerate of unrelated pieces. The explanatory filter has produced a false positive. The very low probability of Sulakhat's emergence by chance is indisputable. A recognized pattern is there for all (from the valley!) to see. "Design!" announces Dembski's filter. "Illusion" is the correct answer.

This example of an illusory pattern can also be viewed as one more manifestation of the subjectivity of Dembski's criterion. From the subjective viewpoint of an observer who looks at Sulakhat from the valley, there seems to be a recognizable pattern. From the subjective viewpoint of an observer who has climbed the mountain, there is no recognizable pattern in the shape of that mountain.

The criterion in question is subjective in regard to both false negatives and false positives.

The nature and role of specification

In Dembski's scheme, specification is explicitly viewed as a category independent of probability. Moreover, he stresses that the side information utilized to establish specification must be conditionally independent of the probability, expressing it as $P(E|H\&I) = P(E|H)$, where P stands for probability, E stands for event, | is a symbol meaning "given that," H is in this case the assumption that the event is due to chance, and I is the side information.

In my view, the described approach is faulty. I submit that the procedure of design inference is essentially an estimation of the *probability* of either design or chance. Therefore the real role of specification is only in enhancing the probability of design compared to the alternatives.

In Dembski's scheme, when we reach the third node of the filter, we first estimate the probability of the event in question, assuming that it occurred by chance. If it turns out to be small, then, according to Dembski, we look for specification, i.e., for a recognizable pattern. If we find such a pattern, it leads to the design inference.

Time and time again, throughout his books and papers, Dembski states that small probability in itself is insufficient to infer design. To infer design, according to Dembski, small probability must be accompanied by a recognizable pattern. In my view, Dembski's formula for inferring design unnecessarily introduces an artificial dichotomy between probability and specification. In fact, specification (i.e., a recognizable pattern) is not a factor independent of probability but rather only one of many factors affecting the estimate of the probability of design as compared with chance.

I think that my approach is rooted in my background in statistical physics. The latter is established as a highly reliable penetration into reality. One of its salient features is that it clarifies the laws of thermodynamics, revealing its actual statistical nature. For example, statistical physics asserts that the predictions of the second law of thermodynamics are not absolute but only determine the most probable outcomes of natural processes. The predictions of this law are highly reliable only because of their overwhelmingly larger probability when compared to any alternatives.

While events may be affected by a multitude of factors, statistical physics incorporates all of them in one ultimate criterion—the probability of the event in question.

Accounting for the success of statistical physics, I see no reason why the same approach should not be utilized in discussing the probability of design versus chance. Like in statistical physics, what really counts is the overall probability of either design or chance, regardless which components it comprises. One of these components may or may not be specification, i.e., a recognizable pattern. In some circumstances specification may be a more important contributor to probability than other factors, while in some other circumstances it may be a less important contributor. In certain situations specification may not be a contributor to the probability of design at all, in that specification (i.e., a recognizable pattern) is not any different from many other factors contributing to the estimate of probability of design versus chance.

If we see a poem or a novel, we unequivocally recognize design because the probability of design is overwhelmingly larger than that of chance. In this particular case, specification (i.e., a detachable pattern) contributes to the large probability of design. However, it is by no means the only possible situation. We may encounter an item which does not look like any familiar image, and hence does not match any detachable pattern, but we may identify it as an obvious artifact, because many factors other than specification combine in an unequivocal indication of the overwhelming likelihood of design as compared with chance. The Voynich manuscript is just one example of such a situation.

Dembski's attribution of a special status to specification as compared with other factors does not seem to be justified by evidence.

The design inference is never absolute. It can only be made in probabilistic terms. If an event has a very small probability of occurring by chance, a hypothesis of design may be highly reasonable. Dembski asserts, however, that small probability itself is not sufficient to infer design. If the event also displays a recognizable pattern, the probability of design becomes even larger. However, in principle it is still a hypothesis, albeit with a higher probability. If, as Dembski states time and time again, low probability in itself is not sufficient to unequivocally infer design, neither is the combination of low probability with specification, because all the latter does is simply decreasing the probability of chance. This in itself does not introduce a new *quality* into the inference procedure, but is only a *quantitative* step toward the hypothesis of design.

Indeed, as several examples discussed in preceding paragraphs have shown, sometimes design can be reliably inferred in the absence of a recognizable pattern, i.e., of specification (as in all cases of false negatives whose possibility Dembski admits), while in other cases specification does not ensure the reliability of the design inference (as in cases of false positives, which have been shown to happen despite Dembski's assertion to the contrary).

Dembski is right when, unlike some of his colleagues in the intelligent design movement, he asserts that low probability is not sufficient to eliminate chance as the source of an event. Indeed, chance events of extremely small probability occur routinely every minute. However, this situation cannot be remedied by mechanically adding to the estimate of probability some other factor, be it specification in Dembski's sense or anything else. Whichever additional factor is taken into consideration, it may or may not change the estimate of the probability in question. Therefore the design inference is doomed to be probabilistic. This does not, though, prevent such an inference from sometimes being highly plausible, but this plausibility may be achieved both in cases when specification according to Dembski is present and in cases when no such specification is discovered. In the example of the Voynich manuscript, no specification in Dembski's sense seems to be found; however, the design inference is quite reasonable, even though it is based solely on the overwhelmingly larger probability of design versus chance.

Another comment in regard to Dembski's definition of specification can be made if we review his example of archery. In that example, Dembski compares two situations. In one an archer hits a wide wall and afterward paints a target around the arrow. In Dembski's terms, this is *fabrication*. In the other situation, an archer hits a small target that was painted on the wall beforehand. This is *specification*, says Dembski.

Whereas the case of fabrication is simply not interesting, it is easy to see that the difference between the two situations is not that in the first case the event was not specified while in the second case it was. The event is obviously specified in both cases. The entire wall is just a larger target (i.e., constitutes a recognizable pattern) than the small round target painted on that wall. The difference is in the size of the target, which, of course, is not a principal difference. The same difference exists between, say, a painted target which is 5 centimeters in diameter and a painted target which is 50 centimeters in diameter. The mere fact that in one case a target is painted while in the other case the entire wall is a target, does not distinguish the events in terms of specification. The difference is only in a *different probability* of a hit.

Likewise, if we see a meaningful text, we conclude that its emergence by chance was highly unlikely. The fact that the text meets a recognizable pattern points to design not because it adds some independent argument in favor of design, but because it enhances the probability of design. Every factor works either for or against probability, and specification is just one more contributor to the estimation of probability rather than a factor independent of probability. It is not combined with low probability, as Dembski maintains, but is incorporated into probability as a part of the necessary background information.

If there is a target painted on a wall and the archer hits that target, this is, according to Dembski, a specified event of small probability which therefore must be attributed to design. This conclusion, according to Dembski, is made because the event (hitting the target) meets the conditions of detachability and delimitation. Detachability, in turn, incorporates conditional independence of the background knowledge and tractability. In relation to the archer's success, let us discuss the conditional independence of the background knowledge which, according to Dembski, is necessary for design inference. From Dembski's explanations and examples, it seems to follow that the background knowledge relevant to the problem at hand boils down to the recognizability of the target. The painted target is specified, i.e., recognizable. The recognizability of the target does not affect the probability of the hit (i.e., conditional independence is satisfied); therefore, when the arrow hits the target, we conclude that the event in question was due to design. Any side information which may affect the probability of the hit is, according to Dembski's scheme, irrelevant for design inference. For the sake of discussion, let us accept Dembski's scheme, which asserts that the successful hit must be unequivocally attributed to design. (In other words, ignore the contribution of laws of mechanics and such chance events as an accidental gust of wind, a small earthquake at the moment of the archer's action, and the like.)

Imagine, though, that prior to the archers' competition, we watched the archers exercising for several days. Imagine that archer A is a world champion with a record of hitting the target 98 percent of the time, while archer B is a beginner who tried many times to hit the target as we watched him and failed to do so in all of his attempts. Thus we acquire knowledge about the two archers in question, which obviously affects our estimate of the probability of success for both archers at the competition. According to Dembski's scheme, this knowledge does not meet the condition of detachability because it is *not conditionally independent*. Hence, according to Dembski's scheme, the knowledge about the skills of the two archers in question has no bearing on design inference. Imagine, though, that on the day of the actual competition both A and B successfully hit the target. According to Dembski's scheme, design inference is equally justified in both cases. I think, though, that our conclusion will be different in the two cases. In the case of archer A with his record of 98 percent success, we will confidently attribute his success to his skill (i.e., to design). In the case of archer B we will justifiably attribute his success to his luck (i.e., to chance). In this example, the choice between chance and design is made based on the background information which, contrary to Dembski's scheme, is not conditionally independent.

Furthermore, in the case of archer A, despite the high probability (98 percent) of a perfect hit, when the archer in question indeed hits the target, we do not conclude that it is due to law (unlike Dembski, who attributes events to law in the case of *high probability*, as his scheme prescribes for the first node of his filter). We conclude that it was due to design, and applaud the archer's skill, while in Dembski's scheme the design inference (see the third node of the explanatory filter) necessarily requires a *very low probability* of the event in question. (As was discussed earlier, the archer's success has to be attributed to a combination of design and law; Dembski's scheme, though, does not recognize such a double attribution, hence we do not need to account for it when discussing that example within the framework of Dembski's theory.)

One more comment about Dembski's analysis of specification/pattern seems to be in order. If the archer hits the target, we identify, according to Dembski, specification. However, when describing this situation, Dembski in some instances mentions just a single hit as being sufficient to qualify as specification—(in particular, such an interpretation follows from the definition of pattern in *The Design Inference* (p. 136). In other instances, however, he mentions an archer hitting the target a hundred times in a row, and that repeated success is viewed as an indication of design, i.e., of the archer's skill (p. 13). He does not seem to see the difference between these two situations.

Indeed, the definition of a pattern given by Dembski in *The Design Inference* seems to make clear that in his scheme the repetition of an event is not considered a factor in determining a pattern.

I believe a repetition of an event is a very important factor in a design inference, and Dembski seems to have missed it. To see why the repetition is important, imagine two situations. In one case, an archer shoots from a certain distance of L meters and hits the target just once. In another case, the archer shoots from the much shorter distance $N < L$ but hits the target ten times in a row. If we adopt Dembski's theory of probability (complexity) difficulty (which we will discuss later), we can assign a certain numerical value of probability to the success in both cases. Since N is shorter than L, it is easier to hit the target from N meters. Assume that the difference between L and N is such that the probability of hitting the target ten times in a row from N meters is exactly the same as hitting it just once from L meters. Despite the equal probabilities in these two situations, I believe our intuitive judgment will be different in the two cases. In the case of just one hit we will be uncertain whether to attribute the archer's success to his skill (i.e., to design) or to chance. Indeed, however small the probability of hitting the target by chance may seem to be, it is not zero. If the archer hit the target just once, it might well be attributed to luck, i.e., to chance. In the second case, we rather confidently attribute the archer's repeated success to his skill (i.e., to design), even though the likelihood of chance in that case is not less than the likelihood of chance in the previous case.

We have to acknowledge the substantial role of a repetition of an event in the procedure of attributing it to either chance or design. Dembski's definition of a pattern and hence of specification fails to recognize that factor.

From yet another angle, there are various classes of events regarding the discrimination between chance and design. On the basis of the same type of background information, some events are readily recognized as having been designed, while others can resist such discrimination. If we find a book of poems, we easily recognize it as a result of design. This is made on the basis of our background knowledge. In particular, our experience tells us that humans write poems and print books, and we have seen many of them and know that they are the products of human design. The same relates to Paley's watch and a myriad of other objects which are familiar to us as a part of our ken. However, there are other classes of objects, and one example of such is any biological structure. A DNA molecule has a very complex structure and carries a lot of information (in the sense of information theory). However, our ken does not include knowledge which would make us conclude that it is the product of deliberate design. The pattern seemingly present in a DNA strand is not "detachable" using Dembski's term, as that pattern is not recognizable.

This would not matter if we had knowledge of some other factors which, if combined, would point to an overwhelming probability of design versus chance. Unfortunately, we have no such knowledge. Biological structures display enormous complexity, but this in itself does not indicate design because structures of unlimited complexity can emerge in a stochastic process. (More about this will be discussed later in this chapter.)

I take the liberty of giving a peculiar example. The famous nineteenth-century Russian anthropologist Nikolai Miklukho-Maklai spent a long time among the aborigines of New Guinea studying the ways of life, language, habits, and mores of those tribes. When the ship that brought Maklai to the island appeared in view, the aborigines, who had never before seen such a big ship, decided that it came from heaven and that the crew members were gods. As a precaution, Maklai never tried to disabuse the villagers of their belief. Moreover, to reinforce their belief, he would sometimes remove his artificial teeth, to the villagers' amazement. Their ken did not include the knowledge of artificial teeth, so they viewed Maklai's action as a miracle performed by supernatural forces. Obviously, such an explanation was unjustified and similar to the interpretation of the DNA structure as the result of supernatural design.

(It can be added that emergence of biological structures could occur without an intelligent designer, but still in a nonrandom way. There are theories offering plausible mechanisms for such occurrences—one such mechanism is described, for example, by mathematician John F. McGowan.)[23]

My conclusion is that Dembski's acclaimed explanatory filter is not a reliable tool to identify design. It often seems to fail even in cases of human design. This appears to be especially true in regard to supposed supernatural design, which seems to be the most interesting case.

Probability According to Dembski

In *The Design Inference,* Dembski devotes the whole of chapter 3 to the discussion of probability and likelihood, and offers various notions related to probability in other chapters throughout his books, as well as in his papers.

Probability and likelihood have been discussed many times before Dembski. (A discussion of the conventional concept of probability, designed for nonexperts, will be given in chapter 13 of this book.)

In probability theory, there are various definitions of probability, such as the classical definition, frequentist definition, geometric definition, statistical definition, and finally the axiomatic definition. There is also the so-called Bayesian approach in which distinction is made between three versions of probability, referred to as *prior probability, posterior probability*, and *likelihood.*

On pages 78 to 79 of *The Design Inference* Dembski briefly discusses the frequentist and classical definitions, and on pages 67 to 69, the Bayesian approach. (With his rather typical lack of excessive modesty, Dembski claims on page 86 that his approach has "a huge advantage" over the Bayesian.) He points to certain limitations of each of those approaches. However, he does not mention any other definitions of probability, including the axiomatic definition by Russian mathematician Andrei N. Kolmogorov.[24] The latter is the most rigorous and the most general; on the one hand, it encompasses all other definitions as particular cases, and on the other hand, it overcomes the limitations of all other definitions.

Having discussed the limitations of the classical and frequentist definitions as well as of the Bayesian approach, Dembski offers his own definitions of probability and likelihood. Here is Dembski's definition of probability from *The Design Inference*:

> **Definition.** *The* probability *of an event* E *with respect to background information* H, *denoted* P(E|H) *and called "the probability of* E *given* H,*" is the best available estimate of how likely* E *is to occur under the assumption that* H *obtains.*

In *The Design Inference* we also find the following definition of likelihood:

> **Definition.** *The* likelihood *of an event* E *with respect to the background information* H, *denoted* Λ(E|H) *and called "the likelihood of* E *given* H,*" is that number, or range of numbers in the unit interval [0,1] denoting how likely* E *is to occur under the assumption that* H *obtains and upon the condition that* H *is as effectively utilized as possible.* (p. 78)

If Dembski offers his own definitions of these concepts, one may expect that such new definitions will shed new light on the matter, or at least reveal some new facets in those concepts, or maybe provide a more convenient or a shorter way to handle these concepts. In my view, they do none of the above.

Let us first look at the definition of probability. On the one hand, this "definition" correctly lists two essential characteristics of probability. Indeed, probability is nothing more than an *estimate* of how likely an event is to occur; this estimate's cognitive value is only as good as the available knowledge about the situation at hand. If that knowledge changes, the estimate of probability changes as well. The more knowledge about the situation available, the better the estimate of probability. However, while the characterization of probability as an estimate which is dependent on the available "information" is correct, it does not tell us anything new because it has been discussed many times before Dembski. Since his definition contains practically nothing beyond the quoted statement, the entire definition does not offer anything new.

On the other hand, Dembski's definition lacks a crucial element. This element may be expressed by the word *quantity*. Probability is a quantity and normally is to be assigned a numerical value. All existing definitions of probability, whatever their limitations, and whatever differences between them, include prescriptions of how to assign numerical value to probability. (Although there are theories of "imprecise probability," which do not assign definite numbers to probabilities, they relate to situations having little relevance to Dembski's case.) Dembski's definition provides no indication that the estimate in question must actually be assigned a numerical value and even less of how to determine the latter.

The reference to "the best available estimate" by no means saves his definition. How do we translate "the best available estimate," even if such has been reliably made, into a numerical value? Moreover, what are the criteria enabling one to distinguish "the best available estimate" from, say, "the second best estimate," or from a poor estimate?

Dembski tries to clarify his idea by indicating that the determination of "the best available estimate" is to be made by "the community of discourse." While this assertion may have a nice sound, it is actually of little substance. What are the criteria enabling one to determine what the agreement within "the community of discourse" is?

In *The Design Inference*, we read: "Within a community of discourse at a particular time probability is therefore uniquely determined" (p. 88). Indeed? I wish this were true. Recall, for example, the ongoing dispute about the probability of appearance of a so-called code in the text of the Bible. (See chapter 14 in this book.) The dispute about this matter within the relevant community of discourse, which included prominent mathematicians, reminded people, in the words of one of its participants, Barry Simon, professor of mathematics at Caltech, of a street fight.[25] Proponents and opponents of the "code" have offered vastly differing estimates of the probability in question, each side offering a host of arguments. While only one side of the dispute was correct, the dispute has not brought about a consensus.

Other examples of disputes among scientists in regard to the calculation of probabilities are given in chapter 13. Quite commonly there is no such thing as an agreement within the appropriate community of discourse regarding the probability of various events.

The vagueness of the concept of an estimate agreed upon by the community of discourse, which may be handy in some philosophical treatise, seems to make it rather out of place in a supposedly rigorous mathematical discourse.

Let us now look at Dembski's definition of likelihood. The concept of likelihood is commonly used in the Bayesian probabilistic approach, where it is rather simply and rigorously defined. (The Bayesian approach has both adherents and critics, but the concept of likelihood, as defined there, is itself not disputed.) This concept has also been used in information theory, the field in which Dembski, according to his admirers, is an expert.

In *The Design Inference* Dembski discusses Bayes's theorem and its constituent concepts, prior probability, posterior probability, and likelihood (pp. 67–69). Then Dembski offers his own definition of likelihood quoted above. Unlike the case of his definition of probability, this time his definition includes a reference to a numerical value ("a number or a range of numbers") which is to be assigned to likelihood. However, like in the case of probability, the definition of likelihood provides no indication of how the numerical value in question has to be chosen, and therefore it also is not a proper mathematical definition.

If we compare Dembski's two definitions—that of probability and that of likelihood—it is hard to see any distinction between them. Both definitions contain as their core the identical phrase "how likely *E* is to occur." Hence, essentially both try to define the same concept expressed by that phrase. One may point to two differences between the two definitions. One is that probability is estimated "under the assumption that *H* obtains," while likelihood is estimated "under the assumption that *H* obtains and upon the condition that *H* is as effectively utilized as possible." The other difference is that probability is determined as "the best available estimate," while the definition of likelihood requires us to assign to that concept a number (or numbers) within a certain range (although it does not provide any prescription of how such number is to be chosen).

It is easy to see, however, that the two mentioned differences are inconsequential. Likelihood is defined with reference to *H* being "as effectively utilized as possible," while probability is defined without referring to that requirement. On the other hand, probability is defined with reference to "the best available estimate." Note that *H* stands for "background information"; i.e., it does not refer to the conditions determining the actual occurrence of the event in question. It is hard to imagine that "the best available estimate" could be made without utilizing all relevant available information *H* "as effectively as possible." If the estimate were not to account for all available information *H* "as effectively as possible," obviously the estimate would not qualify for being "the best available." (The definition of likelihood, if considered in itself, may create an impression that the requirement of *H* being utilized as effectively as possible is a part of the conditions leading to the *occurrence of* E. However, Dembski's elaboration on pages 80 to 81 makes it clear that he actually meant *H* to be the information which must be effectively utilized for the *estimate* of likelihood.) Hence, both definitions seem identical in that respect, while phrased in a slightly different way. Then the only difference between probability and likelihood, according to Dembski's definitions, interpreted literally, seems to be that likelihood is to be assigned a number, while probability is simply some nonnumerical "estimate."

However, Dembski offers the following explanation regarding "the best available estimate": "'Estimate' signifies that an event is being assigned a number between 0 and 1 (or a range of numbers, or no number at all) which, in light of relevant background information, reflects how likely it is for an event to occur" (pp. 87–88).

Hence, probability has to be assigned a number after all. Why, then, is assigning a number mentioned only in the definition of likelihood but not in the definition of probability? If the latter has to be assigned a number between 0 and 1, exactly as prescribed for likelihood by Dembski's definition, what is the difference between his definitions of probability and likelihood? Dembski's two supposedly different definitions actually purport to define exactly the same concept.

Since Dembski's definition of likelihood (as well as his definition of probability) offers no indication of how to actually determine the value (or a range of values), this definition is actually reduced to a pure tautology. It states an obvious platitude—the more likely the event's occurrence, the larger its likelihood; equivalently, the more probable the event's occurrence, the larger its probability.

I believe we would be better off sticking to the concepts of probability and likelihood which have been in use in probability theory for quite a while and which seem to work reasonably well.

Dembski's eagerness to offer his own supposedly innovative concepts of probability and likelihood can be humanly understood, but these concepts in his interpretation do not offer anything new, lack crucial elements necessary for mathematical definitions, seem to have no useful purpose, and at best only introduce notations.

Dembski's colleagues sometimes refer to him as a design theorist, sometimes as an information theorist, and sometimes as a probability theorist. I admit that I am not a probability theorist. I have, though, many times taught a university course on statistical physics, both for undergraduate and graduate students. Statistical physics is based on probability theory, and necessarily includes an introductory chapter where probability theory is discussed. Based on my background in probability theory, I formed the opinion that Dembski's definitions of probability and likelihood are not really useful. However, allowing for the possibility that I did not grasp some hidden meaning in Dembski's discourse, I decided to get a second opinion.

I e-mailed Dembski's definitions of probability and likelihood to a prominent mathematician. This person is an expert in probability and statistics, the author or coauthor of numerous publications in these fields printed in prestigious scientific journals and collections, as well as an editor of a pertinent journal and a member of the Academy of Sciences in his country. This scientist has not read any books or papers by Dembski, nor is he involved in the dispute about intelligent design. In my message to that scientist I did not mention intelligent design, but simply copied verbatim Dembski's definitions of probability and likelihood and asked that scientist to evaluate those definitions. Here is the quotation from that person's reply: "It doesn't work for me, as these definitions sound like mumbo-jumbo."

This reaction from a prominent professional mathematician is easy to understand if we recall that Dembski's definitions of probability and likelihood do not provide any basis for a quantitative estimate of the quantities to be defined, are ambiguous in many respects, and contain some constituent concepts each requiring its own prior definition.

Therefore Dembski's entire chapter devoted to probability seems to be an unnecessary exercise whose removal from his book would hardly cause any harm to the latter.

Complexity According to Dembski

Dembski's Definitions of Complexity/Difficulty

The essence of Dembski's complexity theory can be succinctly expressed by the following two quotations. In *The Design Inference* we read, "Whereas probability theory measures the likelihood of an event, complexity theory measures the difficulty of a problem" (p. 92). Later in the same book we read: "Probability measures are disguised complexity measures" (p. 114). The "disguise," according to Dembski, is in that probability and complexity differ only in direction and scale (p. 115).

If the last statement is true, why is there a need for a complexity theory besides probability theory? Small probability means, according to Dembski, large complexity, and vice versa. The range in one case is [0,1] and in the other [0,∞]. Otherwise, probability values can be mapped into complexity values, and vice versa.

The definition of complexity, according to Dembski, is as follows:

> **Definition.** *The* complexity *of a problem* Q *with respect to resources* R, *denoted by* $\varphi(Q|R)$ *and called "the complexity of* Q *given* R,*" is the best available estimate of how difficult it is to solve* Q *under the assumption that* R *obtains.* (p. 94)

As Dembski indicates, "this definition closely parallels the definition of probability given" in another section of his book.

I believe the above assertion requires an amendment. With all the inadequacy of Dembski's definition of probability, discussed in the previous section, it was at least correct in some limited abstract way, qualitatively defining certain real features of probability. On the other hand, his definition of complexity is rather arbitrary as it seems to unduly generalize a particular class of situations, while being completely off the mark for many other possible situations (I will discuss this in the next section).

Dembski also offers a definition of a quantity which is supposed to stand in the same relation to complexity as likelihood (in Dembski's interpretation) stands in relation to probability. In *The Design Inference*, we read:

> **Definition.** *The* difficulty *of a problem* Q *with respect to resources* R*, denoted by* Δ(Q|R) *and called "the difficulty of* Q *given* R*," is that number, or a range of numbers, in the interval [0,∞] denoting how difficult* Q *is to solve under the assumption that R obtains and upon the condition that* R *is as effectively utilized as possible.* (p. 99)

Both definitions—that of complexity and that of difficulty—define the object of definition in the same words: "how difficult it is to solve *Q*" in the first definition and "how difficult *Q* is to solve," in the second, which is, of course, the same. One seeming difference between the two definitions is that the first one requires "the best available estimate," whereas the second requires assigning a number (or a range of numbers) to the quantity to be defined. However, if we look further at Dembski's discourse, we find that, while using the expression "the best available estimate" in regard to complexity, he actually means a number in the case of difficulty, too. Indeed, in *The Design Inference* we read: "What do the numerical values of complexity signify? So far all I have said about complexities is that they range between 0 and ∞, with 0 signifying minimal difficulty (typically no difficulty at all), ∞ signifying maximal difficulty" (p. 110). Recall that, contrary to the quoted sentence, in his definitions Dembski assigned a number (or a range of numbers) in the interval [0,∞] only to difficulty, but not to complexity. From the last quotation it follows that, in agreement with my interpretation, Dembski himself erases any difference between complexity and difficulty. The difference in wording in the two definitions does not actually represent any real difference between the two concepts being defined.

The second apparent difference between complexity and difficulty is in that the definition of complexity only requires the resources R "to obtain," while the definition of difficulty requires additionally that the resources are "as effectively utilized as possible." However, the definition of complexity requires "the best available estimate." Obviously, if resources R are not assumed to be utilized as effectively as possible, the estimate based on such an incomplete assumption is not the best available. The requirement of the estimate being the best available implies that the resources are utilized as effectively as possible.

Furthermore, like Dembski's definition of probability and likelihood, those of complexity and difficulty provide no indication of how the numerical values are to be assigned to either of the two quantities in question. Instead, Dembski again offers the vague concept of "the best available estimate," which, in turn, is predicated on the concept of "community of discourse." The latter is a nebulous notion and as such hardly has a legitimate place in a supposedly rigorous mathematical definition.

Since Dembski's definition of difficulty provides no information of how to actually determine the number (or range of numbers) which are necessary parts of the definition, the latter does little more than assert that whatever has a larger difficulty is more difficult. Moreover, from these definitions it also follows that, because in Dembski's theory complexity is defined exactly like difficulty, whatever is more complex has a larger complexity. True! It is as true as assertions like whatever has a larger size is larger, whatever is bitter has a larger bitterness, or any other tautology.

The conclusion: Dembski's supposed definitions of complexity and difficulty do not meet requirements for real mathematical definitions and at best only introduce notations. They do not seem to serve any useful purpose.

Other Interpretations of Complexity

As mentioned in the preceding sections, probability theory, with all of its variations and competing concepts, seems to be much better substantiated than Dembski's quasi-rigorous exercise. There is an analogous situation in the case of complexity as well. Although complexity theory is not yet developed to the same extent as probability theory, it has to its credit a number of substantial achievements. In particular, a theory of complexity was developed in the sixties under the name of Algorithmic Theory of Probability/Randomness (ATP). Although some facets of this theory have not yet been fully clarified, it is powerful and largely consistent. This theory will also be briefly discussed in chapter 2. Dembski discusses this theory in

The Design Inference (pp. 167–69), but seems to ignore it when approaching complexity from his own standpoint.

Algorithmic Theory of Probability/Randomness defines complexity (often referred to as Kolmogorov complexity) in connection with the concept of randomness. This theory is often discussed in terms of strings of binary digits, but it is actually applicable to a wide variety of situations. In ATP complexity is assigned a number. Kolmogorov complexity depends on the degree of randomness of a system. The more random the system, the longer in a binary representation becomes the algorithm (or program) which describes the system. The fully random system cannot be described by an algorithm (or program) which is shorter than the system itself. In other words, a fully random system and the algorithm which describes it are almost identical. The less random a system is (i.e., the more its structure is determined by a rule), the shorter the algorithm (or program) describing that system can be. Consequently, ATP defines complexity of a system as the minimal size of an algorithm (or program) which can "describe" the system.

As can be seen, unlike Dembski's exercise in complexity, ATP does not relate that concept to the difficulty of solving a problem.

Regardless of his definitions, Dembski's assertion that complexity is equivalent to the difficulty of solving a problem, and that, in its turn, translates into a small probability of the event, seems rather dubious. At best, such a notion may be true only for certain classes of events.

An example of an event whose complexity indeed parallels the difficulty of solving a problem (which is given in Dembski's writings more than once, but was actually discussed before by Richard Dawkins),[26] is related to a safe combination lock. Such locks have a very large number of possible combinations of digits, of which only one constitutes the code that opens the lock. If the opening combination comprises, say, five two-digit numbers, the opening code can be viewed as quite complex. This complexity translates into a large difficulty of correctly guessing the combination and hence into a very small probability that an attempt to open the lock by randomly choosing some combination of five two-digit numbers will succeed.

However, there are many classes of events for which Dembski's scheme is not only inadequate, but in which the relation between complexity, difficulty, and probability is opposite to that assumed by Dembski. Let us review a few examples.

First, there are situations in which we are not interested at all in solving any problem but may be quite interested in estimating the complexity of a system. In such situations Dembski's definition of complexity (which equalizes it with difficulty of solving a problem) is irrelevant. Such situations are common in psychology, geography, economics, crystallography, and in many other areas of knowledge.

Assume that we wish to compare the *structures* of two objects which we have no intention or need of ever building. In such cases we are not interested in either the available resources for building these structures or in the difficulty in Dembski's sense. The complexity of those structures may be very much of concern, but in this case Dembski's definition is not at all helpful.

Here is a specific example of a situation in which the relation between complexity of a system and difficulty of a problem is opposite to Dembski's scheme. In research aimed at the development of certain types of photovoltaic devices, a need had arisen to deposit electrochemically a thin layer of metallic molybdenum on various substrates. A layer of molybdenum, if successfully deposited, usually contains no more than two different phases. However, because of electrochemical constraints, mainly connected with the low overpotential of hydrogen on a molybdenum surface, the electrochemical deposition of pure molybdenum turned out to be very difficult to achieve. The problem is removed, however, if instead of pure molybdenum, an alloy is deposited containing over 98 percent molybdenum, the rest being metallic nickel and also a small percentage of hydrogen. Such an alloy can be deposited under a rather wide range of conditions and its properties are reasonably close to those of pure molybdenum, thus solving the problem at hand. However, the alloy in question has a much more complex structure than a pure molybdenum layer has. It contains a conglomerate of various phases, such as NiMo, at least three phases of Ni, hydrides of both Ni and Mo, etc. In this case, the difficulty of solving the problem at hand in no way matches the complexity of the system.

Another example is from my experience as a mountain climber. This is a very clear example of complexity being not at all tantamount to difficulty as Dembski assumes. For example, a steep slope covered by smooth ice has a much simpler relief than a rocky slope of a very complex shape. However, the difficulty of climbing over the smooth and steep ice may substantially exceed the difficulty of climbing over a rocky slope (which may be actually quite easy if the rocks, as is often the case, provide multiple cracks and ledges that immensely facilitate the climber's ascent). Also, it is much harder to scale the steep and smooth face of rock whose relief is quite simple than a rock whose face has a complex relief with multiple irregularities. In that example, the complexity of the path to the summit is much larger for a path which is much easier to negotiate and thus to solve the problem at hand.

One more example can be taken from computer science. The concept of *complexity* in this field refers to the efficiency of a program, i.e., the number of computational steps the program needs in order to perform a certain task (i.e., to solve the problem at hand), given a certain amount of data. The most efficient program (or algorithm) is one that requires the minimal number of the steps. Such an algorithm is defined in computer science as having the minimal complexity. However, there is no parallelism between the complexity in the above sense and the complexity of the algorithm's structure. Often the computational algorithm that is the most efficient (i.e., has the minimal complexity in the sense of computer science) is much more complex in its structure than the less-efficient algorithms.

The situations in the above examples, while contradicting Dembski's scheme, are fully compatible with the concept of complexity in ATP.

If we turn now to Dembski's thesis that probability is just disguised complexity, it is easy to provide a multitude of examples to the contrary. I submit that a more common situation is one in which simplicity translates into a larger difficulty and therefore a smaller probability, which is contrary to Dembski's theory.

In chapter 2, an example of pebbles found on a beach will be discussed. In this example, I compare two samples, one a pebble of irregular shape and the other a perfectly spherical piece. Whereas the perfectly spherical piece can be described by a very simple program, the description of an irregularly shaped pebble requires a much more complex program. However, contrary to Dembski's scheme, the spherical piece must be reasonably attributed to design, while the pebble of irregular shape, to chance. Many similar examples can be suggested.

According to Dembski's scheme, complexity is tantamount to low probability, and, hence, points to design, whereas simplicity, which is the lack of complexity, must point to chance. The simple example with two stones shows that more often than not, the actual situation is opposite to Dembski's scheme: simplicity is often a sign of design, while complexity often points to chance.

Dembski's Treatment of Information

Before discussing the particular features of Dembski's treatment of information, let us make some preliminary comments of a general character.

There exists a well-developed science named "information theory." It started with the publications (in 1948) of a paper by U.S. communication engineer and mathematician Claude Shannon.[27] A seminal concept of the theory in question is *information*, which turned out to be very viable and has since become an almost household term.

In a certain sense, the name "information theory" seems to be due more to tradition than to its essence. Perhaps a more appropriate name for it would be "communication theory." (Indeed, Shannon's classical paper of 1948 was titled "A Mathematical Theory of Communication.")

In fact, what information theory studies is the communication process, viewed as the transmission of information, regardless of the presence or absence of a *meaningful message* in that information. This choice of the definition of information was justified for at least two reasons. First, the process of information's transmission is not affected by that information's semantic content. Second, the originators of the information theory did not possess a method for measuring the semantic content. Therefore, for the purposes of communication theory, which is the essence of information theory, the definition of information in that theory was not only adequate but also logical and convenient. However, it becomes inadequate if we wish to use the term *information* in connection with its semantic content.

Information is neither a substance nor a property of some substance. Essentially, information is a *measure* of a system's randomness, as we will discuss in more detail later. Therefore, the assertions that information can exist in some abstract way independent of a material medium or that it is conserved (like energy in physics) seem to be dubious propositions. Information can be unearthed, identified, sent, transmitted, or received, and generally handled in whichever way, only if it is recorded in the structure of a material medium (including electromagnetic waves). For example, this record can be a combination of microscopic magnetic vectors in the layer of some metal oxides deposited on the surface of a disk. These magnetization vectors, each representing the magnetization of a tiny element of the oxide layer, can be oriented in either of two directions, one direction corresponding to a recorded 0, and the opposite direction corresponding to a recorded 1. For all intents and purposes, the described string of zeros and ones represented by alternating directions of magnetization vectors is a *text*.

There are many other ways to record information, but it always requires a material medium in which a certain structure is created by the process of information's recording. Such text can be transmitted from one medium (*source*) through a communication *channel* to another medium (*receiver*). Since the methods of the information's recording (i.e., the physical, chemical, or biological processes utilized for the information's recording) may be different in the source and in the receiver, and since both are usually different from the process in the medium that constitutes a channel, the communication chains usually also include *encoders* and *decoders*. The latter elements of the chain convert the text from one form of recording to another. It must be pointed out that the encoding and/or decoding of information may completely change the appearance of the communicated message without changing the amount of information. For example, a text written, say, in Cyrillic letters can be transmitted in Morse alphabet, which is quite different from the Cyrillic; upon being received, it can be converted into, say, Roman characters. However, the information received is the same as the information sent, except for the unavoidable distortion caused by the *noise* in the channel. It can be said that information is an invariant of transliteration, which means that the set of symbols in which information is recorded can be replaced by another set of symbols without changing the amount of information in the text, provided both sets comprise the same number of available symbols. However, information is not an invariant of translation from one language to another. In such a translation, both the size of the alphabets and the lengths of both the text as a whole as well as the individual words may be different. Therefore, translation changes the amount of information even if the meaningful content of the message is preserved. Information has no relation either to the semantic contents of the message or to the particular appearance of the symbols used to record it.

It is essential for our discussion to note that the more *random* the transmitted text, the larger the amount of information it carries to the receiver.

In many books on information theory one can read direct statements to the effect that information is not a synonym for meaningful content, and that information theory does not provide the means to measure or even to reveal the presence of semantic content in information.

In a set of articles, Brendan McKay and I describe a method of "Letter Serial Correlation" (LSC) and its application to a variety of texts.[28] Meaningful texts display a very distinctive feature of LSC, testifying to special types of order in meaningful texts, common for various languages, styles, authors, etc., but absent in meaningless conglomerates of letters. LSC is a form of statistics. It is not a part of information theory per se (even though it is related to the latter in a certain sense). Unlike the classic information theory, LSC enables one to distinguish between meaningful texts, regardless of language, authorship, style, etc., on the one hand, and gibberish in various forms, on the other. These results show very vividly the principal difference between *information* as defined in information theory and a *meaningful message*. In particular, many meaningless combinations of letters carry much more information than meaningful texts.

In the following discussion I shall use the term *message* to always mean the meaningful contents of information. In some discussions of information, for example, by design theorists, the distinction between information (as it is defined in information theory) and the meaningful message which may or may not be carried by information is overlooked. This confusion is often a source of unsubstantiated assertions. In particular, if the mathematical apparatus of information theory is applied when actually discussing not information but rather a meaningful message, then meaningless conclusions often result.

One of the measures of information, according to information theory, is a quantity named entropy. To all intents and purposes, it behaves like its namesake in thermodynamics. The entropy of a text quantitatively characterizes the level of disorder in that text. The more information carried by a text, the larger the text's entropy. The total entropy of a text as a whole is proportional to the text's length and is therefore an extensive quantity. A more interesting quantity is the specific entropy, which is the entropy of a unit of text, and therefore is an intensive quantity. Usually it is expressed as entropy per character and measured in bits per character. In the following discussion, unless indicated otherwise, the term entropy will mean the specific entropy.

There exists a hierarchy of texts in regard to their entropy. For example, consider a string of the same letter (like *A*) repeated, say, a million times: *AAAAAAAAA* . . . etc. This meaningless text is perfectly ordered. The entropy of that text is practically zero. Now consider a text obtained, for example, by what we can call the *urn technique*. We place into an urn twenty-seven balls, twenty-six of them each bearing a letter of the alphabet, and the twenty-seventh one representing a space. We then extract at random a ball, write down the letter found on it, return the ball to the urn, shuffle the balls, randomly extract another ball, and so on. Let a text in an "urn language" be, say, a million letters long. This string is almost always gibberish. (There is some, extremely small probability that a string of an urn language happens to be a piece of a meaningful message.) If, as is overwhelmingly the case, this string is gibberish, in an overwhelming majority of situations there is no or very little order in that string. We call it a random string. The entropy of that meaningless random string is large, and so is the information carried by that string.

Meaningful texts are located somewhere in the middle of the entropy scale, their entropy being much larger than in perfectly ordered texts of very low entropy (like *AAAAAAAA* . . .) but much smaller than in meaningless random texts.

Here are some typical numbers. The entropy of a normal meaningful text in English (as was estimated already by Shannon) is about 1 bit per character. On the other hand, the entropy of a text written in an urn language, that is, the entropy of a randomized sequence of 27 symbols (26 letters plus space), may be as high as 4.76 bits per character. This means the entropy of a gibberish text which conveys no message may be almost five times larger than that of a meaningful English text of the same length.

For example, in English texts the letter *Q* is almost always followed by *U*. (There may be exceptions; for example, in a paper I published many years ago about some optical phenomena, the abbreviation *QE* was used, which stood for "Quantum Efficiency.") Therefore, if we find the letter *Q* in a meaningful English text, we are pretty confident that the following letter is *U*. Hence, when we indeed discover that the next letter is a *U*, it is not news. In other words, the letter *Q* itself provides as much information as the combination of the letters *QU*. In a random string of characters, however, the letter *Q* can be followed by any letter of the alphabet, including *Q* itself or a space. Hence, when we find which letter follows *Q* in a random text, it is news for us. This means that in a random string, the combination of two letters, the first being *Q* and the next one being whatever it happens to be, supplies more information than the letter *Q* alone. This reflects the fact that *redundancy* is larger for meaningful texts than it is for random strings. The smaller the redundancy, the larger the entropy and, hence, the larger the amount of information.

On the other hand, unlike information, Letter Serial Correlation characteristics of meaningful texts are quite different both from perfectly ordered texts of low entropy and low informational contents, and from random texts of high entropy and high informational contents.[29]

Let us now return to Dembski who discusses information in his books as a tool to support his intelligent design hypothesis.

Let us first discuss what Dembski calls law of conservation of information. (Recall that it is this "law" which Dembski admirer Rob Koons proclaimed to be a "revolutionary breakthrough.") Dembski indicates that the name of his proposed law was used previously by Peter Medawar, whose formulation of the law was, however, in Dembski's opinion "weaker" than Dembski's new version.[30] Dembski's definition of the law of conservation of information is as follows, from *Intelligent Design*: "Natural causes are incapable of generating CSI" (p. 170). The abbreviation CSI stands for "complex specified information." In the further discussion Dembski abbreviates the name of the quoted law to LCI.

Without excessive modesty, Dembski claims that his LCI "*has profound implications for science*" (my italics). A few lines further Dembski lists several "immediate corollaries," of which the first two are: "(1) The CSI in a closed system of natural causes remains constant or decreases, and (2) CSI cannot be generated spontaneously, originate endogenously or organize itself (as these terms are used in origin-of-life research)" (p. 170).

There seem to be several points Dembski left without clarification. Dembski does not provide any definition of a "closed system of natural causes." In particular, does Dembski include intelligent human agents into a "closed system of natural causes"? Human minds are usually viewed as "natural" but can very well generate "complex specified information," including meaningful texts. Hence, if human intelligent agents are to be included in the "closed system of natural causes," Dembski's proposed law seems to be wrong. If, though, Dembski's suggested law implies that the human mind is not natural, it would seem to be contrary to the common interpretation of the word "natural."

Likewise, we find no strict definition of the concept of CSI (complex specified information). The interpretation of that concept is of crucial importance for an analysis of Dembski's LCI. From many remarks scattered all over Dembski's writing it seems that, at least when Dembski discusses *texts*, he uses the expression CSI as a synonym for "meaningful content."

Here is an example illustrating Dembski's interpretation of meaningfulness of a message as a sign of specification: "A random inkblot is unspecified; a message written with ink on paper is specified. The exact message recorded may not be specified, but orthographic, syntactic, and semantic constraints will nonetheless specify it."[31] (Note that in the case of the Voynich manuscript neither orthographic, syntactic, nor semantic data are available, but the artifactual nature of the manuscript is obvious, which is contrary to Dembski's approach.) There are many other statements in Dembski's writing indicating that in his view specification in a case of a text is tantamount to the text's meaningfulness.

The word "text," however, may have a very broad meaning. DNA is a *text*, and so is a novel or a poem, or an infinite string of digits representing the value of *e*, the base of natural logarithms. For example, in the collection *Signs of Intelligence*, of which Dembski is a coeditor, there is a paper by Patrick H. Reardon titled "The World as Text: Science, Letters, and the Recovery of Meaning." In that paper, the word *text* is interpreted in a very broad sense. Since Dembski is a coeditor of that collection, he probably approves of Reardon's thesis, even if only in some general way.

If such an interpretation of the term CSI is indeed correct, then a question arises: what is the word "information" doing in LCI? And if the formulation of LCI includes the word "information," how do we reconcile this "law" with the seminal concepts of information theory?

If Dembski's law indeed is meant to be applicable to information, as its name implies, then it cannot be applied to a text's meaningful content. If, though, the law in question is supposed to be about the *meaningful message* conveyed by the text, then its name is an obvious misnomer.

Reading Dembski's treatment of information, including his alleged law, leaves the impression that when discussing CSI he does not notice how he inconsistently switches back and forth between the concepts of information in the sense of information theory and complex specified information (which actually means a meaningful message and hence is not information in the sense of information theory). The mathematical apparatus of information theory, which Dembski uses in *The Design Inference*, is not applicable to meaningful messages, i.e., to what he refers to as CSI. In my view, this makes Dembski's entire treatment of information largely off the mark.

By calling his assertion the "law of conservation of information," Dembski apparently wished to imbue it with the significance usually associated with laws of conservation, which are an important element of physics.

There are many laws of conservation in physics. All of them are, of course, postulates, but, unlike the "law" suggested by Dembski, the laws of physics are postulates based on the generalization of an immense amount of observational and experimental data.

Despite the name Dembski gave it, the law of conservation of information is rather different from the laws of conservation in physics. Since information is neither a substance nor property, no conservation of information can be asserted. Actually, the outward form of Dembski's suggested law is more similar to the second law of thermodynamics. The latter states that in a closed system a quantity named entropy either remains constant or increases. The term *closed system* has a well-defined meaning in thermodynamics. A system is closed if it does not exchange energy and matter with its surroundings. However, comparing Dembski's proposed LCI with the second law of thermodynamics only further undermines the former.

For example, any text, *left alone,* may be viewed as a closed system as long as no additional texts are added to it nor are any parts of it deleted. The information carried by an isolated text and measured by the text's entropy can only increase or remain constant. Contrary to Dembski's alleged law, entropy of an isolated text cannot spontaneously decrease.

As Dembski himself correctly points out, with time any text left alone can only deteriorate. Certain deterioration also accompanies transmission of a text to another medium. Imagine a manuscript written on paper, papyrus, or cloth, using, say, ink. With time, the material used for the text's recording deteriorates. Aging of ink and paper makes some letters change their shape so their reading becomes uncertain. For example, the original letter *Q* may become indistinguishable from the letter *O*, so while in the original text the following letter *U* was redundant, it is not redundant any longer in the deteriorated text. Hence, the deterioration of the recorded text caused the decrease of its redundancy, i.e., the increase of its entropy. What, then, happens to information associated with that text? Recall that the concept of information is tied to the degree of disorder in the text. Whereas it may seem paradoxical, information (per character) associated with the deteriorated texts actually *increases.*

A similar interpretation is applicable to information's transmission. Because of the unavoidable *noise* in the transmission channel, the meaningful *message,* if any was carried by the signal, enters the receiver partly distorted, i.e., some fraction of the meaningful message is lost. What, however, happens to information carried by the signal? It *increases* (if we mean information carried by a unit of text; the total amount of information can decrease if parts of the text are completely obliterated, thus decreasing the overall length of the text). The disorder in the transmitted message increases because of noise, so its redundancy drops, while its entropy, i.e., the information associated with a unit of that text, increases. This description is true only if the transmitted text is not perfectly random. The specific entropy of a perfectly random text has the maximum possible value—for example, 4.76 bits per character for a random conglomerate of 27 symbols—and cannot increase; if such a random text is transmitted, the noise does not increase its specific entropy any more, but it does not decrease it either. The *specific* entropy of an isolated *random* string is conserved.

When we account for the intrinsic connection between information and entropy, we can see that Dembski's alleged law of conservation of information contradicts the second law of thermodynamics. (An additional discussion of Dembski's alleged law of conservation of information, as it is rendered in his new book *No Free Lunch,* is offered later in this chapter.)

We usually have no problem identifying human design. However, in the case of a supposed supernatural design, our requirements for identifying design must necessarily be much more stringent. Imagine that we generate a random string of letters using the urn technique described earlier. If we create a long string of letters by using this technique, we will confidently attribute the creation of a random string of letters to human design, since we realize that humans had to make the balls and the urn, place the balls in the urn, pull the balls out one by one, and write down the letters; the random string could not be created spontaneously without these deliberate actions by humans. Indeed, the string of gibberish to be compared to Dawkins's example of the phrase "METHINKS IT IS LIKE A WEASEL"[32] was actually deliberately created by Behe (in his foreword to Dembski's *Intelligent Design*; see chapter 2) and by Dembski himself as an example. However, in order to attribute an event to supernatural design, we justifiably demand more—a string of symbols must be specified in a much stronger way to infer supernatural design. The reason for that is the simple fact that stochastic processes which occur without a human or a direct supernatural interference are capable of creating information-rich structures.

As we discussed earlier, Dembski seems to view the *meaningfulness* of a string of letters as an indication of design. To be consistent, he should apply this requirement to biological structures as well as he applies it to strings of letters.

In section 6.5 of *The Design Inference* Dembski estimates the so-called universal probability bound, which he suggests to be at $p_m = \frac{1}{2} \times 10^{150}$. According to Dembski, if the probability of an event, estimated assuming it happened by chance, is less than p_m, then its occurrence necessarily must be attributed to design. In *Intelligent Design*, Dembski again gives the same value for the "universal probability bound," and elaborates by indicating that "the probability bound of 10^{-150} translates into 500 bits of information" (p. 166). Accordingly, "specified information of complexity larger than 500 bits cannot reasonably be attributed to chance" (ibid.).

This is an example of Dembski's inconsistency. In the first of the quoted sentences he speaks about information. He is correct in stating that a probability of 10^{-150} translates into about 500 bits of information. In the next sentence he replaces the word "information" with "specified information," i.e., with a different concept, which actually seems to mean a *meaningful message*.

To be consistent, Dembski had to speak either about "information" or about "specified information" rather than surreptitiously replacing one with the other.

If we want to discuss information, then Dembski's assertion that in cases where its size is above 500 bits, information cannot be attributed to chance, is, in my view, wrong. Indeed, any random string containing, for example, more than 105 letters of English alphabet (26 different letters plus a space) carries over 500 bits of information. It can easily be obtained by chance. For example, recall the urn procedure, described earlier. Instead of pulling out the balls by hand, we can do it in a way utilized in the Keno game played in Las Vegas casinos. In those casinos, a machine is used which constantly shuffles the balls, randomly pushing out ball after ball, each ball bearing a certain number. If instead of numbers, such balls bear letters, the machine can generate a random text of any length. In such a text, the design by a human agent is only limited to creating conditions (designing a machine) which would effect a random procedure. As soon as the text has more than 105 randomly chosen letters, it carries over 500 bits of information, and every particular string obtained this way is the result of a chance procedure. Therefore, Dembski's universal probability bound, or its reincarnation in the universal complexity bound, is irrelevant as long as information per se is in question.

Specified information is a very different animal. This term, in Dembski's usage, seems to denote a meaningful message. The latter is not measured in bits, and cannot be at all treated using methods and mathematical apparatus of information theory. Any random text carries much more information (in the sense of information theory) than any meaningful text of the same length. To judge the probability of a chance emergence of a meaningful text by applying information theory is like measuring the sweetness of ice cream by the brightness of the colors on its label.

If we continue the mechanical procedure of randomly pushing letter-bearing balls out of an urn, the recorded string of letters will become more and more complex. If the machine kicks out balls, say, a little more than 1,000 times, the information in the ensuing random string will soon exceed 5,000 bits, which translates into probability of 10^{-1431}. This value is immensely lower than Dembski's universal probability bound of 10^{-150}.

Again, let us agree that if a not-so-short text carries a message, this points, with overwhelming probability, to an intelligent author (although there is always a very small, though nonzero, probability that the message is an accidental outcome of random events). Information, on the other hand, is indifferent to the message and therefore cannot itself point to an intelligent source. In particular, DNA obviously carries *information* but there is no way to assert that it carries a *message*. Indeed, how can Dembski determine whether or not the information in DNA is *specified*? It is just an unsubstantiated assumption. The available data about the structure of DNA indicate that the DNA strand consists of both pieces which carry a genetic code and segments which do not seem to carry any genetic code.

Here is a quotation from an article by the prominent biologist Kenneth R. Miller (who, by the way, is a Christian believer):

> In fact, the human genome is littered with pseudogenes, gene fragments, "orphaned" genes, "junk" DNA, and so many repeated copies of pointless DNA sequences that it cannot be attributed to anything that resembles intelligent design. If the DNA of a human being or any other organism resembled a carefully constructed computer program, with neatly arranged and logically structured modules, each written to fulfill a specific function, the evidence of intelligent design would be overwhelming. In fact, the genome resembles nothing so much as a hodgepodge of borrowed, copied, mutated, and discarded sequences and commands that has been cobbled together by millions of years of trial and error against the relentless test of survival. It works, and it works brilliantly; not because of intelligent design, but because of the great blind power of natural selection to innovate, to test, and to discard what fails in favor of what succeeds. The organisms that remain alive today, ourselves included, are evolution's great successes.[33]

Let us compare a DNA strand to a string of letters. If such a string is not very short and is a meaningful text (for example, a poem or a novel), the probability of its being the result of chance is indeed exceedingly small. Let us now recall the example discussed both by Dembski and Behe. In that example, two strings of Scrabble letters are compared, one being a piece of gibberish and the other a phrase from *Hamlet*. As Dembski and Behe asserted, the string of gibberish must be attributed to chance because it is not specified (i.e., it is not a recognizable meaningful text). Let us now look at the following string of letters:

> prsdembkreddnpljassddskipqooppppazxkhmainwcloyyrfhlklktain-syuuklscvmwwthatooedflllmqdcompjertfffvaqpurclexitystolm-jdgesetgbdkoqpmzfyhntogetsqoprthjncdeherabpuuerthhhwither-imnaderlthhjkkspecifherrvuiwplkxqcghkricationiiieklodsg . . .

Isn't the above string gibberish? Obviously it is (unless it is meaningful in a language unknown to us, which still does not make it recognizable, i.e., detachable). Hence, according to the criteria suggested by Dembski and Behe, its creation must be attributed to chance. However, if one carefully reads that string letter by letter, one can discover in it, within the gibberish, meaningful segments, which read:

> Dem . . . bski . . . main . . . tains . . . that . . . comp . . . lexity . . . toget . . . her . . . with . . . specif . . . ication . . .

Even if we apply Dembski's criterion (which bases the identification of design in a text on the latter's meaningfulness), discovering the islands of a meaningful text within the above meaningless string will hardly lead us to the conclusion that the string was deliberately designed. A more plausible interpretation seems to be that the meaningful segments happen to occur within gibberish by chance alone. (Of course, I have deliberately created that string in order to provide an illustration of my thesis. However, Behe and Dembski have created their strings of gibberish the same way. Their example implied that the strings of letters were found by accident and their origin was unknown. I have simply followed their way of discussion.) The farthest we may go is to infer that somebody took time to use, say, an urn technique or some alternative method of creating this string; only in this sense can we say the string was designed. However, we will have little reason to see in that string a meaningful message rather than gibberish. The longer the pieces of gibberish and the larger their number in the string, the more it will look like a random string which contains segments of a seemingly meaningful text by sheer chance. As biologists tell us, DNA looks much more like the latter example than like the text of a poem or a novel. Moreover, in the above string the seemingly meaningful segments are readily recognizable as those of English text (in Dembski's terms, their patterns are "detachable"), whereas in a DNA strand even the segments that serve as genes are revealed to be such only via special investigation and are not immediately recognizable, i.e., not "detachable," and hence not really "specified" in Dembski's terms. Moreover, strings of letters are incapable of reproduction on their own; hence, unlike biological structures, they could not have evolved from some other, simpler strings.

Of course, as any analogy, the above example of gibberish with accidental chunks of meaningful English words is not really a representation of a DNA strand. The latter has many features which are absent in the above text. This example only illustrates my thesis and is not intended to serve as a proof of that thesis.

However, I believe that the biological structures seem to better conform to the hypothesis that they emerged as a result of random events than as a result of intelligent design (of course, according to the Darwinian theories, the process of the emergence of biological strands such as DNA has also included nonrandom steps such as natural selection which is not random). The assertion that DNA's complexity is specified and therefore points to design, has so far no foundation in known facts.

Some creationists, both among those openly admitting being creationists and among those protesting against such a label, adhere to a preposterous notion that "chance cannot create information."[34] Since such biological structures as DNA, RNA, or proteins all carry a lot of information, these creationists argue that this very fact testifies that life could not develop without an "intelligent agent" (which usually is just a code word for the biblical God).

Unlike some of his colleagues, Dembski admits that chance *can* create information. Moreover, he even admits that chance can create both complex information and specified information. What chance cannot create, claims Dembski, is information which is both complex and specified. In this claim, Dembski implicitly performs a sleight-of-hand. While saying "complex specified information," he actually seems to mean a meaningful message. Yes, the probability of a meaningful message emerging by chance is extremely small, but this by no means applies to information, however complex and specified the latter may be. Plenty of information can be (and routinely is) generated in stochastic processes. In some sense, such information may be specified and complex, even if it carries no meaningful message.

The enormous amount of information in biological structures does not in itself contradict the hypothesis of a spontaneous emergence of life. To prove otherwise, design theorists would need a different type of argument. So far their theory is little more than an arbitrary hypothesis. Dembski's treatment of information does not, in my view, add anything of substance to the dispute in question.

DEMBSKI'S DESIGN INFERENCE

The main conclusion of Dembski's entire discourse is what he calls "the design inference." This concept has been rendered by Dembski in several versions, both in plain words and in a mathematically symbolic form. Let us look once again at the design inference according to Dembski (in *The Design Inference*):

Premise 1: *E* has occurred.
Premise 2: *E* is specified.
Premise 3: If *E* is due to chance, then *E* has small probability.
Premise 4: Specified events of small probability do not occur by chance.
Premise 5: *E* is not due to regularity.
Premise 6: *E* is due either to a regularity, chance, or design.
Conclusion: *E* is due to design. (p. 48)

A crucial element of the above argument is premise 4, which constitutes what Dembski calls the law of small probability: "Specified events of small probability do not occur by chance."

Without that "law" the entire argument would collapse.

I submit that the law in question is logically deficient. In premise 4, two concepts—specification and small probability—are presented as two independent categories. However, I believe that if an event is judged as having low probability, specification is just one of the factors which contributed to the low estimate of probability rather than an independent factor.

For the sake of further discussion, I will adopt a broader definition of specification, not as a substitute for Dembski's definition when design inference is at stake, but only to identify a certain feature of specification which is relevant to my discussion. My provisional broader definition of specification is based on the common interpretation of that term. I suggest that *selecting* for consideration a particular pattern or event out of the multitude of possible patterns or events makes it specified, and this, in my view, is the most all-embracing definition of specification. Dembski's definition is narrower than the one I offer, but obviously all patterns/events which meet his definition also meet my definition, although the opposite is not true. All events/patterns that meet Dembski's definition constitute a subset of a larger set which comprises all events or patterns that meet my broader definition.

Let us denote specification in a broader sense, according to my definition, as b-specification, and specification according to Dembski as d-specification.

I submit that one of the properties inherent in all events/patterns that meet my broader definition is that b-specification necessarily decreases the estimate of an event's probability. As an example, consider a game in which there are two players. Player *A* has in his pocket a one-dollar bill and player *B* has to guess what the number on that bill is. Let us say the number is L14142983Q. It is specified in that it is a specific number which is unequivocally distinguished from any other number, constitutes a recognizable pattern according to Dembski's criterion, and therefore meets his definition of specification. Of course in this particular case, it meets my broader definition as well. Obviously the probability of player *B* correctly guessing the number in question is very small.

Now, imagine that we have to deal with "fabrication" according to Dembski's concept. It means that player *B* does not guess the number in question before seeing the bill, but first looks at the bill and then announces the number on that bill. Obviously, now the event, which is fabricated rather than specified, has the probability of 100 percent. What made the probability of this event in the first case much smaller was its specification.

The above discussion does not imply that all specified events have low probability in absolute terms. What is viewed as a small probability in a certain situation may be viewed as a not very small probability in some other situation. Many specified events can have a *relatively* large probability. For example, consider a game in which players guess which card has been pulled at random from a deck of fifty-two cards, before the card is actually seen. Since the deck comprises fifty-two cards, the probability that the chosen card will happen to be, say, seven of spades, is 1/52, which is not really small. "Seven of spades" constitutes specification. If the card has not been specified, the probability that the chosen card will happen to be one of the fifty-two possibilities is 100 percent. The probability of choosing at random a *specified* card, while not very low (1/52) is *fifty-two times less* than the case in which the card was not specified. Specification makes probability *relatively small,* as compared with the absence of specification.

Let us take one more look at the example given by Behe in his foreword to Dembski's *Intelligent Design*. Recall that in this example, we find a string of Scrabble letters, spelling a meaningful English phrase in one case and a meaningless sequence in the other. The probability of the appearance of either of these sequences by chance is, according to Dembski, equally low. However, in Dembski's interpretation, the first sequence is specified while the second is not. According to my broader definition of specification, as soon as any of these two sequences has been *chosen*, it has been *specified*, regardless of its being meaningful or meaningless. To see why the latter approach seems to be reasonable, imagine that we obtain the meaningless sequence using the urn technique, continuing the procedure until we have a sequence of N letters written down. If $N > 105$, the probability of a particular sequence to appear is $p < 10^{-150}$. However, the probability that some *unspecified* sequence will appear as a result of the described procedure is of course 100 percent ($p = 1$). Hence, if we b-specify the sequence, this drastically decreases the estimate of probability. It does not matter whether the b-specified sequence is meaningful or not.

Specification according to my definition (b-specification) is not equivalent to meaningfulness.

The difference between the two sequences in Behe's example is not that one is specified while the other is not. Both have been b-specified by the action of Behe who chose them for his example (actually he had deliberately created the meaningless sequence as an example). The probability of the two sequences emerging by chance was equally small, the small probability being due to the b-specification—the choice of those specific sequences for consideration. An unspecified sequence is that which has not been specifically chosen and, hence, is undefined except for being some element of a certain class of objects (like a stack of cards).

If an objection is offered maintaining that the meaningless string does not meet Dembski's definition of specification, it does not matter for our discussion whether or not such an objection is correct. The assertion that specification of a pattern decreases the probability of an event remains valid because any pattern that meets Dembski's condition of specification belongs to a subset of a wider set meeting my definition. Any specification, whether or not it meets Dembski's definition, decreases the probability of the event in question.

A similar situation exists in the case of a raffle. If there are, say, ten million tickets sold, the probability of a *specified* ticket winning is 1/10,000,000. However, the probability of some *unspecified* ticket winning is 100 percent ($p = 1$). What made the probability of a particular ticket winning so much smaller was b-specification—the choice of a particular ticket. Obviously, it has nothing to do with any meaningfulness of the choice.

Therefore the law of small probability in Dembski's rendition, which regards probability and specification as two independent categories, is, in my view, logically flawed.

Dembski's effort to define the features of legitimate specification, such as detachability (and its constituents—conditional independence of side information and tractability) and delimitation, are unnecessary complications which actually seem to be aimed at distinguishing meaningful specified structures from meaningless ones (although Dembski nowhere directly mentions that distinction). However, from the above examples it seems that his convoluted discourse did not achieve its goal. The difference between a meaningful message and a simply information-rich structure remained elusive no less than before Dembski's exercise. (The modern development of information theory, in particular that based on the concepts of the algorithmic theory of probability/complexity, features some substantial advances toward the distinction between noise and "meaningful information."[35] This development goes beyond Dembski's discourse.)

From another angle, if we recall the procedure prescribed by Dembski for the third node of his explanatory filter, we see that the probability of the tested event is estimated in that node assuming that the event occurred by chance. If that is the case, then it seems to me that instead of saying, "Specified events *of low probability* do not occur by chance," Dembski should have said, "Specified events, whose probability *estimated upon the assumption that they occurred by chance turns out to be low,* do not occur by chance."

If the estimation of probability is made assuming chance, and the specification is expected to be found, then the adopted assumption predetermines relatively low probability. We estimate the probability to be low by first assuming that it is relatively low.

By mentioning specified events Dembski actually has already chosen events whose probability is relatively low because of specification, so the phrase "specified event of low probability" really means "events of relatively low probability which have low probability." The formula of Dembski's law seems therefore to actually mean: "Events of relatively low probability whose probability is low do not occur by chance." This is an example of Dembski's "rigorous" logic.

I submit that Dembski's law of small probability makes little sense in that it does not really shed light on the problem of identifying design. No mathematical symbolism, with all of its sophisticated appearance, can save the law in question from being a platitude.

Let us now review the entire design inference offered by Dembski in *The Design Inference*. Look at it once again, this time in its particular form:

Premise 1: LIFE has occurred.
Premise 2: LIFE is specified.
Premise 3: If LIFE is due to chance, then LIFE has small probability.
Premise 4: Specified events of small probability do not occur by chance.

Premise 5: LIFE is not due to regularity.
Premise 6: LIFE is due either to a regularity, chance, or design.
Conclusion: LIFE is due to design. (p. 56)

I have some serious doubts about the validity of some of those premises.

Premise 1 meets no objection since it is simply a statement of fact. However, the rest of the premises sound dubious in some respects.

Start with premise 2—"LIFE is specified." The question here is, What is the meaning of the word "specified" as applied to LIFE? Specification is the choice of a certain event or object from among a number of alternatives, whatever definition of specification one adheres to, including Dembski's own. If he means, for example, specification on a biochemical (molecular) level, then, if we define specification simply as stemming from the rich informational contents of biological structures, premise 2 may be provisionally accepted. However, as I have already discussed, rich informational content is not at all equivalent to a meaningful message. Recall that Dembski views meaningfulness as the feature which defines a string of letters as having resulted from design. To be consistent, he must require at least as much from a biological structure in order to attribute it to intelligent design, even if we may not be able to decipher the meaning stemming from a supernatural intelligence.

Premise 3—if LIFE is due to chance, then LIFE has small probability—seems to be a hypothesis whose validity is not at all clear. First, regardless of the question of LIFE's origin, chance events may have a large probability. If we toss a coin, the probability of tails is 1/2, which is by no means a small probability. The outcome of tails is, though, obviously a chance event. Assigning a small probability to an event simply because it has occurred by chance, as this premise seems to be formulated, is a dubious proposition. Regarding this premise in its specific application to LIFE, it is a hypothesis which may be correct but may be also wrong. There are theories based on the opposite premise, allowing for the possibility that the spontaneous emergence of life had a rather large probability.

Premise 4 is the law of small probability, whose logical deficiency has already been discussed.

Premise 5—LIFE is not due to regularity—is a hypothesis, which may be true but has not been proven. This premise actually is in some sense equivalent to premise 3—if certain regularity was responsible for the emergence of LIFE, this is tantamount to the assertion that the emergence of LIFE was rather likely. The possibility of life having emerged due to a regularity, as some scientists believe may be the case, has not been categorically excluded by any uncontroversial arguments (see, for example, the article by John F. McGowan which offers a plausible mechanism for the spontaneous emergence of life wherein the probability is immensely larger than that for a pure chance and which can possibly be explored under laboratory conditions.)[36]

Premise 6—LIFE can be attributed only to any one of the three possibilities—is also a hypothesis. The origin of LIFE as a result of a superimposition of more than one of the three listed causal factors cannot be excluded.

Hence, the conclusion in the above argument, that LIFE is due to design, is, in my view, based on one statement of fact, three unproved hypotheses, one more or less plausible premise, and one logically deficient statement. Of course, the weakness of the above seven-step argument does not mean that its conclusion is necessarily wrong. It may be true, but it requires a much more consistent and unequivocal set of arguments. As the matter stands now, Dembski's argument, in my view, is far from convincing.

A FREE LUNCH IN A MOUSETRAP

A substantial portion of Dembski's latest book, *No Free Lunch*, reiterates—often verbatim—Dembski's earlier publications. On the other hand, there are also some new elements in this book. Unfortunately, these new elements are mostly characterized by the same penchant for using self-coined terms, making pretentious claims of important insights or discoveries without proper substantiation, and subordinating the discourse to preconceived beliefs in a way that is all too obvious.

I will discuss here only a few selected points from this book. Detailed discussions of other parts, in particular of the author's misuse of the No Free Lunch theorems, as well as his "displacement problem" and what he labels Z-factors, are beyond the scope of this book.[37]

Dembski devotes many pages in *No Free Lunch* to the defense of Behe's concept of irreducible complexity.[38] This is easy to understand: if Behe's thesis, which is allegedly supported by biochemical data, is refuted (as, in my view, it should be), then the design creationists are left with very little that would constitute even a semblance of a genuine scientific discourse.

Although I will review Behe's arguments in chapter 2, I will not devote a substantial space to the critique of Behe's favorite example—that of a mousetrap, which he presents as an example of irreducible complexity and therefore as a model of a biological cell from the viewpoint of that concept. It seems to be a secondary point of his thesis; thus proving the inadequacy of this particular example does not seem as important as unearthing the principal flaws in Behe's concept. I will, though, spend some time on this model here as well as in chapter 12, where I will discuss the question of models in science in general.

Behe's model—his mousetrap—was debunked in a spectacular way by professor of biology John H. McDonald.[39] McDonald demonstrates how Behe's five-part mousetrap can be gradually reduced to a four-part, three-part, two-part, and finally one-part contraption, each preserving the ability to catch mice, albeit not as well as the five-part construction. Therefore, the mousetrap does not seem to be irreducibly complex, which is contrary to Behe's assertion.

Of course, demonstrating the weakness of Behe's particular example does not in itself disprove his thesis. It only shows that he happened to suggest a bad model. Dembski, however, devoted many pages of his new book not only to defending Behe's overall concept, but also to an attempt to overturn McDonald's particular debunking of Behe's mousetrap model.

In *Darwin's Black Box: The Biochemical Challenge to Evolution*, Behe writes that an irreducibly complex system is "a single system composed of several interacting parts that contribute to the basic function, and where the removal of any one of the parts causes the system to cease functioning" (p. 76). He offers several variations of this definition, but none of the others add any essential elements.

Confronted with McDonald's counterexample, Dembski tries to show that McDonald does not prove the inadequacy of Behe's model. When McDonald removes this or that part of the mousetrap, he modifies the shape of the remaining parts, thus preserving the trap's functionality albeit diminishing the trap's quality. The process can be reversed. One can start with the simplest one-part trap, and then gradually add more parts, improving its ability to catch mice at every step of the procedure. In Dembski's view, the modifications of the parts' shape at each stage of McDonald's procedure makes his example irrelevant to Behe's thesis. Behe, though, did not mention the additional condition introduced by Dembski—that the removal of the trap's parts must not be accompanied by any modification of the shape of the remaining parts. Therefore McDonald was not constrained by such a condition at the time he first suggested his counterexample. Furthermore, and this is a more important point, why shouldn't the shape of parts be modified in the "reduced" versions of the trap? McDonald's example satisfied all Behe's originally formulated conditions—it showed that, contrary to Behe's position, parts of the trap can be removed and the remaining contraption can be made to preserve functionality (though at a lower level of fitness). This debunked Behe's assertion as it was originally formulated by Behe.

Is the requirement added by Dembski—that the parts remaining after the removal of a certain other part retain their shape—indeed relevant to the discrimination between irreducible and reducible complexities? Remember that Behe's example of the alleged irreducible complexity of a mousetrap was suggested as an illustration of his concept of irreducible complexity as applied to biological systems. The Darwinian theory of evolution includes as an inseparable part the concept of *gradual* changes resulting in the *slow* accumulation of features advantageous for organisms. The evolution of a mousetrap in McDonald's scheme from a one-part to a five-part contraption comprises steps which are not really small. Actually, each step of McDonald's scheme may be viewed as the sum of many smaller steps wherein the parts gradually change their shape and at a certain stage of the evolution another part is added. At a certain step of that evolutionary process, some of the already existing parts of the system may again change their shape, in particular make it simpler, if such a simplification of shape is not detrimental to the organism's fitness. The mousetrap's parts, if we want to use the trap as a model of biological evolution, can modify their shape at every step of the evolution, preserving the trap's ability to work at a higher level of fitness. Though McDonald's scheme shows only four steps in the path from a one-part to the five-part contraption, actually that path may have comprised many more intermediate steps not shown.[40]

The modification of the parts' shape at every step from a one-part to the five-part contraption in no way diminishes the illustrative power of McDonald's example, which defeats Behe's use of a mousetrap model. Moreover, if the requirement of unchangeable shape of all parts of the mousetrap were included in Behe's original formulation, McDonald, as well as anybody else with some engineering experience, could have shown single-part, two-part, three-part, four-part, and finally five-part mousetraps wherein the shapes of the original parts would not change at all in any of the steps building up to the final five-part contraption. This would result in a five-part trap whose parts would have shapes different from those shown in Behe's picture, but such a five-part trap would be as good as that by

Behe, even if its parts would not have the simplest shape possible. In living organisms, such unnecessarily complex shapes of organs are common, testifying to their evolutionary history.

McDonald states that his example does not represent the actual process of biological evolution. In *No Free Lunch* Dembski grasps at that statement, using it to insist that McDonald's example does not prove that Behe's position is wrong or that Behe's mousetrap model is bad. He says, "The problem is that his progression of mousetraps has little connection to biological reality" (p. 266). You can say that again, Dr. Dembski. What you pretend not to notice is that the reason the progression of mousetraps does not represent biological reality is simply that Behe's example has very little to do with biological reality.

The difference between McDonald and Behe is that the former realizes and freely admits that his scheme does not adequately represent biological evolution, whereas Behe stubbornly tries to unduly use his model as an illustration of biological reality. What McDonald's scheme does very well is show the lack of substantiation in Behe's statement about the irreducible complexity of a mousetrap.

Dembski Salvages Irreducible Complexity

In *No Free Lunch*, Dembski says, "I am not a fan of notation-heavy prose and avoid it whenever possible" (p. xvii). However, leafing through his new book reveals that the quoted statement is contrary to facts. Like his preceding book, this new one is chock-full of mathematical symbols, more often than not adding nothing of substance to his discourse.

A typical example is found in a section titled "The Logic of Invariants" (pp. 271–79). Here Dembski resorts to his favorite method of discussion—presenting a convoluted chain of arguments in a heavily symbolic form, incomprehensible to a general reader. All this mathematical discourse is largely irrelevant to the concept of irreducible complexity.

I suggest that an average reader try to decipher the following passage: ". . . define a function **Invar** on Ω (for definiteness assume **Invar** is real-valued, i.e., **Invar** takes values in the real numbers R). Let $A = \{r \in R \mid$ there exists some natural number n and some X in **Init** such that **Invar** $(\varphi^n)(X) = r\}$ and $B = \{r \in R \mid$ there exists some Y in **Term** such that **Invar** $(Y) = r\}$ . . ." (p. 274).

The quoted passage is just a fraction of a much longer exercise in a heavily symbolic discussion. If it is an example of Dembski's attempts to avoid "notation-heavy prose," I wonder what he considers to be really notation-heavy prose.

The sole purpose of this mathematically convoluted discourse seems to be providing a definition of an invariant.

However, those readers who have enough experience with mathematics to have comprehended Dembski's notation-heavy paragraphs certainly know what an invariant is and need no explanation. Those readers who don't know what an invariant is presumably are also not able to digest the mathematical exercise on page 274, which therefore does not seem to serve a useful purpose.

In any case, the definition of an invariant could be made in one simple sentence. For example, for the purpose of Dembski's discourse it would be sufficient to say that an invariant of a certain procedure is a quantity whose value does not change in that procedure (e.g., entropy is an invariant of a reversible adiabatic process).

After having devoted considerable space and effort to introduce his mathematically loaded definition of an invariant, Dembski makes no use of that definition afterward. Instead, he tries to apply the concept of an invariant as a tool for what he calls *proscriptive generalization.* Proscriptive generalization simply means that certain processes or events are claimed to be impossible based on general considerations rather than on a detailed analysis of the factors preventing the occurrence of these processes or events. What Dembski asserts can essentially be spelled out as the following simple statement: if it were found that in a certain process a quantity which is an invariant of that process would actually change, then such a process must be considered impossible.

When Dembski turns to his defense of Behe's concept of irreducible complexity in *No Free Lunch*, all he says in regard to his preceding lengthy mathematical exercise is that irreducible complexity is "an invariant for the Darwinian process of random variation and natural selection" (p. 279). To make such a statement there was no need for all of the preceding mathematical-looking exercise. More important, though, is that the above statement, which purports to reflect Behe's position, cannot be taken for granted. It requires proof and none has been provided by either Behe or Dembski.

Apparently aware of the objections to Behe's concept from many professional biologists, Dembski admits that Behe's concept is not faultless. He writes, "Behe's idea of irreducible complexity is neither exactly correct nor wrong. . . . Instead it is salvageable" (p. 280). Contrary to his statement, Dembski actually tries to prove that Behe's idea is indeed correct, and asserts that this must become clear if only Behe's original definition of irreducible complexity is slightly fixed. Hence, in order to salvage the concept of irreducible complexity, Dembski, who has so far never admitted a single error or imprecision in his own writings, is even prepared to sacrifice to a certain extent the sterling reputation of his cohort Behe.

Having quoted Behe's definition, Dembski then proceeds to repair it in five consecutive steps. Here is Behe's original definition quoted by Dembski:

> Definition $\mathbf{IC}_{\mathbf{init}}$—A system is *irreducibly complex* if it is "composed of several well-matched, interacting parts that contribute to the basic function, wherein the removal of any one of the parts causes the system to effectively cease functioning." (p. 280)

And here is the final (salvaged) definition of irreducible complexity suggested by Dembski as the result of his five-step salvaging effort:

> Definition $\mathbf{IC}_{\mathbf{final}}$—A system performing a given basic function is *irreducibly complex* if it includes a set of well-matched, mutually interacting parts such that each part in the set is indispensable to maintaining the system's basic, and therefore original, function. The set of these indispensable parts is known as the *irreducible core* of the system. (p. 285)

The comparison of Behe's original definition with the "salvaged" definition by Dembski boils down to, first, the introduction of the concept of "irreducible core" of the system, and second, to the assertion that a system of reduced complexity must retain the "basic (and therefore original)" function.

The "salvaged" definition does not seem to add anything of substance to Behe's original concept, which, according to Dembski, is "neither exactly correct nor wrong." The question of whether all parts of the system are necessary for its original functionality or only those included in an *irreducible core* is of no significance. The boundaries of a system may be set arbitrarily. The irreducible core may as well be viewed as the whole system under consideration. The only purpose of Dembski's modification seems to enable him to reject an argument that points to a system which retains its functionality after the removal of some of its parts by simply asserting that either the removed part did not belong to the "irreducible core," or that the functionality of the reduced system is not equivalent to the "basic (i.e., original)" one.

Dembski's salvaging argument does not really save Behe's concept because it has nothing to do with the actual critique of that concept.

Behe has never proved that even a single protein system he describes was indeed irreducibly complex according to his definition. Dembski's argument that the reduced system must perform exactly the same "basic" function is an arbitrary requirement. Biologists tell us that in the course of evolution many systems changed their functionality. A biochemical system which in modern organisms clots blood, in some of its preceding, simpler forms could have performed a different function or even no function at all, acquiring its ability to do this or that job only at a certain stage of evolution.

The Explanatory Filter Revisited

As I argued in the previous sections of this chapter, the concept of specification, which was presented by Dembski in a heavily symbolic form, actually could be rather simply defined as "a subjectively recognizable pattern." This simple definition follows from all those examples Dembski provided in his previous publications. Dembski, however, chose to cloak this concept in a convoluted mathematical mantle.

To infer design according to Dembski, an event, besides being improbable on a chance hypothesis (the latter not being specified), must also display "specified complexity" (or specification for short).

Specification, according to Dembski's previous renditions of his theory, in turn comprises two *necessary* components, one called *detachability* and the other *delimitation*. Detachability, in turn, *necessarily* comprises two subcomponents, one named *conditional independence* of the background knowledge and the other *tractability*.

This multicomponent scheme has been criticized from various viewpoints by reviewers of Dembski's publications. Apparently the critique has had some effect after all, since in *No Free Lunch* Dembski suggests a discussion of specification rather different from his previous opuses. He introduces two alterations of his earlier discourse.

In the new rendition found in *No Free Lunch*, tractability (denoted TRACT in his work) is no longer a constituent of detachability. Dembski says: ". . . I have retained the conditional independence but removed the tractability condition" (p. 66). In *The Design Inference,* Dembski spent considerable effort justifying the inclusion of tractability into his concept of detachability, using both plain words and mathematical symbolism plus examples illustrating the importance of that alleged insight into the design inference. Although in *No Free Lunch* Dembski does not explicitly admit any faults of his earlier argument, he actually does it implicitly, by removing from the detachability concept the tractability component. Dembski says now that tractability is not really necessary within detachability but rather should be moved to his Generic Chance Elimination Argument (GCEA).

The question of whether tractability is indeed a useful part of the GCEA or is as useless there as it is in detachability is a separate issue. The fact is that Dembski implicitly admits his previous error but is reluctant to say this directly.

Furthermore, there is one more alteration of Dembski's earlier discussion of specification. The delimitation condition, which was prominently discussed in *The Design Inference*, is not used any longer in *No Free Lunch.*

Obviously, if that term is not mentioned any longer when specification is being discussed, it was not really a necessary component of specification, which negates Dembski's convoluted discussion of the delimination condition in his previous book.

As I argued previously, one of the main faults of Dembski's scheme is his attributing to specification the status of a kind of magic. There is, though, nothing magical in that concept. In many examples of false negatives, specification was not discerned, but the event was obviously designed. In many examples of false positives, the specification seemed to be present but the event was due to chance.

Specification, as follows from all Dembski's examples, is nothing more than a *subjectively recognized pattern.* It can be illusory or real, but it has no exclusive status among many factors pointing either to design or to chance. Recognition of a pattern (which necessarily is subjective) affects the estimate of the event's probability, but so do many other factors.

Dembski Suggests a Fourth Law of Thermodynamics

One of the vivid examples of Dembski's propensity to make extraordinary claims is his announcement of a possible discovery of an additional law of thermodynamics. In *No Free Lunch* Dembski writes: "The traditional three laws of thermodynamics are each proscriptive generalizations, that is they each make an assertion about what cannot happen to a physical system" (p. 169). Leaving aside the gist of that statement (which can be disputed), we cannot fail to notice that Dembski seems to have forgotten certain simple facts from the introductory course of thermodynamics: there are not *three* but *four* "traditional" laws of thermodynamics.[41] By a peculiar historical twist they were named the zeroth, the first, the second, and the third laws, so, although there are four of them, none is named the "fourth law."

Why are there four laws of thermodynamics but not three or perhaps two? The reason that the four laws of thermodynamics cannot be reduced to three, two, or one is that these laws are not derivable from each other. The second law is not a consequence of the zeroth law or of the first law, and the first law does not entail the second or the third law, etc. If a new law of thermodynamics is to be discovered, it necessarily must be independent of the four accepted laws.

Of course, another requirement for a supposedly new law of thermodynamics is that it must not contradict any of the four laws of thermodynamics already accepted in that science.

I intend to show that the fourth law of thermodynamics suggested by Dembski fails on both accounts. First, it covers phenomena which have already been covered by the second law of thermodynamics and therefore, even if it were correct in itself, would not constitute a new law. However, the situation with Dembski's alleged new law of thermodynamics is worse because it actually contradicts the second law of thermodynamics.

Let me start with the first point. The fourth law of thermodynamics suggested by Dembski is a generalization of what he calls the law of conservation of information (LCI for short).

In the previous sections of this chapter I pointed to the flaws which, in my view, are present in Dembski's discourse related to information. Dembski's treatment of information was also subjected to critique by several other authors, including Victor Stenger and Matt Young.[42]

In *No Free Lunch* Dembski offers the following definition of *information* $\boldsymbol{I}$ associated with an individual event A:

$$\boldsymbol{I}(A) = -\log_2 P(A), \qquad (1.1)$$

"where $P(A)$ is the probability of event A" (p. 140).

Formula 1.1 as such was not given in Shannon's classical work.[43] However, this formula can be derived from Shannon's formula, which defines information as a change of entropy if we formally apply the latter to an individual event. Formula 1.1 is not peculiar to Dembski's discourse and can be found in textbooks and even in encyclopedias. For example, in *Principles and Practice of Information Theory* by Richard E. Blahut, we find the same formula for information.[44] Likewise, an article on information theory by

Professor George R. Cooper in *Van Nostrand Scientific Encyclopedia* (1976 edition) also contains the same expression as a definition of information of an individual event. It does not contradict Shannon's more general definition of information as a change of entropy, but is rather a consequence of the latter for the case of information associated with an individual event. However, since Shannon's classical work, information theory has undergone a substantial development. In the modern information theory the definition of information by formula 1.1 is often viewed as simplistic. There are various definitions of information, while the quantity expressed by the formula is sometimes referred to as *self-information.*[45] Since Dembski has been acclaimed not just as an *information theorist*, but as an "Isaac Newton of information theory," we are entitled to expect from him a treatment of information on a professional rather than an amateurish level.

For the sake of discussion, let us accept formula 1.1 as a working definition of information.

A completely different question, though, is whether or not Dembski uses this formula properly, and I will show that his treatment of information, including his use of the formula, has a number of faults.

In *The Design Inference*, Dembski does not discuss in detail his theory from the viewpoint of information. In *Intelligent Design*, however, which is of a more popular type, there is a chapter on information (wherein he suggested his law of conservation of information). In *No Free Lunch*, Dembski offers a rather detailed discussion of information and, unlike in the earlier presentation of his views, discusses the concept of Shannon's entropy. The mathematical expression used by Dembski for entropy is

$$H(a_1 \ldots a_n) =_{\text{def}} \Sigma_i \, (-p_i \log_2 p_i).^{46} \qquad (1.2)$$

Various versions of essentially the same expression, all stemming from Shannon's original work, are commonly used. Dembski defines entropy as "the average information per character in a string." This is in agreement with Shannon's definition. I believe, though, that this definition dooms Dembski's attempt to introduce a fourth law of thermodynamics based on his LCI to failure, as I will explain in a few lines.

Like *information **I*** defined by formula 1.1, *entropy* in information theory is also measured in bits. (If natural logarithms are used, the units for entropy and information are called *nats*.)

Sometimes these two terms—*entropy* and *information*—are used interchangeably (Shannon himself was not very stringent in unequivocal usage of these terms). Some writers prefer to use for ***H*** the term *uncertainty* instead of *entropy* (or *Shannon's uncertainty*) and the term *surprisal* instead of *information* for the quantity ***I***. Regardless of the preferred usage of terms and whatever the nuances in interpreting entropy are, the essence of that concept is the same.

Before discussing the concept of entropy, let us look at one of Dembski's examples wherein he estimates the information carried by a certain string of characters. In *No Free Lunch*, Dembski estimates what he labels the complexity of the word METHINKS (p. 166). The formula used by Dembski for what he in this case calls complexity is the same as he previously introduced for *information.*

Note that the term *complexity* is used by Dembski in this example in a different sense than his own definition of complexity found in *The Design Inference*. In that book, Dembski defined *complexity* as "the best available estimate" of how difficult it is to solve a problem at hand. Now, discussing the complexity of the word METHINKS, he uses for complexity formula 1.1, which he introduced a few pages earlier for information ***I***, and which does not seem to have any relation to the difficulty of solving any problem.

Here is how Dembski now calculates the complexity of the word METHINKS. Since this word has 8 characters drawn from the English alphabet (which comprises 26 letters plus a character for space, for a total of 27 characters), Dembski estimates the probability *P* of that word's occurrence as $(1/27)^8$. The logarithm to the base of 2 for this number is 38, so Dembski concludes that the complexity of that word is 38 bits. The actual expression used by Dembski is that "the complexity of METHINKS is *bounded* by $-\log_2 1/27^8$" (p. 166; emphasis added).

Hence in his calculation Dembski assumes the *uniform distribution* of 27 characters in what he calls "reference class of possibilities," so that each of the 8 characters in that word is assumed to have the same probability of appearing in the string. Such an assumption is justified for a string of eight characters randomly drawn from a stock which has an unlimited supply of all twenty-seven possible characters. (The randomized texts obtained in such a way are sometimes referred to as "monkey" texts, because of the famous example of monkeys randomly hitting the keys on a typewriter.) In the case of the urn technique, formula 1.1 is applicable if each letter, after having been drawn from the urn containing one sample for each of the twenty-seven symbols, is returned to the urn.

However, if METHINKS is a part of a message received through a communication channel, then it has to be expected to be a part of a natural language's vocabulary. On page 164, just two pages before calculating the complexity of the word METHINKS, Dembski wrote about *transmission of information* "from one link to another," about "the textual transmission of ancient manuscripts," and about "transmission of texts" (p. 165) and the like. He never indicates that on the following page he discusses the occurrence of the word METHINKS in a different way, as a result of a random selection of letters from an unlimited stock of all 27 symbols, except for using the term "bounded."

The actual distribution of characters in English texts is not uniform. If the letters of the word METHINKS occur within a message in a natural language, the letter *E* has the maximum probability (about 12 percent) of appearing in any location of the word, the letter *T* has a slightly smaller probability, etc. In this case formula 1.1, which is used by Dembski for the calculation of complexity, is not applicable. Formula 1.1, which is legitimate for "self-information" associated with an individual event, can be formally used for a series of events only if their probability distribution is uniform. In the latter case, however, information associated with each individual event (defined by formula 1.1) would formally coincide with entropy defined by formula 1.2, which equals the *average information.* Hence Dembski actually calculated, under the label of complexity, what formally equals the entropy of the word in question *assuming a uniform distribution of letters*. In the case of a nonuniform distribution, a proper calculation of entropy should be done using formula 1.2.

However, even if Dembski used formula 1.2, thus accounting for the nonuniform distribution of symbols, it still would not be sufficient for a correct estimate of the word's entropy (which he calls in this case complexity). Not only is the probability distribution nonuniform, the word in question is also a part of a meaningful English vocabulary, so the probability distributions of digrams, trigrams, etc., are also nonuniform. Furthermore, the natural languages possess redundancy which substantially decreases the entropy of meaningful texts. As this was already shown by Shannon, the first order entropy of a meaningful English text is about one bit per character; hence the first order entropy of the word METHINKS is about eight bits, rather than the thirty-eight bits of Dembski's estimate of the complexity's bound (not to mention the fact that Dembki's estimate ignores the entropies of higher orders than 1, the existence of multiple symbols other than just the 26 letters of the alphabet, etc.).

Now consider a different situation in which the word METHINKS can occur either as a part of a message arriving through a communication channel or through the urn technique wherein, though, the elements of the search space (i.e., of Dembski's "reference class of possibilities") are not individual letters but whole words. For example, imagine that the urn holds every one of the entries from the unabridged Random House dictionary, which contains about 315,000 entries. It means the urn holds about 315,000 words, each as an indivisible unit. Since each word happens only once, every word has the same probability to be randomly pulled out of the urn. The probability that the word METHINKS happens to be the randomly chosen is then $P = 1/(3.15 \times 10^5)$. Then, using formula 1.1, suggested by Dembski for information, we find that information (which he also calls complexity) obtained when the word in question has been pulled out is $-\log_2 P = 18.3$ bits instead of 38 bits of Dembski's estimate. This number, though, again obviously has nothing to do with complexity associated with the word in question.

If the urn holds unequal numbers of words, say (assuming the distribution of words in the urn is determined by the frequencies of their occurrence in English texts), then the probability of the word METHINKS will be different from that estimated for the uniform distribution, and the information calculated by formula 1.1 will be different from either 18.3 bits, 38 bits, or 8 bits, and will also differ from entropy and have nothing to do with complexity.

It is possible to define a procedure wherein a random occurrence of the word METHINKS results in information (as per formula 1.1) much exceeding thirty-eight bits of Dembski's estimate which also has nothing to do with complexity. For example, assume a word is randomly chosen out of all the words found in all the books in the Library of Congress. What is in this case the probability that the randomly chosen word turns out to be METHINKS? If the total number of words in all the books in the library is N, and the word METHINKS happens among them X times, the probability in question is $P = X/N$. Obviously, N is a very large number while X is a relatively small one since the word METHINKS is rare. In such a procedure the information, if defined by formula 1.1, will be much larger than thirty-eight bits of Dembski's suggested "bound."

This shows that the estimate of information associated with a certain word may be different depending on the probability distribution, so defining the information bound requires first defining which probability distribution is considered. Dembski's calculation of the information bound is valid only for a specific (uniform) probability distribution in a specific procedure (random choice of individual letters).

Moreover, information associated with a certain word, if defined by formula 1.1, cannot be simply translated into complexity. In his discourse, Dembski does not clearly define the discussed situation, uses different terms in a haphazard way, and thus creates a mess of concepts and definitions.

Without discussing many nuances of the concept of entropy, its most concise and universal definition seems to be: entropy is a measure of the degree of randomness (disorder) in a system which comprises many constituent elements. The more disordered the conglomerate of whatever elements the system encompasses is, the larger its entropy.

It does not matter what the physical nature of the system's constituent elements is. Initially entropy was meant to characterize thermodynamic systems. For example, it can characterize the degree of disorder in a gas occupying a certain volume. The gas consists of a large number of identical molecules and its entropy measures disorder (randomness) in the distribution of those molecules over the volume. However, entropy is not a property of those molecules. Their physical nature has nothing to do with the degree of randomness characterizing the distribution of their positions (and momenta) in space.

Entropy can be equally applied to estimate the degree of disorder in a gas occupying a certain volume, to a long string of characters, to the DNA strand, to a large gathering of people, and to an infinite number of other systems, each comprising many constituent elements. Regardless of what those constituent elements are, the behavior of entropy is determined by the same laws, of which the second law of thermodynamics is the most widely known.

As quoted above, Dembski defines entropy as the *average information*. Therefore, whatever new law of thermodynamics he may suggest, as long as it deals with information, it deals with entropy.

As I will argue now, however, the fourth law of thermodynamics suggested as a possibility by Dembski in *No Free Lunch* is in fact wrong because it is based on what he calls the law of conservation of information. As I will argue, LCI is neither a law of conservation nor a law about information.

Let us first discuss whether LCI can indeed be named a conservation law. This seems to be a secondary point, but I am forced to discuss it because Dembski seems to attach considerable significance to the name of his supposed new law, devoting many words to justifying the name he gave to it (pp. 161–62).

It seems a platitude to say that a conservation law must necessarily be about something which is *conserved*. There are also many laws in science which are not conservation laws. In particular, out of the four laws of thermodynamics, only the first law is a conservation law. It is a particular form of the energy conservation law applicable to macroscopic systems. The three other laws of thermodynamics are not conservation laws. This includes the second law. It has a number of definitions, but essentially it deals with entropy. This law does not state that entropy is conserved (although it is conserved in the particular case of *reversible adiabatic processes*; the concept of a reversible process is an abstraction since all real processes are irreversible). The second law of thermodynamics states that the entropy of a closed ("isolated") system cannot spontaneously decrease but can either increase or remain constant. Therefore, the second law of thermodynamics is not considered and is not named a conservation law.

Dembski's alleged LCI—law of conservation of information—is not about something that is conserved. It states that a quantity he calls complex specified information (CSI) can either decrease or remain constant in what he calls "a closed system of natural causes" (p. 170). Dembski provides no definition of the "closed system of natural causes." No casuistry can change the fact that Dembski's LCI is not about the conservation of anything. CSI, according to LCI, can decrease; hence there is no reason to name LCI a law of conservation, even if this alleged law made sense.

Now let us discuss whether or not LCI is about information.

In several of his earlier publications Dembski discussed an example (originally suggested by Richard Dawkins) wherein he pointed out the difference between two strings of letters of equal length. One of these strings spells a phrase from *Hamlet*, "METHINKS IT IS LIKE A WEASEL," and the other is a string of gibberish of the same length (28 characters if the space is counted as a character).

Note that the first-order entropy of the above meaningful quotation from *Hamlet*, according to classical information theory, is about twenty-eight bits, while the first-order entropy of a *random* string of twenty-eight characters taken from the English alphabet, which comprises twenty-six characters plus one more for space, is almost five times larger, i.e., about 135 bits. If, though, we also include in the set of available symbols numerals, commas, colons, periods, semicolons, exclamation and question marks, mathematical symbols, etc., the entropy of the string will be larger. Moreover, if we wish to account for the total entropy rather than for only the first-order entropy (which will be discussed later in this section), the numbers will be even larger. Note also that the entropy of a *random* string and, hence, also the amount of information brought by that string to a receiver, is larger than it is for a *meaningful* string of the same length.

According to Dembski, the quotation from *Hamlet* must be attributed to design because the event (in this case the occurrence of the string of characters) has low probability (which in this example is one in so many billions) and is specified (in this case is a recognizable meaningful English phrase). Low probability, according to Dembski, is equivalent to large complexity (statements to this effect are found in many places in Dembski's books and papers). Hence, in the example in question, Dembski obviously finds CSI—his complex specified information—in the quotation from *Hamlet* but not in the string of gibberish of the same length. What is actually the difference between the two strings? It is in that one is a *meaningful* English phrase (i.e., it displays a *recognizable pattern* to those who possess at least some minimal knowledge of English), whereas the other string is meaningless, i.e., displays no recognizable pattern (although, it may, unknown to us, happen to be a meaningful text in some language we are not familiar with). If we rely on the above example, it seems that what Dembski means by CSI is equivalent to the *recognizable meaningfulness* of the string. If we accept this interpretation of CSI, obviously this concept is not what is called information in information theory. The definition of information adopted by Dembski himself (formula 1.1 above) also has nothing to do with the meaningfulness of a string. Hence the interpretation of CSI, as it follows from Dembski's own example, shows that CSI, despite its name, is not information even in Dembski's own interpretation of the latter term. If we were to stick to Dembski's concept of CSI as rendered in his previous publications, we would have to conclude that CSI, despite including the term information, is not information at all and therefore the LCI in its earlier form was not about information just as it was not about conservation.

In *No Free Lunch*, however, Dembski discusses the CSI and LCI in terms sometimes rather different from those which seem to follow from the example of two strings of characters. His new interpretation entails discriminating between what he calls *conceptual* and *physical* information. In *No Free Lunch* we find the following statement: "In practice, there are two sources of information—intelligent agency and physical processes. This is not to say that these sources of information are mutually exclusive—human beings, for instance, are both intelligent agents and physical systems. Nor is this to say that these sources of information exhaust all logically possible sources of information—it is conceivable that there could be nonphysical random processes that generate information" (p. 137).

So, what is Dembski's actual position? The possibility of the existence of mysterious "nonphysical random processes," capable of generating information, which are neither physical systems nor intelligent agents, seems to negate his initial assertion about only two sources of information. He offers definitions of these two kinds of information:

> **Conceptual Information:** *Intelligent agent* S *identifies a pattern and thereby conceptually reduces the reference class of possibilities.*
>
> **Physical Information:** *Event* E *occurs and thereby reduces the reference class of possibilities.* (p. 139)

These definitions allow for various interpretations, but it seems that conceptual information is used by Dembski as a more general concept than just the semantic contents of a string of characters. The meaning of the first definition seems to be that a pattern, if identifiable by an intelligent agent, would point to design. Obviously, though, if a string is a meaningful text in a language familiar to the intelligent agent, it will, according to Dembski's definition, carry conceptual information, although conceptual information is not limited to strings of characters. Therefore, insofar as we deal with texts, the term *conceptual information* seems to coincide with the meaningfulness of that text. The term *text* can have a very wide interpretation. A multitude of systems can be encoded by a string of zeros and ones and hence be represented by a text. If that is so, then *conceptual information* is not what is named *information* in information theory.

Apparently being aware that his term conceptual information is actually often synonymous with the semantically meaningful contents of a message, Dembski tries to salvage that term as allegedly denoting a different concept by devoting a separate section (pp. 145–46) to the discussion of what he calls semantic information. He asserts in this section that semantic information is not a part of CSI. However, the concepts of semantic information and conceptual information, even if not fully synonymous, overlap. (Note that Dembski's discussion of semantic information seems to indicate that he is not familiar with the recent developments in the algorithmic theory of probability which trespassed the boundaries of information theory and have some promising achievements toward distinguishing between noise and useful information.[47] Also, Dembski's assertion on page 147 that what he calls semantic information does not submit to "mathematical and logical analysis" is incorrect. Contrary to Dembski's assertion, semantic contents of texts predetermine statistically discernable patterns which are absent in meaningless strings. In particular, a method of statistical analysis of texts named Letter Serial Correlation, which enables one to distinguish between semantically meaningful texts and gibberish, was developed by Brendan McKay and myself.)[48]

Dembski uses the term conceptual information in his definition of complex specified information (CSI), also referred to as specified complexity (SC):

> **Complex Specified Information**: The coincidence of conceptual and physical information where the conceptual information is both identifiable independently of the physical information and also complex. (p. 141)

We see that Dembski's concept of CSI, as defined above, incorporates, as an inseparable part, conceptual information, which is not information in the sense of information theory. Therefore CSI is not information either. Hence the acclaimed law of conservation of information suggested by Dembski is neither about conservation nor about information.

In *No Free Lunch* Dembski offers interpretations of LCI wherein he introduces a new element absent from his earlier discussions. According to the new interpretation of complex specified information it is indeed information after all, but to qualify for being complex specified information the amount of information must be not less than 500 bits.

The new formulation of LCI seems to become a statement asserting that both stochastic processes and algorithms are capable of generating specified information up to 500 bits but no more than that. The threshold value of 500 bits is based on Dembski's arbitrary probability bound of 10^{-150}.

To provide a feeling for what the seemingly minuscule probability bound of 10^{-150} means, let us note that a random string of characters drawn from the English alphabet (not including numerals, punctuation marks, and spaces) carries over 500 bits of information when its length exceeds only 105 letters. A semantically meaningful English text carries over 500 bits of information if its length exceeds about 500 characters, which is about one-sixth of an average single-spaced typewritten page. (Actually these estimates are good only for the first-order entropy, which, however, constitutes the main portion of the total entropy of a text. The total entropy includes $L - 1$ terms, where L is the text's length expressed in the number of characters. For example, the second-order entropy can be expressed by the same formula 1.2 where, though, p_i denotes the probability of a digram—that is, of a combination of any two characters, rather than of an individual character, and where the sum also has to be divided by 2. If the total entropy is considered, the length of a randomized string of characters drawn from the English alphabet and carrying 500 bits of information is about only 60 letters. For a meaningful English text that number is close to 300 characters.)

Dembski offers his new definition of the law of conservation of information:

> **Law of Conservation of Information.** Given an item of CSI, call it B =

> (T_2, E_2), for which E_2 arose by natural causes, any event E_1 causally upstream from E_2 that under the operation of natural causes is sufficient to produce E_2 belongs to an item of CSI, call it $A(T_1,E_1)$, such that
>
> $$(\text{LCI}_{\text{csi}})\ \boldsymbol{I}(A\&B) = \boldsymbol{I}(A) \text{ mod UCB}, \qquad [(1.3)]$$
>
> where by definition the quantity of information in an item of specified information is the quantity of information in the conceptual component (i.e., $\boldsymbol{I}(A) =_{\text{def}} \boldsymbol{I}(T_1)$ and $\boldsymbol{I}(A\&B) =_{\mathbf{def}} \boldsymbol{I}(T_1 \ \&\ T_2)$). (p. 160)

If the average reader is puzzled by the above definition with its collection of constituent concepts piled upon each other, such a reader can be consoled that he is not alone. First note that the abbreviation UCB stands for universal complexity bound, which, Dembski explains, "throughout this book we take to be 500 bits of information" (ibid.). He also explains that the abbreviation "mod" stands for "modulo," which "refers to the wiggle room within which $\boldsymbol{I}(A)$ can differ from $\boldsymbol{I}(A\&B)$." Dembski elaborates by saying: "To say that these two quantities are equal modulo UCB is to say that they are essentially the same except for a difference no greater than UCB" (ibid.). (Note that this use of the term *modulo* differs from its standard use in mathematics). Regarding notation T, Dembski explains it as follows: "The event E . . . is an outcome that occurred via some physical process. The target T . . . is a pattern identified by the intelligent agent S without recourse to the event . . . both T and E denote events. The ordered pair (T, E) now constitutes specified information provided that the event E is included in the event T and provided that T can be identified independently of E (i.e., is detachable from E)" (pp. 141–42).

Dembski also provides a few more definitions of the same concept in plain words: "If a natural cause produces some event E_2 that exhibits specified complexity, then for every antecedent event E_1 that is causally upstream from E_2 and that under the operation of natural causes is sufficient to produce E_2, E_1 likewise exhibits specified complexity" (pp. 159–60). Strangely, this definition lacks the part of the mathematically symbolic definition concerning the wiggle room. Actually, Dembski permits the natural causes which produce a consequent event E_2 to add information to that already contained in an antecedent event E_1, but only if the additional information does not exceed 500 bits. Dembski says: "Because small amounts of specified information can be produced by chance, this 500-bit tolerance factor needs to be included in the Law of Conservation of Information" (p. 161). This seems to be a small step in right direction on Dembski's part, because his earlier formulation did not allow for any wiggle room as he asserted that a natural cause cannot generate CSI without exception.

It is instructional to look at some passages in *No Free Lunch* which precede the definition of LCI. On pages 151–54 we find a lengthy discussion of whether or not functions can add information, where we read, "Functional relationships at best preserve what information is already there, or else degrade it—they never add to it" (p. 151). Now jump over two pages: "I have just argued that when a function acts to yield information, what the function acts upon has at least as much information as what the function yields. This argument, however, treats functions as mere conduits of information, and does not take seriously the possibility that function might add information" (p. 154). Dembski proceeds with an example of a function which adds information and concludes the passage as follows: "Here we have a function that is adding information. Moreover, it is adding information because the information is embedded in the function itself" (ibid.).

Here is the quintessential Dembski. He makes two important-sounding statements within two pages, with the second statement negating the first. So, which statement is correct—the one asserting that functions "never add" information or the one asserting that there are functions adding information embedded in functions themselves?

Since Dembski has a goal—to prove something he took as true before even considering arguments in favor or against his belief, namely that CSI can only be created by intelligent agent—whereas his own example with functions seems to contradict his thesis, at the end of page 154 and the beginning of page 155 he offers what can only be viewed as a quasi-mathematical trick aimed at allegedly reconciling his two irreconcilable statements.

Here Dembski introduces a new operator U which comprises both the initial information i (source information) and the initial function f (which can add information embedded in it to the initial information i). He insists that, unlike f, U does not add information. In what way the inclusion of f into a composite function U makes f lose its ability to add information is not explained. If f can add information embedded in it, no mathematical trick like making it a part of a composite function can eliminate its ability to add information, regardless of whether it does so as a stand-alone function or as a constituent of composite function U.

Having concluded that functions do not, after all, add information (a conclusion which is necessary to support his thesis) Dembski says: "Formula (*) confirms this as well" (p. 152); formula (*) is as follows:

$$\boldsymbol{I}(A\&B) = \boldsymbol{I}(A) + \boldsymbol{I}(B|A)$$

This formula is a consequence of, first, the formula for the probability of two events *A* and *B* both actually occurring (when *A* and *B* are not independent events) and, second, of the definition of *information* ***I*** as a negative logarithm of probability. It does not in any way confirm or negate Dembski's thesis, according to which CSI can only be created by intelligent agents, or even his narrower thesis that functions do not add information (the latter is actually rejected elsewhere by Dembski himself).

There is one more rather convincing indication of the fallacy of Dembski's LCI. According to Dembski, since *A* entails *B*, therefore the combination of events *A* and *B* carries no more information than was already carried by *A* alone, that is:

$$\boldsymbol{I}(B|A) = 0, \text{ or } \boldsymbol{I}(A\&B) = \boldsymbol{I}(A)$$

I'd like to note that to reconcile formula (*) with Dembski's definition of LCI, it would be necessary to assume not that

$$\boldsymbol{I}(B|A) = 0,$$

but rather that (using Dembski's own notation *mod*)

$$\boldsymbol{I}(B|A) = 0 \text{ mod UCB.}$$

This is just one more example of Dembski's inconsistency.

As mathematician Richard Wein points out, $\boldsymbol{I}(B|A)$ in formula (*) "is *not* Dembski's specified information (SI)! The problem is that $\boldsymbol{I}(B|A)$ is just $\boldsymbol{P}(B|A)$ transformed, and $\boldsymbol{P}(B|A)$ is the *true* conditional probability of the event, which in this case is 1. SI, on the other hand, is based on the assumption of a uniform probability distribution, regardless of the true probability of the event."[49] I agree with Wein.

I believe the arguments listed above effectively lay to rest any claims of legitimacy for Dembski's LCI.

Note that in all of the reviewed discussion Dembski refers to information rather than to complex specified information. As discussed before, CSI is actually not information in the sense of information theory. Since Dembski's law of conservation of information, despite its name, actually asserts something about CSI rather than about information, the whole discussion on pages 151 through 154 does not seem to be related to his subsequent discussion of LCI.

In view of these issues it can be asserted that, whichever of several mutually contradictory interpretations of Dembski's LCI is chosen, this alleged law makes no sense. In a closed system, no spontaneous process may result in a decrease of entropy; if, as Dembski states, entropy is equivalent to average information, obviously information in a closed system can only spontaneously increase or remain constant, which is contrary to Dembski's proposed law. Therefore his suggestion that his LCI can possibly be generalized as the fourth law of thermodynamics contradicts his own thesis. Such a fourth law would contradict the second law of thermodynamics and therefore cannot be taken seriously.

In *No Free Lunch* Dembski continues to adhere to the idea that complexity is inextricably tied to low probability. This idea contradicts a variety of known facts as well as the definitions of both Kolmogorov complexity[50] and computational complexity. Earlier in this chapter I offered examples illustrating that it is simplicity rather than complexity which points to low probability (as in the case of irregularly shaped, i.e., complex pebbles versus a stone which is perfectly spherical, i.e., quite simple in shape).

Dembski and the No Free Lunch Theorems: The "Displacement Problem"

The title of *No Free Lunch* was borrowed from the name of a set of mathematical theorems proven a few years ago by David Wolpert and William Macready (the "No Free Lunch" or NFL theorems).[51] Actually, although Dembski's book is named after the NFL theorems, he discusses these theorems in only one of the chapters where he attempts to utilize these theorems to support his thesis about the necessity of an intelligent agent for the generation of complex specified information (CSI).

However, Dembski applies the NFL theorems to evolutionary algorithms where these theorems, though correct, are irrelevant.

The NFL theorems establish that performance of all algorithms is the same *if averaged over all possible fitness functions*. Dembski illegitimately applies this result to the algorithms' performance on *specific fitness functions* where different algorithms can (and do) perform very differently. Dembski's assertion that no evolutionary algorithm can outperform a random search because of the NFL theorems and that therefore Darwinian evolution is impossible is absurd. The NFL theorems in no way prohibit Darwinian evolution.

David Wolpert rejected Dembski's use of these theorems as lacking sufficient rigor.[52] A detailed analysis showing that the NFL theorems are irrelevant to evolutionary algorithms in general and Darwinian evolution in particular can be found elsewhere.[53]

In *No Free Lunch*, Dembski also suggests that Darwinian evolution is impossible because of what he calls a "displacement problem" (pp. 203–206). In fact, it has been shown that the alleged displacement problem is a phantom and has no bearing whatsoever on the question of the feasibility of Darwinian evolution.[54]

Overall, the book seems to be a hodgepodge of unsubstantiated but quite pretentious claims and unnecessary quasi-mathematical exercises shedding little light on the issues discussed.

A critique of *No Free Lunch* is also found in other publications,[55] whose authors' views regarding Dembski's literary production I largely share. A discussion of various shortcomings of *No Free Lunch* will also be found in the collection *Why Intelligent Design Fails: A Scientific Critique of the New Creationism* (forthcoming from Rutgers University Press).

CONCLUSION

As I have stated before, there is no way for me to offer a comprehensive review of Dembski's three books and his many papers and Internet postings in one chapter of a reasonable length. For example, I have left out of consideration those two parts of *Intelligent Design* in which Dembski discusses the historical development of the design concept (part 1) and "bridging science and theology" (part 3), as well as some other parts of his books and papers. I have concentrated on the most salient points of what Dembski calls his "theory of design."

Let me briefly summarize the main elements of my critical review:

- Formulating his concept of design, Dembski suggests that design is not a causal but a logical category, and that design does not necessarily entail a designer. However, contrary to that notion, he often refers to design as a synonym for agency, having defined the latter as a conscious activity of an intelligent agent.
- Having defined design as simply the exclusion of regularity and chance, he often forgets about his own definition and refers to design as an independently defined category.
- The law of small probability, suggested by Dembski as one of the

main pillars of his theory of design, is intrinsically contradictory in that it fails to recognize the probabilistic character of design inference and artificially separates probability from specification.

- Of the three nodes of Dembski's explanatory filter, the first two play no useful role, as they imply an unrealistic procedure of estimating probability prior to assuming chance or regularity as the event's causes.
- The explanatory filter as a whole, which seems to be the heart of Dembski's theory, produces both false negatives (as Dembski admits) and false positives (which is contrary to Dembski's assertion).
- Dembski's categorical demarcation between law, chance, and design as the three independent causes does not seem to be realistic either, as it ignores multiple situations wherein either two or all three causes may be at play simultaneously. His scheme ignores many situations where the causal history of events is complicated by feedbacks, conditional causes, etc.
- The versions of probability theory and complexity theory suggested by Dembski do not seem to make much sense either. His definitions of probability, likelihood, complexity, and difficulty lack any indications of how to choose the quantities which are necessary parts of those definitions. Besides, his definitions of complexity and difficulty, while they may be relevant for some particular situations, are unduly generalized as supposedly having universal applicability. There are classes of events wherein his definitions are contrary to the actual situations.
- Dembski's theory does not recognize the differences between human, alien, and supernatural types of design, which is arguably the most interesting problem (not to mention the problem of design by artificial intelligence).
- Contrary to Dembski's scheme, I have demonstrated that sometimes design can be reliably inferred in the absence of a recognizable pattern, i.e., of specification, while in some other cases specification does not ensure the reliability of design inference. He elevated specification, i.e., a recognizable pattern, to a unique status among all possible factors which affect the design inference, without any justification for such a preference.
- The design inference argument, utilized by Dembski to conclude that life is due to design, includes a number of arbitrary assumptions and logically deficient assertions, and therefore lacks evidentiary value.
- The law of conservation of information suggested by Dembski

makes no sense; it contradicts his own thesis. Its extension to the level of an allegedly new law of thermodynamics in fact contradicts the second law of thermodynamics and cannot be taken seriously.

- Dembski's use of the No Free Lunch theorems (which allegedly make Darwinian evolution impossible) is based on his misinterpretation of these fine mathematical results; in fact the NFL theorems in no way prohibit Darwinian evolution.
- The "displacement problem," which, according to Dembski, is an insurmountable obstacle to Darwinian evolution, is not actually a problem at all, insofar as it concerns the existing biological reality.

This list of inconsistencies and weaknesses can be extended beyond the discussion detailed in this chapter.

While I am of the opinion that Dembski's effort to create a consistent theory of design fails, I cannot assert that the hypothesis of intelligent design itself is wrong, but only that neither Dembski nor any of his cobelievers have so far succeeded in proving it.

The Design Inference contains a short introduction by the mathematician David Berlinski, known as an uncompromising adversary of Darwinism. Berlinski writes, among other things, about Dembski's book, that "It is a fine contribution to analysis, clear, sober, informed, mathematically sophisticated and modest."[56]

The first ("fine contribution") and the second ("clear") points in Berlinski's six-point evaluation are a matter of personal opinion, so I will not argue against them.

Regarding "sober," I don't know what Berlinski means. If he uses it as a synonym for "reasonable," I would only accept it with reservation. As can be seen from the preceding sections, in my view, many parts of Dembski's discourse make no sense. Apparently, this view is shared, at least to a certain extent, by mathematicians Eli Chiprout, Jeffrey Shallit, Richard Wein, John S. Wilkins, and David Wolpert; philosophers Ellery Eells, Branden Fitelson, Philip Kitcher, Robert T. Pennock, Del Ratzsch, Elliott Sober, and Christopher Stephens; biologists Wesley R. Elsberry, Gert Korthof, H. Allen Orr, Massimo Pigliucci, and Robert D. Schneider; physicists Taner Edis, Victor J. Stenger, Howard J. Van Till, and Matt Young; and other highly qualified individuals as well.

Regarding "informed," I agree if it means that Dembski is well educated and possesses knowledge of various subjects and topics.

Regarding Berlinski's fifth point ("mathematically sophisticated"), again, the question is, What is the meaning of that term? If Dembski's mathematical sophistication is supposed to mean that he is familiar with various fields of mathematics and comfortable with mathematical symbolism, I would agree with Berlinski's assessment. However, if the term in question implies that Dembski's work constitutes an innovative contribution to mathematics or to its application in some field, in my view this assertion would be hardly justified. Some parts of Dembski's mathematical exercise, while possibly interesting by themselves (for example, all the material on pages 122 to 135 or 209 to 213 in *The Design Inference*), seem to be not germane to his theme. On the other hand, some parts of his mathematical exercise (for example, his treatment of probability and likelihood) in my view do not meet the requirements for a rigorous mathematical discourse.

Berlinski's final descriptor, "modest" seems to be quite off the mark. Dembski's style reveals his inflated feelings of self-importance, which is obvious not only from his penchant for introducing pompously named "laws," but also from his categorically claimed conclusions and such estimates of his own results as calling some of them "crucial insight," "profoundly important for science," or "having a huge advantage" over existing concepts.

Then, in his replies to his critics, Dembski's main thesis seems to be that his critics simply do not understand his fine discourse. This method of discussion hardly seems to meet the concept of modesty. It reminds me of the story about a detachment of soldiers in a boot camp. When a sergeant reproached a soldier for walking out of step with the rest of the platoon, the soldier replied that it was he who walked in step, while all the rest of the soldiers walked out of step.

Starting already from Dembski's early publications, where he compares himself to Kant and Copernicus, he often acclaims his own supposed breakthroughs and great contributions to science.[57] In his newest book (forthcoming from InterVarsity Press), Dembski in all seriousness claims that his ID theory is a revolution, after which science will never be the same.[58] This self-aggrandizing puffery is in fact typical of purveyors of crank science.

A maxim usually attributed to the great Russian writer Leo Tolstoy says that the actual value of a person is like a fraction wherein the numerator indicates the person's achievements and talents, while the denominator indicates what the person thinks of himself. If that denominator is very large, the fraction's value is nearly zero. In Dembski's case, while the estimates of the numerator may vary depending on the level of acceptance of his work, the denominator is obviously enormous.

A good scientific work includes as a necessary part relentless attempts by the scientist himself to find the most powerful arguments which would disprove his conclusions. Dembski seems to have failed to perform such a self-check. The result is a seemingly very sophisticated theory, which may be impressive to those who find in it a confirmation of their preconceived convictions, but which nevertheless has so many holes and inconsistencies that it is overall meaningless.

NOTES

1. William A. Dembski, *The Design Inference: Eliminating Chance through Small Probabilities* (Cambridge: Cambridge University Press, 1998); *Intelligent Design: The Bridge between Science and Theology* (Downers Grove, Ill.: InterVarsity Press, 1999); *No Free Lunch: Why Specified Complexity Cannot Be Purchased without Intelligence* (Lanham, Md.: Rowman and Littlefield, 2002).

2. Michael J. Behe, foreword to Dembski, *Intelligent Design.*

3. Robert T. Pennock, *Tower of Babel: The Evidence against the New Creationism* (Cambridge: MIT Press, 2000).

4. Del Ratzsch, *Nature, Design, and Science: The Status of Design in Natural Science* (New York: State University of New York Press, 2001).

5. Ellery Eells, review of *The Design Inference*, by William Dembski [online], philosophy.wisc.edu/eells/papers/direv.pdf [August 6, 2003]; also published in *Philosophical Books* 40, no. 4 (1999).

6. Brandon Fitelson, Christopher Stephens, and Elliott Sober, "How Not to Detect Design—Critical Notice: William A. Dembski, *The Design Inference*," *Philosophy of Science* 66 (September 1999):472–488.

7. Massimo Pigliucci, "Design Yes, Intelligent No: A Critique of Intelligent Design Theory and Neo-Creationism," *Skeptical Inquirer* 25, no. 5 (2001):34–39.

8. Philip Kitcher, "Born Again Creationism," in *Intelligent Design and Its Critics*, ed. Robert T. Pennock (Cambridge: MIT Press, 2001).

9. Eli Chiprout, "A Critique of *The Design Inference*," Talk Reason [online], www.talkreason.org/articles/Chiprout.cfm [August 14, 2003]; Taner Edis, "Darwin in Mind: Intelligent Design Meets Artificial Intelligence," Committee for the Scientific Investigation of Claims of the Paranormal [online], www.csicop.org/si/2001-03/intelligent-design.html [January 17, 2001]; Wesley R. Elsberry, "Review of W. A. Dembski's *The Design Inference*," Talk Reason [online], www.talkreason.org/articles/inference.cfm [November 22, 2001]; Gert Korthof, "On the Origin of Information by Means of Intelligent Design: A Review of William Dembski's *Intelligent Design*," Was Darwin Wrong? [online], home.wxs.nl/~gkorthof/kortho44.htm [November 22, 2001]; Thomas D. Schneider, "Effect of Ties on the Evolution of Information by the EV Program" [online], www.lecb.ncifcrf.gov/~toms/paper/ev/dembski/claimtest.html [November 22, 2001]; Jeffrey Shallit,

review of *No Free Lunch*, by William Dembski, www.math.uwaterloo.ca/~shallit/nflr3.txt [August 7, 2003]; Howard J. Van Till, "*E. Coli* at the No Free Lunch Room: Bacterial Flagella and Dembski's Case for Intelligent Design" [online], www.aaas.org/spp/dser/evolution/perspectives/vantillecoli.pdf [August 6, 2003]; Richard Wein, "What's Wrong with the Design Inference?" [online], website.lineone.net/~rwein/skeptic/whatswrong.htm [November 22, 2001]; Matt Young, "Intelligent Design Is Neither" [online], www.mines.edu/~mmyoung/DesnConf.pdf [January 17, 2002].

10. Dembski, *The Design Inference*, p. 9.

11. Ibid., p. 8.

12. Ibid., p. 9.

13. Pigliucci, "Design Yes, Intelligent No"; Elsberry, "Review of W. A. Dembski's *The Design Inference*."

14. Dembski, *The Design Inference*, p. 11.

15. See Dembski, *The Design Inference*; *Intelligent Design*; "Redesigning Science," in *Mere Creation*, ed. William Dembski (Downers Grove, Ill.: InterVarsity Press, 1998); "The Third Mode of Explanation," in *Science and Evidence for Design in the Universe* (San Francisco: Ignatius Press, 2000); "Signs of Intelligence: A Primer on the Discernment of Intelligent Design," in *Signs of Intelligence: Understanding Intelligent Design*, ed. William A. Dembski and J. M. Kushiner (Grand Rapids, Mich.: Brazos Press, 2001); *No Free Lunch*.

16. Dembski, *The Design Inference*, p. 36.

17. Readers who are skeptical of assertions not supported by direct references may refer to two of my published articles, both of which are chock-full of formulas: "Slot-Type Field-Shaping Cell: Theory, Experiment, and Application," *Surface and Coatings Technology* 31 (1987):409–26; and "Calculation of Spontaneous Macrostress in Deposits from Deformation of Substrates and Restoring (or Restraining) Factors," *Surface Technology* 8 (1979):265–309.

18. Ratzsch, *Nature, Design, and Science*, pp. 27–78.

19. A. Zakharenkov, "A Poem" (in Russian), in *Strofy Veka*, ed. Evgeniy Evtushenko (Moscow: Polyfact, 1997), p. 985.

20. William A. Dembski, "Redesigning Science," p. 107.

21. Ibid., p. 111.

22. Ratzsch, *Nature, Design, and Science*, p. 166.

23. John F. McGowan III, "Jigsaw Model of the Origin of Life" [online], www.jmcgowan.com/JigsawPreprint.pdf [August 6, 2003]; also published in *Instruments, Methods, and Missions for Astrobiology*, vol. 4, Proceedings of the SPIE 4495 (2001).

24. See Nikolai B. Tikhomirov, *Theory of Probability and Mathematical Statistics* (in Russian) (Kalinin, USSR: Kalinin State Institute Press, 1971).

25. Barry Simon, personal communication with author, 1998.

26. Richard Dawkins, *The Blind Watchmaker* (New York: W. W. Norton, 1996), pp. 7–8.

27. Claude E. Shannon, "A Mathematical Theory of Communication," parts 1 and 2, *Bell System Technology Journal* (July 1948):379–90; (October 1948):623–27.

28. For the articles on Letter Serial Correlation by Mark Perakh and Brendan McKay, see the following Web site: www.nctimes.net/~mark/Texts.

29. Ibid.

30. Dembski, *Intelligent Design*, p. 170.

31. William A. Dembski, "Signs of Intelligence: A Primer on the Discernment of Intelligent Design," p. 189.

32. Dawkins, *The Blind Watchmaker*, pp. 46–49.

33. Kenneth R. Miller, "Life's Grand Design," *Technology Review* 97, no. 2 (1994): 28–29.

34. See, for example, Phillip E. Johnson, *Defeating Darwinism by Opening Minds* (Downers Grove, Ill.: InterVarsity Press, 1997); Lee Spetner, *Not by Chance: Shattering the Modern Theory of Evolution* (New York: Judaica Press, 1998).

35. Paul Vitanyi, "Meaningful Information," *Proceedings of the Thirteenth International Symposium on Algorithms and Computation (ISAAC): Lecture Notes in Computer Science* (Berlin: Springer Verlag, 2002).

36. McGowan, "Jigsaw Model of the Origin of Life."

37. But for discussion on these issues see Mark Perakh, "There Is a Free Lunch after All: Dembski's Wrong Answers to Irrelevant Questions," in *Why Intelligent Design Fails: A Scientific Critique of the New Creationism*, ed. Matt Young and Taner Edis (Piscataway, N.J.: Rutgers University Press, forthcoming); Mark Perakh and Matt Young, "Is Intelligent Design Science?" in *Why Intelligent Design Fails.*

38. Presented in Michael J. Behe, *Darwin's Black Box: The Biochemical Challenge to Evolution* (New York: Simon and Schuster, 1996).

39. John H. McDonald, "A Reducibly Complex Mousetrap" [online], udel.edu/~mcdonald/oldmousetrap.html [April 9, 2002].

40. In fact, McDonald has recently improved his model; in his new version there are more, smaller steps of evolution that are depicted with animation. See the Web site udel.edu/~mcdonald/mousetrap.html [June 12, 2003].

41. For details, see Thomas Espinola, *Introduction to Thermophysics* (Dubuque, Iowa: William C. Brown, 1994).

42. Victor J. Stenger, "Intelligent Design: The New Stealth Creationism," Talk Reason [online], www.talkreason.org/articles/Stealth.pdf [June 12, 2003]; Matt Young, "How to Evolve Specified Complexity by Natural Means," *Pacific Coast Theological Society Journal* [online], www.pcts.org/journal/young2002a.html [August 6, 2003].

43. Shannon, "A Mathematical Theory of Communication."

44. Richard E. Blahut, *Principles and Practice of Information Theory* (New York: Addison-Wesley, 1990), p. 55.

45. See, for example, Robert M. Gray, *Entropy and Information Theory* (Berlin: Springer Verlag, 1991).

46. Dembski, *No Free Lunch*, p. 131.

47. See, for example, Vitanyi, "Meaningful Information."

48. Mark Perakh and Brendan McKay, "Study of Letter Serial Correlation (LSC) in Some Hebrew, Aramaic, Russian, and English Texts" [online], www.nctimes.net/~mark/Texts.

49. Richard Wein, "Not a Free Lunch but a Box of Chocolates: A Critique of William Dembski's book *No Free Lunch*," Talk Reason [online], www.talkreason.org/articles/choc_nfl.cfm [August 6, 2003].

50. Andrei N. Kolmogorov, "Three Approaches to the Quantitative Definition of Information" (in Russian), in *Problemy Peredachi Informatsii* 1, no. 1 (1965):3–11; English translation in *Problems in Information Transmission* 1 (1965):1–7, and *International Journal of Computational Mathematics* 2 (1968):157–68; Gregory J. Chaitin, "Randomness and Mathematical Proof," *Scientific American* 232 (May 1975):47–52; reprinted in *From Complexity to Life: On the Emergence of Life and Meaning*, ed. Niels Henrik Gregersen (New York: Oxford University Press, 2003), pp. 19–33.

51. David H. Wolpert and William G. Macready, "No Free Lunch Theorems for Optimization," *IEEE Transactions on Evolutionary Computation* 1, no. 1 (1997):67.

52. David Wolpert, "Dembski's Treatment of the No Free Lunch Theorems Is Written in Jello," Talk Reason [online], www.talkreason.org/articles/jello.cfm [January 7, 2003].

53. See, for example, Perakh, "There Is a Free Lunch after All."

54. Ibid.

55. See, for example, John S. Wilkins and Wesley R. Elsberry, "The Advantages of Theft over Toil: The Design Inference and Arguing from Ignorance," *Biology and Philosophy* 16 (2001):711–24; Jeffrey Shallit and Wesley R. Elsberry, "Playing Games with Probability: Dembski's 'Complex Specified Information,'" in *Why Intelligent Design Fails*; Jeffrey Shallit, review of *No Free Lunch*, by William Dembski [online], www.math.uwaterloo.ca/~shallit/nflr3.txt [August 7, 2003].

56. David Berlinski, "The Design Inference," introductory note to Dembski, *The Design Inference*.

57. William A. Dembski, "Randomness by Design," Design Inference Website [online], www.designinference.com/documents/2002.09.rndmnsbydes.pdf [June 3, 2003].

58. William A. Dembski, introduction to *The Design Revolution: Answering the Toughest Questions about Intelligent Design*, Access Research Network [online], www.arn.org/docs2/news/designrev061003.htm [June 14, 2003].

2

IRREDUCIBLE CONTRADICTION

Michael J. Behe's book *Darwin's Black Box: The Biochemical Challenge to Evolution*[1] is one of the most popular polemic publications arguing against Darwin's theory of evolution.

The purpose of Behe's book is to provide a new type of argument in favor of the so-called intelligent design theory. As we have already discussed, intelligent design theory states that the universe, and, more specifically, life, is not the accidental outcome of a spontaneous chain of random events but the result of a deliberate design by an intelligent mind. Usually the proponents of the intelligent design theory do not discuss the question of who the designer is. Sometimes they indicate that clarifying the identity of the designer is a task for theology (see, for example, part 3 of William Dembski's book *Intelligent Design*). However, there is little doubt that the designer implied by the theory in question is a supernatural mind, i.e., God.

A manifestation of the strong splash made by Behe's book can be seen, for example, in a voluminous collection of articles titled *Mere Creation*[2] in which almost every paper contains a reference to Behe's book. The level of discourse in that collection is uneven, but it includes some fairly sophisticated articles in which numerous references are made to Behe's book as an allegedly revolutionary step toward proving intelligent design.

We read on the cover of Behe's book opinions of some prominent supporters of the intelligent design concept, who extol the virtues of Behe's breakthrough on the way to the complete defeat of Darwinism and neo-Darwinism. For example, David Berlinski, a mathematician known as an outspoken adversary of Darwin's theory of evolution says, "Mike Behe makes an overwhelming case against Darwin on the biochemical level. No one has done it before. It is an argument of great originality, elegance, and intellectual power."

A similar view of Behe's book is evident from some of the references made to *Darwin's Black Box* by mathematician and philosopher William Dembski in his books *Intelligent Design* and *The Design Inference*. Professor of law Phillip E. Johnson, one of the most prolific propagandists for the intelligent design theory, touts Behe's book in equally enthusiastic terms in several of his books (for example, in *Defeating Darwinism by Opening Minds*).[3] Hence, there seems to be a view—widely shared by people of various professional backgrounds who are all believers in intelligent design and opponents of Darwin's theory of evolution—that Behe's book provides an indisputable argument in favor of intelligent design and thus against all the versions of Darwinian hypotheses and theories.

Behe's book, while highly acclaimed by many proponents of intelligent design, was criticized by opponents of his thesis, including prominent biologists such as Professors Kenneth Miller, Russell Doolittle, H. Allen Orr, and David W. Ussery.[4] This criticism does not seem, though, to have impressed Behe, who continues to publish papers in which he shows no intention to modify his views. He repeats the same arguments time and time again, despite the objections to those views from many critics.

Behe is a biochemist and his book reveals his knowledge of that subject. Since I am not a biochemist, I will not try to delve into Behe's detailed descriptions of biochemical systems; he is in his domain there while I am not. I shall accept Behe's biochemical discussion as flawless, even assuming that some other critics, possessing a wider ken in biochemistry and related fields, could possibly argue about some details of those biochemical descriptions (as, for example, was done by Kenneth Miller). While the biochemical descriptions take up a considerable part of his book, Behe ventures outside biochemistry in his determination to offer a strong rebuttal of the evolution theory, invoking certain mathematical and philosophical concepts—it is these excursions beyond biochemistry that are the target of this critical review. My intention is to show that Behe's principal concept is poorly substantiated and his discourse in no way proves his thesis.

Before discussing in detail Behe's main concept, it seems proper to point out that when Behe ventures beyond biochemistry, he sometimes looks more like a dilettante than an expert.

One of the examples of Behe's dilettantism is how he discusses probabilities. Calculations of probabilities, such as those for the spontaneous origin of life, are very common in books aimed at disproving the theory of a natural emergence of life. More often than not, these calculations produce exceedingly small probabilities which lead to the conclusion that the spontaneous emergence of life was too improbable to be taken seriously. There are also some opponents of the natural emergence of life who realize that the small probabilities in question, if viewed alone, are irrelevant. For example, Dembski, who is proficient in probabilities, correctly points out more than once that a small probability in itself is not a proof (see *Intelligent Design* and *The Design Inference*). Therefore Dembski suggests more elaborate criteria to decide whether an event was the result of chance or of design (see chapter 1).

BEHE CALCULATES PROBABILITIES

Unfortunately, Behe's discussion of probabilities is just a recital of many other similar calculations, based on an insufficient understanding of probabilities. On pages 93 through 97 of his book, Behe criticizes Professor Doolittle's explanation of the blood clotting sequence. The discussion is about the probability that Tissue Plasminogen Activator (TPA) could have been produced by chance rather than by design. Behe suggests some calculations: "Consider that animals with blood-clotting cascades have roughly 10,000 genes, each of which is divided into an average of three pieces. This gives a total of about 30,000 gene pieces. TPA has four different types of domains. By 'variously shuffling,' the odds of getting those four domains together is 30,000 to the fourth power, which is approximately one-tenth to the eighteenth power."[5]

First, let us note the imprecision of Behe's statement. Obviously, 30,000 to the fourth power is a very large number, while one-tenth to the eighteenth power is a very small number, so these two numbers are not even "approximately" close to each other.

Probably Behe meant to say *one in 30,000 to the fourth power.* This is a small mistake, but it hints at Behe's possible discomfort with mathematics. Indeed, he continues as follows: ". . . if the Irish sweepstakes had odds of winning of one-tenth to the eighteenth power, and if a million people played the lottery each year, it would take an average of about thousand billion years before *anyone* (not just a particular person) won the lottery."[6]

Behe's statement is flawed in several respects.

First, Behe's example is contrived, artificially decreasing the chance of winning, and this makes his example irrelevant. On the one hand, he estimates the probability of an event in question (winning the Irish lottery) as one in ten to the eighteenth power. On the other hand, he assumes that only one million people play that lottery. One million is ten to the sixth power, which is just a tiny fraction of ten to the eighteenth power. In this way Behe drastically decreases the chance of winning for anyone (not just a particular person). To clarify that observation, let us discuss a small raffle where only 100 tickets are available for sale. The probability of winning is the same 1/100 for each ticket. If all tickets are sold, one of the tickets (we do not know in advance which one) will necessarily win. Therefore, if all tickets are sold, the probability that some of the tickets will win is 100 percent. Now assume that out of 100 available tickets, only ten have been sold. Each ticket has the same chance to win, regardless of its being sold or not. Therefore, the probability that at least one of the *sold* tickets will win is now 10 percent instead of 100 percent (which it would be if all tickets were sold). This example shows that the exceedingly small probability of anyone (not just a particular person) winning in the imaginary lottery described by Behe is due to his deliberate choice of numbers—only one million tickets sold while the number of potentially available tickets is immensely larger.

In Behe's example, he discusses not a raffle but an Irish-type lottery where players themselves choose the sets of numbers on their tickets. In this case, it is possible, unlike in a raffle, that more than one player happen to choose the same set of numbers on their tickets. This decreases the probability of someone (not a particular player) winning, making it less than 100 percent. As shown in chapter 13, in such a lottery the probability of someone winning is at least about 37 percent, which is still immensely larger than in Behe's contrived example.

Obviously, Behe's example has nothing to do with a real lottery. In every real lottery the number of tickets sold is usually close to the total number of available tickets and the coincidental choice of the same numbers by more than one player is very rare. Therefore the probability that someone (not just a particular person) wins is at least about 37 percent. (This question is discussed in detail in chapter 13.)

Second, Behe's discussion is irrelevant if the probability of a *particular person* winning is considered. This probability does not depend on the number of tickets sold. If the total number of available tickets is 100, then each ticket, whether sold or not, has the same probability of winning, namely 1/100. If, as in Behe's example, the total number of possible events is ten to the eighteenth power, then the probability of a particular event happening is one in ten to the eighteenth power. It is exceedingly small. However, it is *equally small for all possible events*. One event (one set of numbers winning) must necessarily happen in at least 37 percent of the games, despite its individual probability being exceedingly small. Therefore, the exceedingly small probability calculated by Behe for the case of TPA in no way proves his thesis and in no way rebuts Professor Doolittle's discourse.

Third, Behe seems to assume that an event, whose probability is $1/N$, where N is a very large number, would practically never happen. This is absurd. If the probability of an event is $1/N$, it usually means that there are N equally probable events, of which some event must necessarily happen. If event A, whose probability is very low ($1/N$), does not happen, it simply means that some other event B, whose probability is also very low, has happened instead. According to Behe, though, we have to conclude that, if the probability of an event is $1/N$, none of the N possible events would occur (because they all have the same extremely low probability). The absurdity of such a conclusion requires no proof.

The assertion that events having very low probabilities do not occur was suggested by the prominent French mathematician Emile Borel.[7] Borel suggested what he labeled the single law of chance, which stated that "phenomena with very small probabilities do not occur." He estimated that the events that cannot reasonably be attributed to chance are those whose probability does not exceed one divided by ten to the power of fifty. Since Borel was an influential mathematician who contributed fruitful ideas to the field of probabilities, the single law of chance gained wide acceptance, often extending its meaning beyond its legitimate implications. A little later we will discuss some facts illustrating that Borel's law, if interpreted literally, is absurd (see more in chapter 13).

These paragraphs in Behe's book may seem to be of minor importance, as they appear to be beyond the main theme of his discourse. However, this item is actually closely connected to the core of Behe's main idea, his "irreducible complexity." The concept in question comprises two elements, complexity and irreducibility, both being necessary parts of Behe's idea. The question of exceedingly small probabilities calculated for the emergence of biological structures by chance is just another facet of the concept of complexity. The complexity of a biological system is a necessary component of Behe's scheme because, as Behe's idea implies, a system of low complexity has a much better chance of emerging spontaneously as a result of a chain of random events. In order to build a bridge from irreducible complexity to intelligent design, Behe must assume that the probability of his system being the result of a random unguided process is exceedingly small. Therefore, I had to discuss the faults in Behe's treatment of probabilities.

ADDITIONAL REMARKS ON BEHE'S USE OF PROBABILITIES[8]

Suppose that, according to Behe's assertion, there is indeed only one sequence of proteins that can perform a specific function (for example, to clot blood). Suppose further, again according to Behe's approach, that there are no other simpler biological processes that could perform those functions. Also suppose that it can be somehow proved that higher organisms could not evolve without those particular mechanisms (like blood clotting).

Following Behe's discourse further, suppose also that a spontaneous emergence of the protein's sequence which is necessary to perform the function in question, by randomly joining individual proteins, is extremely unlikely (i.e., assuming that the probability of such an outcome of random events is too low to expect that it could have occurred during the time of the Earth's existence).

In other words, grant Behe all his assumptions.

The conclusion that seems to follow from all of Behe's assumptions is that the "protein machines" were not created by joining proteins at random. This is the first part of Behe's conclusion. However, even if we accepted this highly disputable part, the next part of his discourse—which asserts that therefore those "machines" must be the products of intelligent design—would pose very serious problems.

One of the problems in question is that Behe has not eliminated other actions of *randomness* besides the simple random joining of proteins. There are many alternative possibilities. A few examples follow.

1. There could exist stable protein sequences quite similar to clotting (or any other) sequence. Assume, for example, that there is such a stable sequence which differs from the one necessary to perform the clotting by, say, 5 percent only. If the sequence in question is stable, it tends to hang around a long time. Perhaps it is biologically useful, or perhaps it resulted along a common chemical pathway from something biologically useful. In such a case, we have to attribute the emergence of a sequence differing only by 5 percent from an existing one to random combinations, which is immensely easier. Those "sequences correct by 95 percent" could, in their turn, have earlier evolved in similar fashion from "sequences correct by 90 percent."

There is no need to assume that the whole edifice was created in a single step from the primordial soup.

2. Perhaps clotting or any other useful sequence can be decomposed into relatively few fragments ("bricks") which are also parts of other bio-

logically useful sequences. Then the clotting sequence could have resulted from the random combination of bricks broken off sequences that already existed. Again, this is immensely easier probability-wise.

Suppose that we have 4 types of blocks. Consider a particular sequence which is 100 blocks long. Estimate the likelihood of its emergence by a random joining of blocks. There are 803,469,022,129,495,137,770,981,-046,171,215,126,561,215,611,592,144,769,253,376 100-block sequences of 4 different blocks. Of course, this number is extremely large, hence the probability of the spontaneous emergence of the particular sequence is exceedingly small. So, having accepted Behe's approach, we don't expect to see the right sequence to emerge any time soon. Now suppose that each group (brick) of ten blocks is a stable configuration. All of the ten bricks we need can be "made" in parallel, each by joining ten blocks at random, which is immensely easier because there are only 524,800 ways to join ten blocks in a sequence. Given the required ten bricks, there are approximately 1,858 million ways to join ten of them together, which is also a reasonably small number for nature. So, overall, the expected time for the first appearance of the required 100-block sequence is reasonably modest.

3. There can be an adaptive search process. Consider a field that is 100×100 meters and contains a single pit somewhere in it, with the overall surface of the field slightly sloped toward the pit. We define the pit as an area of one square meter. If we want to find the pit by randomly probing points in the field, we will not be surprised if it takes a long time. However, we can find it much faster by a certain, still random process. Put a drunken man at a random place in the field and allow him a "random walk." At each time unit (say, every second) the man takes a step in a random direction. However, downhill steps are on average a little bit longer than uphill steps. Eventually, the man will reach the pit. It might still take a long time, but the expectation of the time will be much less than the expectation of the time required using random probing at the field. (In computer science there is a number of optimization methods that rely on this type of random process. There are even some directly modeled on evolutionary processes and using the same terminology. They are often successful at optimizing functions over search spaces much too complicated for traditional methods.)

4. It might be true that sequences which are very close to the clotting sequence are of no use for modern organisms, but maybe they were useful to earlier organisms. It could be that in early times there was some primitive organism with some primitive (but useful) protein sequence, and that the organism and sequence evolved together into gradually more complex

forms. Changes in either the organism or the sequence could help direct the evolution of the other, so there is no real surprise if the sequence is useful to the organism at each point of time.

Even if options (1) through (4) can be ruled out somehow, what about some mechanisms (5), (6), etc., that we didn't think of yet? To assume that everything in nature happens only according to *known* mechanisms would unduly limit the path to the scientific elucidation of the unknown.

I can foresee a common creationist counterargument to the notions presented above—an assertion that these scenarios are "just so stories" that do not prove anything, because there is no direct empirical evidence of their actual occurrence. Indeed, all these scenarios are speculative. They prove, though, one thing—that the claims of creationists asserting the alleged impossibility of evolution because of the extremely small probability of its individual steps are invalid insofar as they consider only the exceedingly unlikely combinations of chance events as if such purely random chains of events were the only option. In fact, nature has in its arsenal a host of other options ignored by the creationist scenario. Unless all these options are shown to be impossible (which is not the case, by a long shot) the creationist claims remain much more speculative than those listed above or than many other "natural" scenarios we have so far not even imagined. This fully applies to Behe's naive attempt to prove the impossibility of a "natural" emergence of a TPA molecule by assuming that of all its constituent parts encountered one another by chance and that it emerged in one fell swoop.

In the following sections I shall concentrate on the main thrust of Behe's book, namely his attempts to prove the so-called intelligent design based on his concept of "irreducible complexity."

INTELLIGENT DESIGN ACCORDING TO BEHE

Of course, the claim that intelligent design is responsible for the existing universe's structure in general, and for the existing forms of living organisms in particular, was not invented by Behe. In various forms, this concept has been discussed many times before Behe. Behe's contribution to this discussion is in evoking the images of immensely complex biochemical systems and claiming that the complexity of those systems is "irreducible" and therefore points to intelligent design.

There are many descriptions of those fascinating, exceedingly complex biochemical systems in Behe's book. Among those examples are the mechanism of blood clotting, the device used by bacteria for moving (the cilium mechanism), the structure of the human eye, etc. All these systems look like real miracles and it is fun to read Behe's well-written discussions of those immensely complex combinations of proteins, each performing a specific function.

The complexity of the biochemical systems has been demonstrated by Behe in a spectacular way.

To lead the way to his conclusion about intelligent design, Behe claims that the complexity in question is "irreducible." This term means that the removal of even a single protein from the convoluted chain of proteins' interaction would render the entire chain nonoperational. For example, removing even one protein from the process of blood clotting would make blood either not clot, causing the organism to hemorrhage, or totally coagulate, also leading to the organism's demise. From that statement Behe proceeds to claim that the irreducible complexity could not be the result of an evolutionary process and therefore can only be attributed to intelligent design.

Let us discuss all three steps in Behe's reasoning, namely (a) complexity, (b) irreducibility, and (c) attribution to intelligent design.

COMPLEXITY AS A FACADE FOR PROBABILITY

Complexity is one of the two components of Behe's *irreducible complexity* concept. Behe does not offer any definition of what he means by complexity. Therefore, to analyze the actual meaning of his overall idea of irreducible complexity, we have to find clues in his descriptions of those biochemical systems he views as being *complex*. When discussing the concept of complexity, we can turn to the writings of some supporters of Behe who have invested considerable effort in solidifying Behe's assertions by plugging certain obvious holes in them, including his failure to define complexity.

In particular, it is interesting to look at a definition of complexity offered by Dembski in *The Design Inference*. To explain why Dembski's explanation should be considered as a legitimate elaboration on Behe's discourse, let us review certain quotations. In his foreword to Dembski's book *Intelligent Design*, Behe wrote, "Although it is difficult to predict (frequently nonlinear) advance of science, the arrow of progress indicates that the more we know, the deeper design is seen to extend. I expect that in the decades ahead we will see the contingent aspect of nature steadily shrink.

And through all of this work we will make our judgments about design and contingency on the theoretical foundation of Bill Dembski's work."[9] This quotation allows us to confidently say that Behe accepts Dembski's ideas, including his treatment of complexity. Indeed, nowhere can we find a single instance of Behe's disagreement with any of Dembski's arguments.

It seems obvious that Dembski enjoys a great deal of authority amongst the proponents of intelligent design, and therefore when he writes about subjects related to Behe's work, his opinions can be viewed as authoritative expressions of that camp's position.

For our discussion of Behe's book it seems appropriate to point out that complexity as defined by Dembski is essentially a concept quite different from complexity in Behe's interpretation. Complexity as defined by Dembski is practically synonymous with a "difficulty of solving a problem" (see chapter 1). On the other hand, in Behe's scheme, the complexity is in the structure of the biochemical system. It is determined by the number of components of the system and the number of links and interconnections between those components. The more components the system includes, and the more interconnections between those components, the more complex the system. The two concepts of complexity are essentially different. However, there is a link between the two concepts. It is probability. According to Dembski, the harder it is to solve a problem, the smaller the probability that it will be solved by some unguided random action. The more complex a system, insist Dembski and Behe, the smaller the probability that it could have emerged as a result of unguided random events. I submit that more often than not the relation between the complexity *of a system* (as implied by Behe) and the probability of its spontaneous emergence is opposite to the relation assumed by Dembski.

The only aspect of Dembski's complexity that has a bearing on Behe's line of arguments is the suggestion that the complexity of a system translates into the very small probability of its emergence via unguided random events. All other aspects of complexity (of which there are many) are irrelevant for Behe's discourse. Later in this chapter I will return to discussing complexity in general, and of biochemical systems in particular, from a viewpoint ignored by Behe. But now let us take a closer look at the very facet of complexity which is at the core of Behe's use of it, namely, its probability aspect.

First, recall our discussion of Behe's calculation of probabilities. Can a very small probability serve as a decisive argument against the possible occurrence of an event? As Dembski admits, it cannot. Events whose probability is exceedingly small occur every day.

Imagine tossing a die with the letters *A*, *B*, *C*, *D*, *E*, and *F* on its six faces. Assume it has been tossed one hundred times. After each trial we write down the letter facing up. The combination of 100 letters obtained after 100 trials constitutes an event. There are six to the power of 100 possible events, i.e., of possible combinations of 100 letters comprising six letters listed above. This is an enormously large number, about 6.5 times ten to the power of seventy-seven. Only one particular set of letters out of that vast number of possible sets has occurred. Whatever combination has actually resulted from the test, it has an *exceedingly small probability*, close to one divided by more than ten to the power of seventy-seven. The denominator of that fraction is by forty-three orders of magnitude larger than the number Behe calls "horrendously large."[10] This fraction is by twenty-eight orders of magnitude smaller than the lowest limit of probability for a random event as suggested by Borel (ten to the power of minus fifty). Nevertheless, the event in question, whose probability was so exceedingly small, actually occurred. Nobody would be surprised by the occurrence of that exceedingly improbable event because some combination of 100 letters must unavoidably happen with all of the possible combinations having an *equally* minuscule probability. Hence when one of those exceedingly improbable events occurred, there was no reason for surprise.

Unfortunately, in many publications aimed at supporting intelligent design theory, including Behe's book, very small *calculated* probabilities of events, such as the spontaneous emergence of proteins, are offered as alleged proof that such events could not happen. An often repeated statement in such publications is that events, whose probability is so exceedingly small, just do not happen. That statement is tantamount to the absurd assertion that nothing happened, i.e., that no set of letters resulted from 100 tests. The indisputable fact is that exceedingly improbable events happen all the time (see more on that in chapter 13).

Behe's use of biological system's complexity to posit the improbability of their spontaneous emergence without intelligent effort is hardly convincing.

While Behe shares this misconception about the impossibility of events whose calculated probability is very small with many other proponents of intelligent design, Dembski is one of the few design theorists who realize the falsity of that assertion. Indeed, in Dembski's *The Design Inference* we read: "Sheer improbability by itself is not enough to eliminate chance."[11] This runs contrary to Behe's interpretation of low probabilities. In Dembski's *Intelligent Design*, we read a similar assertion, again contrary to Behe's understanding of probabilities: "Complexity (or improbability) isn't enough to eliminate chance and establish design."[12] These statements are especially telltale since they are written by a man highly regarded by proponents of intelligent design, including Behe, and who himself is one of the staunchest design theorists (see chapter 1).

Since, unlike Behe, Dembski asserts that very small probabilities do not prove design or disprove chance, he suggests a more elaborate criterion, which, in his view, enables one to empirically discover design. Dembski's idea is expressed in the concept of what he calls explanatory filter (see chapter 1). This term denotes a three-step scheme for choosing one of the three causes of events, which, according to Dembski, are regularity, chance, and design.

Although a detailed discussion of Dembski's theory was offered in chapter 1, let us briefly summarize his approach. Like many other proponents of intelligent design theory, Dembski maintains that a very low probability of an event (which he views as tantamount to its complexity) is a necessary condition to infer design. However, unlike many other adherents of intelligent design theory, Dembski asserts that an extremely small probability (high complexity) of an event, even if *necessary*, is not in itself a *sufficient* condition to infer design. He perceives the additional condition that will make up for the missing sufficiency in what he calls either "specification" or "pattern." Therefore, according to Dembski, if an event is (a) highly improbable and (b) specified, this points to design.

I find many points in Dembski's argument highly disputable. In chapter 1, a detailed discussion is offered disproving Dembski's assertions. To see how Behe relates to Dembski's theory, let us review a quotation from Behe's foreword to Dembski's book *Intelligent Design*. Behe writes: "For example, if we turned a corner and saw a couple of Scrabble letters on a table that spelled AN, we would not, just on that basis, be able to decide if they were purposely arranged. Even though they spelled a word, the probability of getting a short word by chance is not prohibitive. On the other hand, the probability of seeing some particular long sequence of Scrabble letters, such as NDEIRUABFDMOJHRINKE, is quite small (around one in a billion billion billion). Nonetheless, if we saw that sequence lined up on a table, we would think little of it because it is not specified—it matches no recognizable pattern. But if we saw a sequence of letters that read, say, METHINKSITISLIKEAWEASEL, we would easily conclude that the letters were intentionally arranged that way. The sequence of letters is not only highly improbable, but also matches an intelligible English sentence. It is a product of intelligent design."[13]

The above quotation is a concise representation of Dembski's idea stripped of its mathematical and sophisticated embellishments. Note that this quotation from Behe's foreword shows how he has abandoned the assertion made in his own book that events with a very low probability just do not happen. Instead, he now adopts Dembski's more sophisticated approach, asserting that design must be inferred only when there is a combination of a very small probability with a recognizable pattern.

As I argued in my review of Dembski's work in chapter 1, while Dembski correctly denies the evidentiary power of low probability alone, adding specification does not eliminate the probabilistic nature of design inference.

In view of the above, how can Dembski's filter help Behe in proving the irreducible complexity? The answer does not seem to be very encouraging for Behe and his supporters. There are no distinguishable "recognizable patterns" in the biochemical systems so beautifully described by Behe. Looking at those immensely complex biochemical machines, we do not see recognizable patterns as defined by Dembski, but rather patterns that are, according to his definition, *not recognizable* (in Dembski's terms, "not detachable") as we have no independent background knowledge enabling us to match the observed pattern to any sample known a priori.

After reviewing Dembski's concept of complexity, we have no choice but to conclude that the only input this concept provides is that biochemical machines are highly improbable because they are very complex. We will see later that even this statement is highly questionable. However, even if it were true, it would provide no new insight into Behe's concept of irreducible complexity. Indeed, whatever advantages Dembski's definition of complexity may possibly supply, it contains no facet which would elucidate what makes complexity *irreducible*. Therefore, to discuss irreducibility, we have to first talk about complexity from a viewpoint different from that chosen by Dembski.

COMPLEXITY FROM A LAYPERSON'S VIEWPOINT

I will later return to complexity as a mathematical concept, limiting the discussion in this section to some intuitively understood meaning of complexity of a system.

In Behe's view, enormous complexity (combined with its alleged irreducibility) is a sign that the system in question must have been designed by some unnamed intelligent mind.

Is complexity indeed an attribute of an intelligent design? Human experience points in the opposite direction. The simpler the solution to a problem, the more intelligence and ingenuity it requires. The entire history of technological progress proves that the best designs are always the simplest. Let us look at a few examples.

Remember the electronic circuits which appeared at the beginning of the twentieth century? They were based on vacuum tubes. The simplest vacuum tube, a diode, had a number of fragile parts, soldered together in a vacuumed vessel. A triode, which was a necessary part of an amplifier, had several electrodes of complex shape soldered into a glass or metallic body with a bunch of contacts penetrating the walls of the vessel.

Recall the first electronic computer, named ENIAC, which was built in 1946 by J. Presper Eckert and John Mauchly. This wonderful achievement of the human mind from today's point of view seems to be a monster. It was a huge contraption containing 18,000 vacuum tubes and 3,000 switches.

If we accepted Behe's concept, improvements in electronics and computer design should have proceeded via increased complexity of both vacuum tubes and circuitry. Indeed, for a while, the increased ability of electronic circuits to perform various tasks was achieved by increasing the complexity of both the tubes and the circuits. Vacuum tubes were designed with four electrodes, then five, six, seven, and so on. The number of tubes in a circuit increased. The more complex the tubes and the circuits became, the slower their performance improved, hitting a wall when the cost of the systems became prohibitive without a substantial improvement in performance, and was accompanied by a drop in reliability. Then, in 1948, J. Bardeen, W. H. Brattain, and W. Shockley of Bell Telephone Laboratories invented the transistor. A transistor is much simpler than a vacuum tube. Its introduction led to the enormous simplification of electronic circuits and thus to the vastly increased ability to perform more complex tasks. While modern computers are much more complex than that built by Eckert and Mauchly, were they to perform only the same tasks as the ENIAC, they could be immensely simpler than the ENIAC. This simplification has enabled the enormous progress in computations, communications, and automatics we witness today.

Remember another example, from a completely different field. In the nineteenth century, various inventors tried to design a sewing machine. A number of patents were granted. All those machines were unreliable and heavy, and inventors tried to solve the problem by adding more parts, each designed to get rid of some of the shortcomings of the machine but at the same time making it more complex.

In 1851, a man by the name of I. M. Singer invented a simple shuttling hook and a simple needle of a special shape. These two elements immediately made all the complex devices used by Singer's predecessors unnecessary. His machine was much simpler than any before him, while also much more reliable and easier to use. It became the model for further improvements by A. B. Wilson, who introduced a swinging hook, thus further simplifying the design.

Would anybody say that Singer's and Wilson's predecessors were more intelligent than these two inventors because the predecessors' designs were more complex?

It has been agreed among experts in warfare that the Russian-made tank T-34, designed by Joseph Kotin, was the best tank of World War II. It was also the simplest in design.

The best submachine guns in their categories are considered to be the Russian piece designed by Kalashnikov (AK-47) and the Israeli-made Uzi. Both are also the simplest in design among all the submachine guns ever produced. The Uzi has only seven parts, and is easily assembled and dissembled.

Many more such examples could be listed. Now recall Dembski's assertion, obviously supported by Behe, which maintains that complexity is equivalent to low probability. I submit that the factual situation is opposite to Dembski's idea.

Imagine that you have embarked on an excursion from Rome into the Italian countryside and have lost your way. Of course, everybody knows that all roads lead to Rome. Since you wish to be back in Rome as soon as possible, you would like to choose the shortest road. There are many different roads to choose from, but only one of them is the shortest (i.e., the simplest); let us denote it *S*. There are many other roads which all are more complex than *S*. However, you do not know which of the roads is your most desired *S*. Imagine that you decide to rely on chance—let's say you assign a number to every possible road, write those numbers on pieces of paper, and then randomly pull one of the numbers out of your hat. Of course, the probability that the randomly chosen road turns out to be *S* is much smaller than the probability that the randomly chosen road turns out to be one of the more convoluted ones, simply because there are so many convoluted roads but only one shortest road. Imagine now that you do not rely on chance but rather decide to approach the problem in an intelligent way. For example, you buy a map in the nearest village and determine the shortest road to Rome. In this case you have a good chance of selecting the shortest of available roads.

Hence, if you learned that your friend who lost his way in the countryside chose the shortest (i.e., the simplest) road to Rome, you would have a good reason to assume that he made an intelligent decision, choosing the road by design rather than by chance. If, though, you learned that your friend chose some convoluted, complex way, it might indicate that he relied on chance.

Similarly, any task in either a mechanical or a biological system can be performed in many ways. There are always much more complicated, convoluted ways of performing a task than simple ways to do the same job. If a machine, be it mechanical or biochemical, is very complex, it points to its unintelligent origin. Nothing prevents a system of any degree of complexity from emerging as a result of unguided random events. If, though, a task is performed in a very simple way, there is a good chance a design may be inferred. The simple reason for that is that there are many convoluted ways to do a job but only a few simple ways.

In view of this discussion, I submit that Dembski's assertion that complexity equals small probability can reasonably be turned upside down. The simpler the system that successfully performs a job, the smaller the probability that it is a result of spontaneous random events. The more complex the system is, the less probable is its origin in an intelligent design. Of course, if the latter statement is accepted, it undermines the very core of Behe's concept.

Behe has convincingly showed that biochemical systems are extremely complex. That complexity, according to Behe, is one of two necessary facets pointing to intelligent design (the other is irreducibility). Why this complexity *in itself* should point to intelligent design, remains Behe's (and his supporters') mystery.

Of course, Behe considers complexity together with irreducibility, and when combined, they provide, in his view, a strong argument in favor of intelligent design. In the following sections I will discuss the role of the alleged irreducibility of the systems described by Behe.

WHAT IS THE REAL MEANING OF IRREDUCIBILITY?

The discussion of Behe's concept of irreducible complexity must be twofold. On the one hand, we will have to discuss the following question: If the biochemical systems described by Behe are, as he asserts, indeed irreducibly complex, does this indeed point to a deliberate design? On the other hand, we will have to address the question of whether or not these systems are not just complex, but indeed *irreducibly* complex.

Let us discuss the definition of irreducible complexity.

There exists a quite rigorous and quite general definition of irreducible complexity, of which Behe was apparently unaware, and which actually defines something quite different from what Behe means by his term. This definition is given in the algorithmic theory of probability (ATP). That chapter of statistical science was developed in the 1960s. Its main creators were U.S. mathematician Ray J. Solomonoff of Zator Co., Russian mathematician Andrei N. Kolmogorov of the Russian Academy of Sciences, and U.S. mathematician Gregory J. Chaitin of the IBM research center.[14] ATP makes use of elements of mathematical statistics, information theory, and computer science.

The definition of irreducible complexity developed in ATP, while being rigorously mathematical, is quite universal and applicable to any system, regardless of its particular nature. It is based on the concept of randomness, for which ATP provides a strict definition as well.

It seems easiest to explain irreducible complexity according to ATP by using a mathematical example and a computer analogy, although this specific example and analogy in no way limit the applicability of the concept in question to any system, including the biochemical systems discussed by Behe.

Consider the following set of digits: 01 01 01 01 01 . . . and so on. It is obvious that this sequence is highly ordered. It is constructed by the repetition of zeros and ones in pairs. The size of this sequence, depending on the number of repetitions, can be any number, for example, one billion bits. How can we program a computer to reproduce this sequence? It is obvious that there is no need to tell the computer all the numbers of which this sequence consists. It is sufficient to tell the computer the rule which determines the sequence. The program in question can be written in a very simple and short form, essentially as the following instruction: *Print 0,1* n *times*, where *n* can be any number. The length of the program is much shorter than the length of the sequence itself. No matter how we increase the size of the sequence, the size of the program will always remain much shorter than the sequence itself.

Now imagine the following sequence: 10110110001011001110100111 101 . . . etc. Such a sequence can be obtained, for example, by flipping a coin many times and writing *1* each time the result is heads and *0* when it is tails.

Viewing this sequence, we cannot see any specific order in it. This set of numbers corresponds to our intuitive concept of a *random sequence*. How can we program a computer to reproduce this sequence? Since there is no evident rule determining which digit must follow any digits already known, there is no way to produce this sequence by using any program shorter than the sequence itself. To program a random sequence, we need to feed into the computer the entire sequence, which serves as its own program. Hence, the size of a program that produces a *random sequence* necessarily equals the size of the sequence itself. (We can also discuss the problem in more general terms of *algorithms* instead of *programs*.) Again, a random sequence cannot be encoded by a program of a reduced size, hence a random sequence is *irreducible*. Any ordered sequence, on the contrary, can be encoded (at least in principle) by a program (or an algorithm) which is shorter than the sequence itself. Therefore, an ordered sequence is reducible.

While the above discussion is a simplified presentation of some seminal concepts of ATP, it can, hopefully, help us comprehend the definition of irreducible complexity given in ATP. We will discuss this definition after a few preliminary remarks.

Every system, including the biochemical ones described by Behe, can be represented by a certain algorithm, or, if we prefer a computer-related parlance, it can be represented by a program which encodes the system. The code can be represented by a sequence of binary symbols. If a system is not random, and hence obeys a certain rule, the encoding program in question (or algorithm) can be compressed by using the rule in question, i.e., made shorter (in number of bits) than the size of the system itself.

The complexity of a system (often referred to as *Kolmogorov complexity*) is defined in ATP as the minimum size of a program (or of an algorithm) that is capable of encoding the system. The more complex a system, the larger the size of the *minimal* program that can encode that system. If the size of the minimal encoding program cannot be reduced below the size of the system itself, i.e., if the minimum size of the encoding program (or algorithm) approximately equals the size of the system itself, the complexity of such a system is defined as *irreducible*.

Note that the definition of complexity in ATP is very different from the definition given by Dembski, who defines complexity in terms of the difficulty in solving a problem and also identifies complexity as low probability (see chapter 1). Dembski's definition provides no clue as to what can make complexity irreducible. The definition of complexity in ATP is a definition of the complexity *of a system* per se rather than of the difficulty in its reproduction. We will see later that ATP complexity has a relationship to probability which is opposite to that of Dembski's complexity.

The basic definition relevant to our situation is then as follows: a system is irreducibly complex if the minimum size of a program that is capable of encoding the system approximately equals the size of the system itself. On the other hand, if a system is not random, there exists (at least in principle) a rule prescribing the structure of that system. Using that rule, an encoding program can be designed (at least in principle) which is much shorter than the system itself.

Hence, a very important consequence of the basic theorems of ATP is as follows: *if a system is indeed irreducibly complex, it is necessarily random.* In other words, ATP has established that irreducible complexity is just a synonym for randomness.

Whatever examples of biochemical systems Behe can come up with, he cannot eschew the mathematical fact: if a system is indeed irreducibly complex, it is random. Of course, a system that is the result of intelligent design is, by definition, not random. The conclusion: if a system is irreducibly complex, it is not a product of design.

Of course, the proponents of intelligent design theory may insist that the alleged intelligent Creator is not constrained in his choice of design and can, if he wishes so, create systems which appear random despite having been designed. This argument would essentially make the entire dispute meaningless by erasing any discernable difference between objects or events that are designed and those that are not.

We see that if Behe wished to stick to his term of irreducible complexity, his entire explanation, which suggests that biochemical systems are irreducibly complex and therefore must be products of a design, would make no sense.

In terms of ATP, however, biological systems are never irreducibly complex. Indeed, a tiny seed contains the entire information necessary to grow an oak. The entire complex structure of an oak is encoded by the much smaller program contained in a seed.

While Behe's term is a misnomer, and no biological organism is irreducibly complex in terms of ATP, we can just say that Behe simply has not chosen his term well. Does his concept, mislabeled irreducible complexity, nevertheless have some meaning different from the term of ATP?

Reviewing Behe's multiple examples of biochemical machines, we can see that what he actually implies by his term is the interdependence of all the components of a biochemical machine, such that the removal of any element of it renders it dysfunctional. We will have to discuss whether or not biochemical systems are indeed characterized by such a tight interdependence of all of their constituents, as suggested by Behe, and if they are, whether or not this indeed points to intelligent design.

MAXIMAL SIMPLICITY PLUS FUNCTIONALITY VERSUS IRREDUCIBLE COMPLEXITY

Let us approach the problem of the connection between complexity and design utilizing an analogy to William Paley's famous watchmaker argument. In that argument, one is asked to answer the following question: If you found a watch, would you believe that it was a result of a spontaneous natural process or that it was designed by a watchmaker? Of course, the answer is unequivocal, and everyone agrees that a contraption which performs a well-defined function could only be the product of intelligent design. Let us analyze: What feature of that watch led to the conclusion that it was a product of design? Was it the watch's complexity?

To answer the last question, let us formulate the problem a little differently. Suppose you sit at a beach and pick various pebbles. Most of them have an irregular shape, with a rough surface, with their color varying from spot to spot, and their density also varying over their volume. Suppose that you come across one particular piece which, unlike all other pebbles, is of a perfectly spherical shape, its color and density perfectly uniform all over its volume and its surface polished mirrorlike. Obviously, the rational conclusion is that the perfectly spherical piece is an artifact, a result of an intelligent effort, including design, planning, and a set of actions aimed at achieving the goal of producing that perfectly uniform ideal sphere. While we don't know the purpose of the designer of that spherical artifact, we have to admit that its spontaneous appearance is unlikely. Any other piece of pebble with an irregular shape is more likely a result of some spontaneous natural process. Now, the spherical piece is extremely simple and can be described by its color as well as a very simple formula requiring only two numbers, the diameter and the constant density. The full description of that spherical artifact requires a simple program of a small size. Any other piece of pebble with its complex shape and a nonuniform distribution of density, color, and surface roughness cannot be described by a simple program, but rather by a much more complex one containing many numbers.

This example again illustrates that complexity in itself is more likely to point to a spontaneous process of random events while simplicity (low complexity) more likely points to intelligent design. This is in full agreement with the definition of complexity given in ATP but contrary to the definition of complexity given by Dembski. Regarding the ideal sphere, its complexity in terms of ATP (that is, its complexity as a system) is very small. However, the probability of its spontaneous emergence is also very small, which is opposite to the relation between Dembski's complexity and probability. In Dembski's terms, the simpler a system is, the larger its probability. On the other hand, a system which is simple in ATP's sense, but fully functional, must be complex in Dembski's terms, since its probability is small.

A system that is simple in ATP's sense and also fully functional more likely points to design than to chance. This conclusion is contrary to Behe's concept which attributes large complexity to design.

In fact, our conclusion about the likely origin of Paley's watch was based not on its complexity, but rather on its *functionality*. The watch performs a definite task, and that gives rise to our conclusion.

Nothing prohibits a very complex system from emerging as a result of random events. Functionality is what seems to point to intelligent design. In the case of an ideal sphere, we inferred design not because of the sphere's complexity, but because of its obvious *artifactuality*, a term introduced by Del Ratzsch.[15]

If we analyze the examples given by Behe, we have to conclude that his thesis was not about *irreducible complexity* but rather about the *functionality* of biochemical systems, or, more specifically, about a strict interdependence of the system's components, each of them being necessary for the system to properly function. In Behe's often discussed example of a mousetrap, the feature relevant to the discussion was not the trap's complexity or irreducibility. The indication of design was the trap's functionality, its ability to perform a certain task by means of a *simple* combination of parts.

It is easy to provide examples of systems which are even much simpler but nevertheless meet Behe's actual rather than proclaimed formula. One such example was given by mathematician David Berlinski, who strongly supports the concept of intelligent design, but is obviously aware of weaknesses in Behe's position.[16] Imagine a regular chair with four legs. Of course, it is a very simple system. Cutting off a leg renders the chair unusable. Hence, according to Behe's actual formulation, which he strangely seemed not to realize himself, that chair meets the requirement of what Behe unduly labeled "irreducible complexity."

Now, since complexity in itself more likely points to a spontaneous chain of random steps rather than to intelligent design, what are the features of a system which would point to intelligent design? They are simplicity and functionality. Hence, Behe's definition, first, should have been turned upside down (simplicity instead of complexity), and second, complemented by one more necessary component—functionality.

If we discover that a system performs a certain task, then the simpler the system in question and the better it performs a certain task, the more likely is its origin in intelligent design. The more complex the system performing a certain task, the less likely is the suggestion of intelligent design. Hence, if Behe and his followers want to test whether a system was likely created via intelligent design, they must have in their possession criteria which would enable them to determine whether the complexity of the system performing a certain task is close to the minimally possible while preserving its functionality. Behe did not suggest nor did he apply such criteria to the biochemical systems he described. Therefore his assertion that those systems were irreducibly complex (which actually should be redefined as "most simple but functional") was not substantiated. The sheer complexity of the biochemical system is rather an argument against the probability of intelligent design, especially if it is not shown that their complexity is close to the minimum possible while preserving functionality.

As an example of an irreducibly complex system, Behe persistently refers to a simple five-part mousetrap. A mousetrap can be constructed in various ways. The simpler its design, the more intelligent that design is. What is missing in Behe's discourse is proof that the particular design of a trap he described is indeed close to being as simple as possible. Moreover, it is easy to demonstrate that the five-part mousetrap described by Behe can be reduced to a four-part, three-part, two-part, and finally a one-part contraption still preserving the ability to catch mice.[17] Behe should have thought more carefully about the example he uses to illustrate his thesis.

TWO FACETS OF INTELLIGENT DESIGN

Note that the concept of intelligent design is twofold. It comprises, first, the idea of *design*, and second, the idea of that design being *intelligent*.

For the sake of illustration, imagine a situation which is intentionally simplified to the extreme, and not intended to be viewed as realistic. Hopefully readers will forgive the obvious frivolity of such an example. Assume that on a certain planet X civilization developed without inventing chairs, so the inhabitants of that planet, if they wished to sit down, had to do so by sitting on the ground. Imagine further that the idea of a chair took hold and a competition was announced for inventing a comfortable chair. Now assume that among the submitted proposals were chairs with various numbers of legs. Of course, all of those chairs, once made, would be the results of design. However, not all of them would qualify to be viewed as designed intelligently. For example, chairs with only one or two legs attached at the corners of the seat would be impractical and therefore their design would be viewed rather as imbecilic. A chair with three legs would be the most intelligently designed since it would combine a reasonable level of comfort with the best stability, since it is least dependent on the flatness of a floor. The design of a three-legged chair would deserve the definition of intelligent design. A four-legged chair would have the drawback of being less stable on an imperfectly flat floor. However, a four-legged chair may win as a little more comfortable. Hence, both the three-legged and four-legged designs could reasonably be viewed as intelligent. (Since this example is somewhat on the facetious side, we ignore the multiple variations of possible seats, such as chairs without legs, chairs hung from ceilings, sofas, etc.) Assume now that among the submitted designs were five-, six-, and seven-legged chairs. Now imagine that the inhabitants of X were not among the smartest people in the universe, so they chose to opt for seven-legged chairs (maybe in their religion the number seven was also supposed to have a special meaning). Assume that a visitor from another planet, Y, where still no chairs had been in use, came to X and saw the seven-legged chairs. The visitor, who never saw any other type of chairs, might admire the seven-legged chairs as an amazing invention. Since he never saw four- or three-legged chairs, he might believe that all seven legs were necessary. If that visitor happened to be a disciple of Behe, and had never had a chance to try a four- or a three-legged chair, he could conclude that he saw in a seven-legged chair that famous irreducible complexity. This, in its turn, might lead him to the conclusion that a seven-legged chair was a result of intelligent design. He might never suspect that the design of that chair was in reality not very intelligent, but rather based on *excessive complexity.* (We will additionally discuss excessive complexity in a following section.)

Similarly, many biochemical systems described by Behe could very well be excessively complex. If they are excessively complex, it can be ascribed either to an unintelligent design, or to the blind evolution. I don't think anybody would entertain as reasonable the idea of an unintelligent design on a cosmic scale. Hence, unless there are proofs that the complexity of a system is not excessive, its complexity more likely points to randomness than to intelligent design.

At the beginning of this chapter I quoted Berlinski, a mathematician who highly praised Behe's book. It is of interest to look at another quote where Berlinski relates to specific points in Behe's book, using a mathematical approach. In *Mere Creation*, Berlinski writes, "The definition of irreducible complexity makes strong empirical claims. It is foolish to deny this as well to suggest that these claims have been met. The argument having been forged in analogy, it remains possible that the analogy may collapse at just the crucial joint. The mammalian eye seems irreducibly complex; so, too, Eucariotic replication and countlessly many biochemical systems, but who knows?"[18]

Indeed, who knows? This statement comes from a man who is regarded by the proponents of intelligent design as an accomplished mathematician and a staunch anti-Darwinist. His statement reveals one of the weakest points of Behe's position—the absence of proof that the biochemical systems he describes indeed possess what he mislabeled *irreducible complexity*, but which actually is a tight interdependence of components and an accompanying lack of compensatory mechanisms (see the next sections).

While Behe's discourse provides no proof that irreducible complexity (in his sense of the term) is indeed present in cells, it provides even less indications that the systems in question are *irreplaceable*, that is, they cannot be replaced by other systems, which would perform the same function, possibly in a more efficient way.

At least two features, if indeed present in a system, speak against the hypothesis of intelligent design. One is *excessive complexity*, and the other, the *absence of self-compensatory mechanisms*.

EXCESSIVE COMPLEXITY

We have established so far that if the irreducible complexity (either in ATP or in Behe's sense) is indeed present, it rather points (for two different reasons) to the absence of design. However, this does not mean that if complexity is not irreducible, it must point to design. The complexity that is reducible (in Behe's sense) can be justifiably viewed as *excessive*, and as such it may also point to absence of design. Another term for it can be *redundant complexity*.[19]

The proposition of the excessive (or redundant) complexity is not just a logical conclusion. There exists direct experimental evidence pointing to the excessive complexity of some biochemical systems. Moreover, one such evidence relates precisely to that mechanism of blood clotting which Behe had chosen as an example of what he named irreducible complexity. In the nineties, biochemists learned how to "knock out" individual genes from an animal's genome. In a paper by T. H. Bugge an his colleagues, the results of an ingenious experiment have been reported.[20] These researchers succeeded in removing from a group of mice the gene which was instrumental in producing fibrinogen, a protein necessary for blood clotting. Naturally, these mice lost the ability to clot blood and suffered from hemorrhage. In another group of mice, the researchers "knocked out" the gene responsible for production of plasminogen, the protein ensuring a timely cessation of blood clotting and hence preventing trombosis. As it could be expected, the mice without plasminogen had serious trombotic problems. However, when both groups of mice were crossed, the issuing generation, which had neither fibrinogen nor plasminogen, turned out to be normal. This experiment has shown that what Behe described as irreducible complexity of the blood clotting system was actually excessive complexity, since the removal of two proteins from the system resulted in some alternative mechanism taking over.

As Professor Russell Doolittle, who is a prominent microbiologist and an expert on blood clotting, wrote in reference to Bugge et al., "Music and harmony can be achieved also with a smaller orchestra."[21]

In "Answering Scientific Criticisms of Intelligent Design," Behe responded to Doolittle and argued that the experiment described by Bugge et al. did not actually prove what I call excessive complexity.[22] Since I am not a biologist, I will not delve into the arguments offered by Doolittle or Behe. We see here a dispute between two microbiologists, so the problem for us laypersons is whose view to trust. However, I could not fail to notice a peculiar feature of Behe's argumentation. In the above-mentioned paper, he claims that, after having considered Behe's rebuttal, Doolittle conceded that he was wrong in his interpretation of the results by Bugge et al. To my inquiry, Professor Doolittle denied having ever given a reason for Behe's claim. Contrary to Behe's claim, Doolittle adheres to his original view. When I see a discussion in which one attributes to one's opponent something the latter never said, as seems to have been done by Behe in this case, I am inclined to doubt the rest of Behe's assertions as well.

Other examples of experimentally confirmed redundant complexity are known.[23]

This discourse shows that the enormous complexity of events in a cell, so amply demonstrated by Behe, can often be suspected to be *excessive* (or *redundant*) *complexity*. Therefore the conclusion that the complexity in question can only be attributed to intelligent design is unfounded. On the contrary, the complexity in question is quite often an *excessive* complexity, which points to its being a result of random events rather than that of a deliberate design.

ABSENCE OF SELF-COMPENSATORY MECHANISMS

Excessive complexity is not the only argument against intelligent design. While the strict interdependence of all elements of a biochemical system, even when it is not excessive, does not contradict the possibility of *design*, it clearly speaks against the design being *intelligent*. Indeed, *intelligently* designed machines are expected to have built-in self-compensatory resources. If unforeseen circumstances render some elements of the machine dysfunctional, the self-compensatory mechanism automatically takes over the damaged function. If after buying a car we discover that its designer failed to provide space and a holder for a spare tire, we hardly would praise the designer's intelligence. Without a spare tire, each time a tire blew, it would render the entire vehicle dysfunctional, precisely like the removal of or damage to a single protein, which allegedly renders the entire biochemical machine dysfunctional according to Behe's concept of irreducible complexity. The very essence of Behe's mislabeled irreducible complexity implies the absence of self-compensatory mechanisms in biochemical machines. If the removal (or damage) of a single protein indeed makes the entire machine dysfunctional, as Behe asserts, it is a very serious fault of the alleged designer, whose intelligence immediately appears suspicious. Since, again, the hypothesis of a stupid designer acting on a cosmic scale is hardly satisfactory for most of the design theorists, the constitution of biochemical machines, if they indeed are as described by Behe, is an argument against intelligent design.

It seems appropriate to point out that biologists tell us something about biochemical systems that does not seem to jibe with Behe's notion of their irreducibility. Actually, many such systems possess redundancy rather than irreducibility, thus again pointing to evolution rater than to creation.[24]

I can envision a counterargument accusing me of not leaving room for intelligent design at all. Indeed, if, in my view, either the absence of self-compensatory mechanism or the irreducible complexity (in the ATP sense) both point to a random chain of events rather than to intelligent design, isn't this self-contradictory? My answer to such an argument is as follows: First, my task is not to suggest a criterion of intelligent design but rather to test the validity of Behe's arguments. In my view (see also chapter 1), the design inference is necessarily probabilistic. If we see a poem or a novel, we have no difficulty in attributing it to design because design is in this case overwhelmingly more likely than emergence of a long meaningful text as a result of random unguided events. Our inference in this case is based on our ken, as we have extensive experience with texts written by people and can easily recognize them. On the other hand, when we deal with a biological system, our probabilistic estimate cannot be based on our ken because we don't know in advance what the system in question must look like to be attributed to design. If a biological system is indeed irreducibly complex, either in Behe's or in the ATP's sense, then in both cases this is compatible with the assumption of its origin in a random process but (probabilistically) speaks against the design inference. If a biological system possesses redundancy which serves as a self-compensatory mechanism, this is equally compatible with both design inference and the absence of design, but in this case Behe's assertion of irreducible complexity is contrary to the facts. If a biological system has no built-in self-compensatory mechanisms (as Behe suggests), then this is an argument against the *intelligent* design (although not necessarily against *design* as such).

Overall, Behe's concept does not seem to jibe with reality. It does not seem to provide a reasonable argument in the design versus random variations controversy.

One more comment, although of secondary importance, seems to be appropriate. Besides being a faculty member of the department of biological sciences at a U.S. university, Behe is also a fellow of the Discovery Institute's Center for Science and Culture. Members of this well-financed organization are actively promoting intelligent design, irreducible complexity, and other similar concepts. There is a feature typical of the numerous publications by the members of the center in question which justifies referring to it as a club of mutual admiration. The center's fellows often praise each other's literary production and views in superlative terms. Some of them go even further, not shying away from self-praise. Behe seems to be one of the practitioners of that genre. In *Darwin's Black Box*, Behe claims that his thesis of irreducible complexity "must be ranked as one of the greatest achievements in the history of science. The discovery rivals those of Newton and Einstein, Lavoisier and Schroedinger, Pasteur, and Darwin."[25] This display of amusing self-aggrandizing puffery seems to run contrary to the actual scientific significance of Behe's thesis which, in my view—obviously shared by many highly qualified scientists such as Doolittle, Miller, Orr, and Ussery, among others—adds no valid argument in favor of intelligent design.

CONCLUSION

In this chapter, I have omitted discussion of many secondary points and details in Behe's book, concentrating only on his main argument in favor of intelligent design, the concept he calls by a misappropriated term of irreducible complexity of biochemical machines.

Let me briefly summarize the main points of the preceding discussion.

- The very term "irreducible complexity" has been misappropriated by Behe, probably inadvertently since before Behe used it, the term has been rigorously defined mathematically but has denoted a very different concept. If any system reviewed by Behe happened to indeed be irreducibly complex, according to the proper definition of that concept, it would mean that the system is random, and, hence hardly the product of design.
- An inseparable part of Behe's concept is the complexity of biochemical systems. That complexity itself, however, points not to an intelligent designer but rather to a chain of unguided, largely random events. The probability of a spontaneous emergence of a *complex* system performing a certain function is much larger than the probability of the spontaneous emergence of a system which performs the same function in a *simpler* way. (The simpler the system capable of performing a certain function, the more difficult it is to create such a system, and hence its spontaneous emergence is less probable.)
- Biological systems are never irreducibly complex in the mathematical sense of the term. Their programs are intrinsically reducible to shorter sets of instructions contained in embryos, seeds, the combinations of spermatozoids and eggs, etc.
- Since there is no proof that any of the systems described by Behe is indeed irreducibly complex (in Behe's terms), many of them may be excessively complex, which is an argument against intelligent design.
- If any biochemical machine is indeed irreducibly complex (in Behe's terms), it means it lacks compensatory mechanisms and is highly vulnerable to any accidental damage to a single protein which would render the entire system dysfunctional. Such a structure of a biochemical machine, if it is indeed as described by Behe, points to a lack of intelligence of the alleged designer, and, hence, rather to the absence of a designer.

In view of the above, I submit that Behe's book and his theory of irreducible complexity add nothing useful to the discussion of the evolution versus intelligent design controversy.

It seems worthwhile to notice that Behe's (and his supporters') rejection of Darwin's theory is limited to pointing to aspects of that theory which, as they think, have not yet been sufficiently explained or understood. Every scientific theory is incomplete and fails to explain some facts. This does not negate the theory's positive features. Newton's mechanics fails to explain, for example, the behavior of elementary particles. This does not mean that Newton's theory must be rejected. Indeed, this theory is extremely useful, for example, in planning the flights of spacecrafts where its precision is amazingly good. Darwin's theory (or neo-Darwinism in its various forms), like any scientific theory, may be correct in some respects and weak in others. Moreover, the progress of science may indeed reveal that Darwin's theory contains more weaknesses than truths. This does not seem very likely, though, since the evolution theory certainly contains many empirically verified elements and to dismiss it, as some creationists actively try to do, would be a regrettable loss. Therefore, attempts to overthrow Darwin's theory, as Behe tries to do, on the basis of often dubious and sometimes even obviously incorrect notions is not a fruitful way to search for truth.

NOTES

1. Michael J. Behe, *Darwin's Black Box: The Biochemical Challenge to Evolution* (New York: Simon and Schuster, 1996).
2. William A. Dembski, ed., *Mere Creation* (Downers Grove, Ill.: InterVarsity Press, 1998).
3. Phillip E. Johnson, *Defeating Darwinism by Opening Minds* (Downers Grove, Ill.: InterVarsity Press, 1997).
4. See Kenneth Miller, "Life's Grand Designs," *Technology Review* 97, no. 2 (1994):24–32; Russell F. Doolittle, "A Delicate Balance," *Boston Review* 22, no. 1 (1997):28–29; H. Allen Orr, "Darwin v. Intelligent Design (Again): The Latest Attack on Evolution Is Cleverly Argued, Biologically Informed—and Wrong," *Boston Review* [online], www.bostonreview.net/BR21.6/orr.nclk [August 6, 2003]; also published in *Boston Review* 21, no. 6 (1996–97); David W. Ussery, "A Biochemist's Response to 'The Biochemical Challenge to Evolution,'" *Bios* 70 (1999):40–45.
5. Behe, *Darwin's Black Box*, pp. 93–94.
6. Ibid.
7. See, for example, Emile Borel, *Probability and Life* (New York: Dover, 1962).

8. This section, which is meant to complement the previous discussion of the treatment of probabilities by Behe, is based on comments made by Brendan McKay in personal communication with the author, July 2001.

9. Michael J. Behe, foreword to *Intelligent Design: The Bridge between Science and Theology*, by William A. Dembski (Downers Grove, Ill.: InterVarsity Press, 1999).

10. Behe, *Darwin's Black Box*, p. 96.

11. William A. Dembski, *The Design Inference: Eliminating Chance through Small Probabilities* (Cambridge: Cambridge University Press, 1998), p. 3.

12. Dembski, *Intelligent Design*, p. 130.

13. Behe, foreword to *Intelligent Design*.

14. See, for example, Andrei N. Kolmogorov, "Three Approaches to the Quantitative Definition of Information" (in Russian), in *Problemy Peredachi Informatsii* 1, no. 1 (1965):3–11; English translation in *Problems in Information Transmission* 1 (1965):1–7, and *International Journal of Computational Mathematics* 2 (1968):157–68; Gregory J. Chaitin, "Randomness and Mathematical Proof," *Scientific American* 232 (May 1975):47–52, reprinted in *From Complexity to Life*, ed. Niels Henrik Gregersen (New York: Oxford University Press, 2003), pp. 19–33.

15. Del Ratzsch, "Design, Chance, and Theistic Evolution," in *Mere Creation*, p. 289.

16. David Berlinski, "Gödel's Question," in *Mere Creation*, pp. 402–26.

17. Matt Young, "Intelligent Design Is Neither," [online], www.mines.edu/~mmyoung/DesnConf.pdf [January 17, 2002]; John H. McDonald, "A Reducibly Complex Mousetrap" [online], udel.edu/~mcdonald/oldmousetrap.html [April 9, 2002].

18. Berlinski, "Gödel's Question," p. 406.

19. Niall Shanks and Karl H. Joplin, "Redundant Complexity: A Critical Analysis of Intelligent Design in Biochemistry," *Philosophy of Science* 66, no. 2 (June 1999): 268–82.

20. T. H. Bugge et al., "Loss of Fibrinogen Rescues Mice from the Pleiotropic Effect of Plasminogen Deficiency," *Cell* 87 (1996):709–19.

21. Doolittle, "A Delicate Balance," p. 29.

22. Michael J. Behe, "Answering Scientific Criticisms of Intelligent Design," in *Science and Evidence for Design in the Universe*, ed. Michael J. Behe, William A. Dembski, and Stephen C. Meyer (San Francisco: Ignatius Press, 2000), p. 133.

23. See Shanks and Joplin, "Redundant Complexity."

24. Gert Korthof, "On the Origin of Information by Means of Intelligent Design," Was Darwin Wrong? [online], home.wxs.nl/~gkorthof/korthof44.htm [August 6, 2003].

25. Behe, *Darwin's Black Box*, p. 233.

3

A MILITANT DILETTANTE IN JUDGMENT OF SCIENCE

Phillip E. Johnson is one of the most prolific writers and debaters vigorously promoting the intelligent design theory. Whereas there are several variations of that view, Johnson suggests an approach wherein the question of the designer's identity is left out. Also, Johnson avoids a discussion of the Bible's literal inerrancy, aiming at uniting all species and subspecies of anti-Darwinism under one banner. Johnson's writings include a number of books and papers in various periodicals and collections.[1]

In *Darwin on Trial*, Johnson proclaims his goal and outlines his attitude in the pursuit of that goal. He writes, "My purpose is to examine the scientific evidence on its own terms, being careful to distinguish evidence itself from any religious or philosophical bias that might distort our interpretation of that evidence."[2]

Of course, such an attitude can only be greeted with approval. Unfortunately, a few sentences down the road Johnson writes, "Given the emphatic endorsement of naturalistic evolution by the scientific community, can outsiders even contemplate the possibility that this officially established doctrine might be false? Well, come along and see."

The words "officially established" in that sentence betray Johnson's bias. He surreptitiously squeezes into his sentence a ready-made conclusion that the prevailing view in the scientific community has been "officially established." This assertion pictures the scientific community as a kind of Orwellian organization, such as the one that existed in the former USSR. Within the framework of the monstrous Soviet system the validity of scientific theories was officially determined by decree of the ruling party. Even within that system, scores of scientists managed to preserve their scientific integrity and achieve important results in science. To do so, they sometimes had to risk their freedom and even life, not to mention their academic positions.

The actual situation in the world scientific community has nothing in common with the caricature painted by Johnson. Nothing has been "officially established" in science and no scientific theory has been protected from criticisms, be it quantum mechanics or the theory of evolution. When mentioning the "emphatic endorsement" of naturalistic evolution by the scientific community, if he has indeed detected such, Johnson should have paused and analyzed without prejudice why that endorsement has been given, rather than attributing it to an alleged "official establishment."

In *Darwin on Trial* we also read, "I am a philosophical theist and a Christian. I believe that God exists who could create out of nothing if He wanted to do so, but who might have chosen to work through a natural evolutionary process instead. I am not a defender of creation-science and in fact I am not concerned in this book with addressing any conflicts between the Biblical accounts and the scientific evidence."[3] This statement meets no objections (a critical comment regarding the logic in the quoted sentence will be discussed a little later). If Johnson proclaims a certain *belief,* it is fine as long as he has no intention to impose his belief on others by force. Johnson may exercise his right to believe in God's existence without providing rational reasons for that belief, just as an atheist has the same right to adhere to the belief—also without providing a rational basis—that there is no God.

Unfortunately, Johnson's books and papers reveal that his innocent-sounding claim is actually not so innocent. We learn from Johnson's writings that in reality he views himself as a crusader against materialistic science and specifically against one of its most perfidious offspring, the Darwinian theory of evolution.

As biologist Brian Spitzer, himself a devout Christian, has shown in a well-documented review of Johnson's literary output, the way Johnson promotes his thesis is characterized by a consistent use of quotations out of context, frequent ad hominem attacks, and often deliberate, crude distortions of the views and arguments of his opponents.[4]

JOHNSON AS THE LEADER OF DESIGN PROPONENTS

Johnson is often referred to as the most prominent representative of the intelligent design movement. In Johnson's *The Wedge of Truth* we read,

"The Wedge of my title is an informal movement of like-minded thinkers in which I have taken a leading role."[5]

In her paper entitled "You Guys Lost," theologian Nancy R. Pearcey writes, "It would appear that the latter-day design theorists have caught on. The movement has capable leadership—such as that provided by Phillip Johnson."[6]

Similar statements can be found in books and papers of other proponents of intelligent design and opponents of Darwinism. We see that Johnson has proclaimed himself the leader of the antinaturalist assault force, and has been acknowledged as such by others.

In order to form an opinion regarding Johnson's qualifications to be a leader of a movement whose proclaimed goal is to prove intelligent design by purely scientific arguments, it may be useful to look at a story Johnson tells in his book *Defeating Darwinism by Opening Minds*. In this book Johnson tells about a conversation he had with a colleague at the University of California, Berkeley. According to Johnson, he remarked to his colleague "that the scientific community was baffled at its failure to convince the general public to believe in evolution." The colleague replied that "the people don't understand the theory." To that, Johnson tells us, he "blurted: 'Oh, no. The people understand the theory better than the scientists do.'"[7]

This story seems to be telltale evidence as to what Johnson's real attitude to science is. In Johnson's view, people who are uneducated in science understand scientific problems better than the scientists who are experts in the field. This frank statement is in contrast to Johnson's multiple assertions of his respect for science as long as it is based on evidence.

Of course, each group is free to choose as its leader whomever they view fit. Recall, though, that this is a group which supposedly intends to win the battle by presenting scientific evidence. The fact that they have for a leader a man who has neither conducted a single scientific experiment, nor derived a single formula, nor proved a theorem, nor suggested any scientific theory in any field, and, moreover, frankly admitted his contempt for scientists, testifies to the actual status of that group in relation to a genuinely scientific discourse.

On the other hand it is telltale that the people who offer supposedly scientific arguments in favor of their views and beliefs have chosen as their leader a man who is a lawyer. Such a choice is easy to understand if we account for a simple fact: the entire set of arguments in favor of the intelligent design theory is actually far from being scientific, since it is based not on any empirical proofs but rather on convoluted casuistry. Naturally, a lawyer's skill is a suitable tool if the discussion is based on casuistry rather than on an empirical foundation.

Every line in Johnson's writings reveals a lawyer's approach to the subject of discussion.

In a court of law a lawyer has to confront the other side and convince the jury. In his writing, Johnson is free from the constraints of a legal procedure. He is the prosecutor, defense lawyer, and judge all in one. He prosecutes Darwinism and the materialistic worldview, he defends the intelligent design theory, and he pronounces the verdict.

A good example of Johnson's lawyer technique—using to his advantage any small piece of evidence, however insignificant or even contrary to his case—is found in *Darwin on Trial*. Johnson discusses Stephen Jay Gould's review of the first edition of *Darwin on Trial*. That review, Johnson tells the reader, was "a hatchet job." Nevertheless, Johnson tells us that he was "elated" by the review, because its very appearance testified to the strength of Johnson's position, which forced his adversary to write a lengthy review of Johnson's book. There is little doubt that if such a review did not appear, Johnson would be "elated" as well, and would interpret the absence of a review as proof that his adversaries just have no good counterarguments and therefore chose to keep silent about it.[8]

A statement on the back cover of his book *Darwin on Trial* asserts that the book "rocked the scientific establishment." I have news for Johnson and the authors of the above statement: his books have made very little impression on scientists. Its popularity among his cobelievers, who have already been convinced of the validity of his arguments, in no way translated into even a ripple in the minds of experts in the field he has amateurishly attacked.

Of course, a lawyer's logic would also provide a ready explanation if one pointed out the minimal impact on scientists of Johnson's writings. For example, Johnson's comrade-in-arms, philosopher Alvin Plantinga, asserts on the same book's back cover that the book in question "shows just how Darwinian evolution has become an idol." This implies that the defense of Darwinism may only be due to the atheistic religion of the scientific "establishment," which prohibits criticism of an "idol" blindly worshiped by scientists. Of course, the real reason for the absence of any noticeable effect of Johnson's writing on the scientific community is that scientists know the facts whereas Johnson, as is typical of dilettantes, only thinks he knows them.

Johnson also assumes the role of a prophet, foretelling the imminent demise of Darwinism and of materialism/naturalism. Here is a quotation from Johnson's paper "How to Sink a Battleship": "I believe that at some time well before 2059, the bicentennial year of Darwin's 'Origin of Species,' perhaps as early as 2009 or 2019, there will be another celebration that will mark the demise of the Darwinist ideology that was so triumphant in 1959."[9]

Well, let us wait and see. Darwinism and neo-Darwinism are scientific theories suggesting certain interpretations of experimental and observational evidence. As any scientific theory, they may be correct in some respects and wrong or uncertain in some others. Their survival or demise will result from the further progress of science but not from the beliefs and emotions of Johnson and his cohorts.

Johnson's entire literary production is valueless as far as a legitimate scientific dispute goes. However, his lawyer skills and eloquence can be persuasive to laymen; because of the popularity of his books and papers, it seems desirable to provide some rebuttals to his unfounded claims.

JOHNSON'S QUALIFICATIONS AS A NEMESIS OF DARWINISM

In *Darwin on Trial*, Johnson offers two reasons for being viewed as a legitimate participant in the dispute about Darwinism despite not being a biologist. One of the reasons is, as Johnson says, "Practicing scientists are of necessity highly specialized, and a scientist outside his field of expertise is just another layman."[10]

Of course, a biologist usually is not an authority on physics and an astronomer is not an expert in biology. Therefore astronomers usually do not endeavor to discuss biology and biologists normally keep clear of disputing the problems of theoretical physics. Why does the fact that scientists are usually specialized in relatively narrow fields qualify a lawyer to discuss problems obviously far from a lawyer's background and expertise? Isn't something not quite right with this logic?

Moreover, although a scientist is indeed usually an expert only in his own narrow field, every practicing scientist has a general understanding of how science works. Such an understanding is acquired through experience in designing experiments, collecting and sorting data, discerning regularity in the experimental results, and interpreting the data and distilling the grains of truth from the chaos of measured numbers. This experience is similar for a physicist, biologist, or chemist, but is alien to a lawyer. Therefore, a scientist endeavoring to discuss problems of a field in which he has no specific experience is still better equipped to do so than a lawyer is.

To provide one more reason why he should be considered an authority in biology, Johnson says, "I should say something about my qualifications and purpose. I am not a scientist but an academic lawyer by profession, with a specialty in analyzing the logic of arguments and identifying the assumptions that lie behind those arguments."[11]

Hence, we may expect that in his books and papers Johnson will shower the readers with strictly logical arguments, void of any personal bias, proving his thesis with a mathematical certainty. Is this indeed the case?

There are various kinds of logic. One is that of a scientist, whose goal is to establish facts and to distinguish clearly between the facts and their interpretation. When a scientist starts to research a problem, he or she is not supposed to know in advance what he will find. He or she is equally interested in every fact regardless what theory or hypothesis that fact jibes with. A scientist uses logic to separate the chaff from the wheat and thus to arrive at the most reasonable and plausible interpretation of his findings.

A lawyer's logic is different. The lawyer's goal is not to find the truth but to win in an adversarial process. A lawyer's arguments tend to stress everything that supports his already adopted position and to downplay or disregard everything that contradicts it. Therefore, Johnson's assertion that he is perfectly qualified to attack a biological theory does not seem to be convincing.

Johnson certainly has the right to discuss any problem of his choice, be it in biology or the mathematical group theory. However, when evaluating his argumentation, there is no reason to attribute to him the status of an expert in any fields beyond his legal expertise. His discussion, no matter how popular his writings are, is that of an amateur.

Like Johnson, I am not a biologist. Unlike Johnson, I will not pretend to have mastered biology. Also unlike Johnson, I am a practicing scientist with a half-century of experience in scientific research. Hence, while both Johnson and I are dilettantes in biology, I have at least a general expertise in scientific research, and thus in applying scientific logic to the problems to be discussed. Johnson lacks such expertise.

Not being a biologist, I have no intention to delve into details of Johnson's discussion of particular features of the theory of evolution. There is also no need for that because, as mentioned earlier, Johnson's arguments have failed to persuade a single biologist who supported the theory of evolution to change his views. Johnson has a ready explanation for his lack of success in fighting the views of scientists. He wants us to believe that scientists' minds are "closed" because their views are based not on evidence but on an "atheistic religion" or on materialistic philosophy. This assertion is implausible. It distorts the actual position of scientists and is nothing short of slander. In reality, Johnson's particular arguments against Darwinism have been rebuffed more than once. Darwinists have responded to Johnson's predecessors, who used similar arguments and examples, before Johnson ever offered his views. Therefore, there is no need for me to repeat those counterarguments, even if I were well qualified to do so.[12]

As mentioned before, in *Darwin on Trial*, Johnson writes, "I am a philosophical theist and a Christian." He continues saying that he is "not concerned with addressing any conflicts between the Biblical account and the scientific evidence." What a convenient position! Before even starting the discussion, Johnson demonstrates the peculiarity of his alleged logic. If he is a believing Christian, does that not mean he is supposed to believe in the biblical story? And if that story contradicts scientific evidence, how come he is not concerned about it?

This statement alone portends very peculiar twists of alleged logic a reader may expect to see in the rest of Johnson's writings.

JOHNSON SINKS A BATTLESHIP

In his article "How to Sink a Battleship," Johnson writes, "What went wrong is that scientists committed the original sin, which in science means believing what you want to believe instead of what your experiments and observations show you."[13]

The reply to that assertion is twofold. First, one may wonder whether or not it is actually Johnson himself and his cohorts who have committed the described original sin. Indeed, do not Johnson and his colleagues believe what they want to believe instead of what experiments and observations show? Can Johnson point to a single experiment or observation confirming his religious beliefs? The accusation against scientists (seemingly all scientists) in something Johnson is certainly himself guilty of sounds rather amusing.

Second, Johnson's accusation is plain slander. Scientists who conduct their research properly do not believe what they want to believe but only what their experiments and observations show them (otherwise, they would hardly deserve the title of scientist). When I say "scientists," I do not imply particular individuals but rather the community of scientists. Individual scientists are human and, like Johnson, scientists may be prone to err and sometimes to believe what they want to believe. Such beliefs, though, as a Russian adage says, have short legs. They do not survive scrutiny by the scientific community and are swiftly dismissed if they are not supported by independent verifications.

The quoted statement at the beginning of this section portends the overall level of Johnson's subsequent rebuttal of naturalism. Since his article is a call to arms, he needs to frighten his cobelievers with the image of an alleged monopolistic bunch of vile antireligious scientists all in cahoots to suppress the free exploration of the truth and of any theories that contradict their atheistic religion. Of course, this picture is a caricature having no basis in reality.

Here is another quotation from Johnson's article: "What went wrong in the wake of the Darwinian triumph was that the authority of science was captured by an ideology. . . ." He continues with a question, "What are we going to do to correct this deplorable situation?"[14]

No, Mr. Johnson, science is not, and never has been, in the claws of any ideology, while you and your colleagues have obviously been captured by your religious beliefs. Among scientists, there are men and women of various ideological and religious persuasions, from believing Christians, Jews, Muslims, Hindus, or Buddhists, to agnostics and atheists, who manage to reconcile their beliefs with proper scientific research. There is no conspiracy by scientists to suppress whatever ideas and theories one may devise. There are no scientific police that try to suppress the views of Johnson and his cohorts and their right to express them, however preposterous those views may seem from a scientist's viewpoint.

Why, then, is Johnson trying to present this false picture of science and of the allegedly reigning "deplorable situation"? Because his real agenda seems to be not "opening the minds," as he claims,[15] but imposing his ideology on society.

Just look at the list of publications by InterVarsity Press, which printed all of Johnson's books. It includes dozens of titles, none of which reports results of any scientific research. Instead, most of these books are polemic escapades often disparaging legitimate science. So far nobody has tried to suppress the activity of that publishing house, or of the numerous periodicals which print articles by the adherents of intelligent design. Many scientists view the abundance of such publications as a really "deplorable situation," but it has never occurred to scientists to do anything to suppress those publications, regardless of the level of ignorance or the distortion of facts which, as many scientists think, is not uncommon in these publications.

Of course, Johnson's article is what could be expected—a speech by a lawyer whose task is to prove his point regardless of the facts.

JOHNSON OPENS MINDS—OR DOES HE?

On the back cover of Johnson's book *Defeating Darwinism by Opening Minds*, we read accolades by Johnson's cohorts, such as Michael Behe, Charles Colson, and Dallas Willard. According to Behe, Johnson is "our age's clearest thinker on the issue of evolution." Colson tells us that Johnson's analysis is "brilliant." Willard asserts that Johnson's production must be "closely studied by all who wish to understand the forces that actually govern the intellectual world in the United States today."

It is nice to know that Johnson has found full approval of his ideas and discourse among his cobelievers who seem to belong to a mutual admiration society. However, for an unbiased reader the book in question looks rather different. In this book Johnson openly shows his contempt for scientists. He spares no sarcasm or derisive labels when referring to professors of zoology such as Tim Berra, Nobel laureates such as Francis Crick, and renowned astronomers and authors of popular books on science such as Carl Sagan.

Arguing against Berra, Johnson suggests the derisive term "Berra blunder" as a general definition of logically untenable propositions. Even if we disregard the rudeness of this way of conducting a discussion, the very substance of Johnson's rebuttal of Berra's discourse is a display of Johnson's own blunder. Berra illustrates the process of evolution using the example of how an automobile's design led from an original model, through a number of modifications, to the most recent model. In Johnson's view, this example is a blunder because automobiles are designed by engineers and therefore the example in question does not testify against the concept of intelligent design, but rather supports it.

This argument is preposterous. Berra's example was just an *analogy*. It is a ridiculous assumption on Johnson's part that Berra, a university professor, did not realize the difference between natural selection and the deliberate design of an automobile. Berra gave his example to *illustrate* the process of gradual modification, which, as any analogy, has both similarity with and difference from the process of natural selection. The similarity, legitimately noted by Berra, is in the sequential chain of modifications, common to both the deliberate design of automobiles and evolution via natural selection. The difference is that the automobile design was indeed a product of an intelligent agent, an engineer (or engineers), while evolution via natural selection, according to Darwinian theory, is an unguided process. The difference between the two situations does not make the use of such an example a "blunder" and is justified, as long as nobody tries, as Johnson does, to represent it in a light different from Berra's intent. Since Berra did not use his example as a *proof* of evolution theory, Johnson's obvious fallacy meets the definition of fighting a straw man.

Johnson does not seem to distinguish between a replica and an analogy. An analogy is a method of illustrating a situation A by considering another situation, B, which has some features common with A. Situation B is chosen because it is easier to comprehend than A, often simply because B is more familiar to a wider audience.

Analogies are similar to *models*, which are the mainstay of physics. Models will be discussed at length in chapter 12.

An analogy is also a model, just not of an object but of a situation. In order for an analogy to be useful, the real situation has to be replaced by another, usually simpler, situation, which has certain features in common with the situation under discussion. A good analogy preserves the essential features of the situation and ignores the nonessential ones, thus making it easier to comprehend the essential core of the situation.

Berra's discourse has met this rule, because his goal was to illustrate how the evolution process passes consecutive stages, and for that illustration the source of changes—either intelligent agent or random events—was inconsequential.

Johnson's attack on Berra displays the smugness of the dilettante, which Johnson appears to be even in the area he claims to be his field of expertise—the analysis of the logic of arguments.

Another example of Johnson's contempt of those who do not share his views is seen in *Defeating Darwinism by Opening Minds*. Here Johnson discusses a computer modeling of a process of descent with modification: "I am amused by self-styled 'skeptics' who invariably seem able to believe the wildest nonsense if it supports Darwinism."[16]

Remember that the "self-styled skeptics" Johnson refers to in this passage include professional biologists and computer scientists, experienced in scientific research and the interpretation of experimental data. "The wildest nonsense" mentioned by Johnson is the extensive experimental evidence in favor of Darwinian theories.

Derogatory remarks about science and scientists are scattered all over

Johnson's opuses. For example, in *The Wedge of Truth*, Johnson asserts that Darwinists "do not understand the difference between intelligent and unintelligent causes."[17] Really, Mr. Johnson? Are Darwinists such fools that they need a lawyer to teach them the meaning of common terms?

In the same book Johnson informs readers that Einstein "was apparently unaware that his own statement was both immodest and self-contradictory" (p. 92).

What a travesty—a self-appointed arbiter teaching Einstein modesty and logic!

On the other hand, Johnson's writings are indeed full of real blunders, some of which will be discussed in the next section.

HOW A DILETTANTE DISCUSSES INFORMATION

One of the examples of Johnson's dilettantism and of the feebleness of his logic is his discussion of information. In the section titled "A Book Isn't Just Ink and Paper" from his book *Defeating Darwinism by Opening Minds*, Johnson discusses the difference between information and the material medium used to record this information. One of his main theses is that information is a separate entity unconnected to the material medium on which it is recorded.

Johnson writes about his book, "The information in each chapter was exactly the same whether it was recorded on paper or on computer disk or in some fragmented and disembodied form as it moved over the links of the internet."[18] He then provides an example: "If all copies of Shakespeare's plays were destroyed, nothing would be permanently lost. Actors who had learned the roles could easily re-create the texts from memory" (ibid.) In Johnson's view, this example illustrates the independence of information from a material medium which can be used for its storage.

The statement about "disembodied" information floating over the Internet as well as the assertion allegedly illustrated by the example of actors memorizing plays are so blatantly lacking logic that I am tempted to give it a generic name of a "Johnson blunder." All his examples show is that information can be transmitted from one material medium (for example, paper and ink) to another material medium (for example, the actor's brains).

Of course, information and the medium carrying it are not the same. The medium can carry or not carry information. However, information cannot exist without some material medium in which it is recorded.

Recording information entails a certain change in the medium's structure. This change is reversible. Information can be erased as well as recorded. It does not exist unless it is recorded and hence cannot exist without a material medium.

Moreover, Johnson's example fails to mention that in the process of information transmission it is impossible to eliminate noise. The information received by another medium (in his example, the actors' brains) is never identical with the original information (in his example, the written text).

However, the most essential fact disproving Johnson's assertions is that, unless information has been transmitted to another material medium, it is destroyed as soon as the medium that contained that information has been destroyed. Information does not exist independently of the material medium on which it is recorded.

Another fault of Johnson's treatment of information is that he evidently confuses *information* with a *meaningful message*, which is a rather common misinterpretation.

Since Johnson concentrates on the information that is carried by biological macromolecules, mainly DNA, we can limit our discussion only to information contained in *texts* (a term meaning any string of symbols). These symbols can be letters of an alphabet, numbers (as in the infinitely long string representing π), or the elements of the DNA strand.

Information has been discussed in detail in chapter 1, so I will not repeat it here. That discussion shows how poorly Johnson understands the topic—information—he so bravely endeavored to discuss.

Indeed, in *The Wedge of Truth* Johnson discusses information once again, this time "teaching" his readers the real meaning of that term. Johnson writes: "Information theory is a complex subject far beyond the purview of this book."[19] This sentence may seem to imply that Johnson simply wants to spare his readers the travails involved in a discussion of information theory, which he himself has mastered. Actually, he betrays his ignorance of even the seminal concepts of that theory as soon as he tries to write about it. He writes: "By information I mean a message that conveys meaning, such as a book of instructions" (ibid.). It looks as if Johnson offers his own information theory, quite different from the one commonly accepted in science.

If Johnson had spent some time acquainting himself with the seminal concepts of the real information theory, he might have learned that according to this theory the concept of information is defined quite differently. Rather than referring Johnson to some high-level monograph, let us quote from a commonly available source, *Van Nostrand's Encyclopedia of Science* written for nonexperts. In the article titled "Information Theory" we read, "The term information as used in the context of information theory is not related to the meaning."[20]

Since Johnson obviously uses the term *information* in a way different from information theory, we have to carefully look at his statements each time he uses the word *information* to see whether or not that statement makes any sense. When he borrows in his discussion some concepts related to information from other sources, it very well may not be applicable to what he means by that term. On the other hand, when he makes use of that term as *he understands it*, it may behave very differently from the behavior of the real *information* as it is defined in information theory.

Johnson is not alone in the confusion of information with "meaning." Some other design theorists often make a similar mistake. Usually, however, it is done in an implicit way, when, in the course of a discussion, the subject is imperceptibly changed, switching from information to a meaningful message and vice versa. Johnson seems to stand alone in that he makes his mistake obvious by an explicit statement.

Johnson's main thesis, which he shares with other adherents of intelligent design, is that biological structures carry vast amounts of information and the latter cannot be generated by chance or by a combination of chance and natural law. Therefore, insists Johnson, living organisms must have been created by a supernatural intelligent agent.

The fallacy of this argument is due to the mix-up of two concepts—*information* and *meaning*.

Biological structures do indeed carry a lot of information. However, it is not what Johnson defines as information. The information carried by, say, a DNA strand, can very well be created by chance. Nothing prevents the generation of information, as it is defined in information theory, in a stochastic process. In particular, Johnson seems not to know that meaningless strings of symbols often carry much more *information*, in the proper sense of that term, than meaningful messages. This point was discussed in chapter 1. If, though, the term *information* is used, as Johnson suggests, to denote a *meaningful message*, then the probability of its being generated by chance is indeed exceedingly small. This statement, however, has little to do with biological structures. While the latter indeed carry a large amount of *information*, there is no evidence that biological structures carry a *meaningful message*. Actually there is plenty of evidence to the contrary.

In *The Wedge of Truth* we read, "A random assortment of letters also contains no significant information unless the sequence is also *specified* by some independent requirement."[21] If Johnson read at least an introductory text on information, he might learn, again, that a random sequence of letters carries, as a rule, more information than a meaningful text. To learn that, Johnson would need to familiarize himself with the concept of a text's entropy (see chapter 1). Regretfully, it is not being taught in law schools.

For example, the DNA strand, which is the primary example of an information-rich biological structure, is known to have a composition whose origin is very hard to explain by design but is better explained by a process of random unguided events. The DNA strand includes a large number of genes, each performing a certain function. If this statement were the whole story, it might give at least some foundation to the hypothesis of DNA carrying a "message." However, the whole story is different. Besides genes, the DNA strand also includes a very large number of so-called pseudo genes which do not perform any function: gene fragments, "orphaned" genes, "junk" DNA, and "so many repeated copies of pointless DNA sequences that it cannot be attributed to anything that resembles intelligent design," according to Kenneth Miller.[22] Of course, unlike Johnson, Miller, a professor of biology, is a real expert in the matter and has expressed in the above quotation the view of the overwhelming majority of biologists.

We see that Johnson is not only a dilettante in biology, he is also largely ignorant of *information* which he has the gall to "explain" to his readers.

The fact that Johnson, who may be an expert in law but is an obvious dilettante in biology and is ignorant of information theory, has acquired the status of a leader of the intelligent design theorists, shows the abject paucity of substance in that "theory."

JOHNSON PROSECUTES DARWIN

As mentioned before, I have no intention of engaging in a detailed discussion of biological problems because in my view such discussions should be left to experts. However, I think I may point out some quite obvious faults in Johnson's attack on Darwinism, in particular in his book *Darwin on Trial*.

The very title of that book betrays Johnson's real agenda—to prosecute Darwin and his followers as a bunch of criminals.

Being apparently confident, as is usual for a dilettante, of having mastered the nuances of biological science, Johnson has the gall to accuse scientists of lies, maintaining that scientists deliberately conceal information that may reveal a weakness in their position. Scientists, according to Johnson, do it for selfish reasons—to preserve their dominance in academia. In *Darwin on Trial* Johnson refers to evolution theory as a "subject that has for too long been protected from critical thinking by law and academic custom."[23] Indeed, Mr. Prosecutor? Which law prohibits a critique of Darwinism? If we believe Johnson, for a long time there has been no critical discussion of Darwinism among scientists, but now a lawyer named Johnson has arrived to finally clear up the matter. To remedy the sad situation, i.e., the alleged shortage of critical discussions of Darwinism, Johnson suggests a prescription: "If you are a scientist you can follow the path set by Michael Behe and others and bring out the crucial information that is not widely reported because it does not fit materialist preconceptions" (p. 452). It seems Johnson is talking here not about contemporary science but rather about the medieval church which indeed made it quite unsafe for a scientist to report information which contradicted the Bible. It remains Johnson's secret—the identity of those vile scientists who hide crucial information contradicting materialist preconceptions. As for his example of Michael Behe, the latter is indeed a scientist, a biochemist. However, Behe's reports which Johnson has mentioned are not a part of Behe's biochemical research. Johnson refers here to Behe's controversial book *Darwin's Black Box*, discussed in chapter 2. This book is a popular tale in which many interesting facts of molecular biology are mixed with arbitrary creationist interpretations of them, rejected by many experts in biology. It would be a very sad situation if other scientists followed Behe's path, so enthusiastically endorsed by Johnson.

Johnson, having read some material in biology, imagined that he mastered all the intricacies of that marvelous science. Actually, according to the reaction by professional biologists, Johnson's anti-Darwinian diatribes are like a blind man disputing the quality of Rembrandt's paintings.

Johnson's subterfuge is in attributing to scientists an attitude which actually belongs to Johnson and his cohorts. His main thesis is not a discussion of the intricacies of biological theories but rather his insistence that Darwinism is based not on evidence, but only on a purely philosophical premise, *metaphysical naturalism* (which he alternatively refers to as "metaphysical atheism," or "materialism"). Never mind that it is actually Johnson and his cobelievers who base their views on faith rather than on empirical evidence. In Johnson's picture, scientists who support Darwinism are depicted as members of a sect, who have a very narrow vision because of their doctrinaire adherence to atheism. On the other hand, Johnson depicts himself and his cohorts as open-minded and tolerant people, prepared to listen to and reasonably discuss any arguments, both supporting and contradicting their views. If only biological (and any other science) were freed from the shackles of the perfidious materialism, maintains Johnson, Darwinism would have no chance against creationism.

To support such a view, Johnson offers some critical comments on the Darwinian theory of evolution. Like his predecessors in the attack on Darwinism, Johnson concentrates on some aspects of Darwinism which have not yet been sufficiently clarified or understood. His argument is purely negative, since he does not provide any arguments which would indeed show the advantages of the supernatural explanation of the origin of life and of species. Johnson's negative argumentation does not offer anything that has not been heard before from other creationists. Most of these critical comments have been shown by many Darwinists to be superficial and usually unconvincing if the real details of biological science are taken into account.

Let us review some examples of Johnson's rebuttal of certain arguments in favor of Darwin's theory of natural selection.

Chapter 2 in *Darwin on Trial* is titled "Natural Selection." Johnson's main thesis in this chapter can be briefly expressed as the assertion that the principle of natural selection as one of the driving forces of evolution has no empirical confirmation. Johnson discusses several approaches used to substantiate the principle of natural selection and finds all of them unsatisfactory.

In a section titled "Natural Selection as a Tautology" Johnson misrepresents some arguments by Darwinists when he formulates that principle in a derisive form. He writes, for example, that the principle of natural selection is nothing more than a tautology of the type: "organisms that leave the most offspring are the ones that leave the most offspring" (p. 22). Does Johnson indeed believe that biologists who accept Darwinism are such dimwits that they do not understand the obvious emptiness of the quoted assertion?

Without delving into the meaning of the principle of natural selection, we can state an obvious fact: if this principle is supported by evidence (as it is indeed), it has considerable explanatory power. It shows the path of evolution as controlled by the easily understood force of predominant survival of species better adapted to the environment. If Johnson wishes, he may present this principle in a tautological form, although such a presentation is misleading. Presenting something in a tautological form does not at all mean disproving it.

Actually, it seems that, indeed, in Johnson's view too many scientists are fools who do not understand elementary logic and stubbornly disregard facts in order to preserve their philosophical prejudices. Many statements to that effect are scattered all over his books, with the frequent use of the term "nonsense" in regard to views that differ from Johnson's.

Apparently realizing that derision of the seemingly tautological character of the principle of natural selection does little to demolish that principle, Johnson proceeds to show that the idea of natural selection has no basis in any empirical evidence.

The two most important points in that discussion can be found in the sections "Natural Selection as a Scientific Hypothesis" (pp. 24–28) and "Natural Selection as a Philosophical Necessity" (pp. 28–31) in *Darwin on Trial*.

Discussing natural selection as a scientific hypothesis, Johnson insists that the evidence offered by the Darwinists is not at all convincing. He writes, "None of the 'proofs' provides any persuasive reason for believing that natural selection can produce new species, new organs, or other major changes, or even minor changes that are permanent" (p. 27).

Of course, finding some arguments to be not persuasive enough is everybody's prerogative. However, experts in biology overwhelmingly view the evidence as persuasive enough to accept it as a good scientific theory. I find it more reasonable to rely on the opinion of experts than on that of a lawyer who is obviously prejudiced against the theory in question because of his proclaimed philosophical and religious views. This is more so because of Johnson's obvious blunders when he discusses questions which are closer to my own area of expertise (for example, his discussion of information). Why should I expect his discussion of biological problems to be any better? He is no more a biologist than he is an expert in information theory.

Biologists overwhelmingly agree that there is very strong evidence in favor of the natural selection at work in the process of evolution. They overwhelmingly believe that evolution results in the emergence of new species. While Johnson is not persuaded by the evidence in question, he does not offer a single argument showing that the emergence of new species or organs via natural selection is impossible. He admits that evolution as a simple change within some narrow limits takes place. Where are those limits? Johnson provides no arguments to support his contention that those limits are narrow indeed to the extent making the emergence of new species impossible.

It seems amusing that Johnson's quasi-logical rebuttal of the arguments by evolutionists in the chapter in question has been demonstrated to be baseless, this demonstration having come from his own camp. I am referring to the writing of Del Ratzsch.

Ratzsch participated in the conference of design theorists at Biola University, where he gave a talk. His paper titled "Design, Chance, and Theistic Evolution"[24] appears in the collection *Mere Creation*, edited by William Dembski, along with the papers by Michael Behe, William Dembski, Phillip Johnson, Stephen Meyer, Paul Nelson, and other principal proponents of intelligent design. No opponent of that theory had access to either the conference at Biola or to the collection in question. His book *The Battle of Beginnings*[25] was published by the same InterVarsity Press which has published a long list of pro-design books but not a single one arguing against intelligent design. Hence, he obviously belongs to the same camp as Dembski, Johnson, and their cohorts.

However, Ratzsch stands alone in one respect: he is logical. This logic is the foundation of Ratzsch's relative impartiality. In *The Battle of Beginnings*, Ratzsch offers an analysis of the false arguments which are often the staple of the mutual attacks by the adherents of the two opposing views.

Ratzsch's analysis actually shows that the position of evolutionists is much stronger than that of creationists. His critical autopsy of false arguments by creationists depicts a pitiful picture of either ignorance or of a deliberate disregard for the real arguments of evolutionists, thus making the creationists' critique hopelessly inadequate. On the other hand, when Ratzsch turns to debunking the critique of creationism by evolutionists, he finds mostly secondary erroneous points and misinterpretations by evolutionists. Therefore, while Ratzsch evidently aimed at demonstrating that both sides of the dispute are equally guilty of inaccuracies (which may be true), his analysis pictures a rather unequal stand of the two sides in the dispute, with evolutionists looking much better than their creationist opponents.

In Ratzsch's book we find something not found in any other writings of design theorists and their cohorts. Ratzsch reveals the errors in Johnson's arguments. He did not shy away from debunking the "leader" of the design theorists. Unlike admirers of Johnson, the only praise Ratzsch is willing to bestow on Johnson is by saying that Johnson's book *Darwin on Trial* along with Denton's *Evolution: A Theory in Crisis* served as "catalysts" of the trend to elevate the sophistication of discussion from the primitive level of "early creationists" to the more recent "higher tier" of participants (p. 84). On the other hand, Ratzsch shows the fallacy of Johnson's arguments. First he debunks the creationists' argument that the theory of natural selection is a tautology (pp. 144–45). Ratzsch does not mention Johnson by name in this section, but since Johnson uses, as we discussed, the argument alleging the tautological character of the natural selection concept, Ratzsch's debunking tool shows the absurdity of Johnson's stand on that question. A few pages further, in a section titled "Circular Reasoning," Ratzsch directly names Johnson and shows again the fallacy of the latter's argument against evolutionists. Ratzsch shows here that Johnson (as well as Denton) does not understand the nature of scientific reasoning and the relationship between experimental data and theories. The fact that this analysis comes from Johnson's own camp makes it unnecessary for me to delve into the details of Johnson's faulty arguments; I refer instead to Ratzsch's book.

THE WEDGE OF ARROGANCE

It is educational to look at how Johnson reacts to critical comments about his previous publications. In particular, in *The Wedge of Truth* he responds to professors Robert Pennock and Kenneth Miller.

In Johnson's view, both his opponents simply don't know what they are talking about. Both writers, Johnson tells us, resort to a caricature instead of honestly reporting on Johnson's strong arguments. Pennock is "naïve" (p. 60), and his errors are "elementary" (p. 135). According to Johnson, Miller "grotesquely distorts the design concept" (p. 131), and so on.

However, Johnson's reasoning shows signs of his own, often elementary, misunderstanding of the subjects he so bravely argues. Here is an example. Miller compares an explanation that life emerged because of a supernatural intelligence to an imaginary situation in which some adherent of intelligent design would refuse to explain in natural terms the spectrum of solar radiation, resorting instead to the hypothesis of a supernatural origin of that spectrum. Johnson derides Miller's example as allegedly showing Miller's egregious misunderstanding of the difference between the two situations. He writes, "Miller either does not know, or chooses to ignore, that the argument for intelligent design rests primarily on the existence of complex genetic information and the absence of a natural mechanism for creating it" (p. 130). In fact, Miller simply illustrates the difference of approaches to the explanation of phenomena between science and the so-called intelligent design theory. While the scientific approach, based on a search for a natural explanation, is capable of providing a strong explanatory power, a reference to a supernatural cause lacks such a power. In Johnson's view, Miller's example is a caricature, because the sun's radiation follows "a regularity produced by a known physical process" (ibid.). On the other hand, Johnson and his cobelievers in design "are concerned with the much more important question of the origin of irreducibly complex systems or new complex genetic information" (p. 131).

Why this interest in a much more important question makes Miller's example a caricature is Johnson's secret. I will give a slightly different example. Johnson probably has heard about the laws of thermodynamics. There are four of them. Thermodynamics itself does not explain these four laws but simply postulates them as a generalization of immense observational and experimental evidence. If we wish to understand deeper the nature of these laws, we can choose one of two approaches. One is to assert that we do not know the intrinsic meaning of these laws and attribute them to a supernatural origin. This is similar to the attitude adopted by adherents of intelligent design when discussing the origin of genetic information. Science, however, chose another path. As a result of an intensive scientific work, the nature of the laws of thermodynamics found a very transparent explanation in a purely natural way. This explanation is given in statistical physics, which makes clear the underlying molecular mechanism of the laws of thermodynamics which are a manifestation of an enormous difference between the probabilities of various microscopic configurations of the host of particles constituting macroscopic bodies. If physicists chose the prescription for solving problems given by Johnson and his cohorts, the laws of thermodynamics would remain mysterious products of the supposed intelligent agent (i.e., completely unexplained).

To refute each statement in Johnson's books would require a few more books of about the same size. Of course, I have no intention to do that, as I have already spent more time and effort on discussing Johnson's quasi-scientific exercise than the latter deserves. Let me quote one more time from *The Wedge of Truth.* According to Johnson, scientists who happen to disagree with his views on science maintain that "scientific evidence is not really needed to prove the theory true any more than scientific evidence is needed to prove that two plus two equals four" (p. 45). Recall that it was Johnson who accused some other writers of making caricature of the views of design theorists.

The quoted statement is indeed a caricature, moreover, a caricature which does not have even a remote semblance to facts. It is Johnson who offers assertions without any evidence to support them. When Johnson claims to be a Christian believer, he does not suggest any evidence which could support his beliefs. In his view, only scientists are required to provide evidence, while Johnson and his fellow believers are not. In fact, contrary to his caricature, scientific theories are inevitably based on evidence, and no scientist has ever suggested that evidence is not necessary. Johnson tries to attribute his own attitude to scientists.

CONCLUSION

Throughout his books, Johnson repeatedly emphasizes that he is not a "creationist." Moreover, while rebuffing some of his opponents, for example Robert Pennock, Johnson refers to their use of the term "creationism" for him and his cohorts as an attempt on the part of his opponents to smear him. However, as we proceed chronologically through Johnson's publications, we see how this camouflage gradually becomes more and more transparent. Johnson is as hard-boiled a creationist as they come, no matter how fervently he protests such a label. This follows from the essence of his arguments throughout his books and papers, and comes into the open in his most recent book *The Wedge of Truth.* In that book Johnson largely sheds the mask of an objective analyst concerned only with facts and openly proclaims the superiority of religious faith over science.

Arguing against the critique by Robert Pennock in the latter's *Tower of Babel*,[26] Johnson writes ". . . intelligent design advocates [the "new creationists" targeted in Pennock's subtitle] do not bring Genesis into the discussion at all" (p. 136). Here Johnson again claims that his and his cohorts' arguments have no religious foundation. In fact, the opposite is true. Chapter 7 in *The Wedge of Truth*, titled "Building a New Foundation for Reason," is replete with statements to this effect. Here are a few examples of such statements: ". . . science means a very partisan adherence to a philosophy variously called naturalism, materialism or physicalism" (pp. 145–46). "By any realistic definition naturalism is a religion, and an extremely dogmatic one" (p. 148). "Naturalism is ultimately incompatible with the existence of reason" (p. 149). "At a fundamental level we know the reality of God . . ." (p. 152). "The materialist story counterfeits the authority of science in order to masquerade as an inference from scientific evidence, but it is actually based on idolatrous fantasy" (pp. 155–56). This comes from a man who claimed to make conclusions based solely on evidence but not on philosophical or religious grounds. In Johnson's view scientists who disagree with him are "claiming to be wise, but become fools" (p. 156). "Once we learn that nature does not really do its own creating, and we are not really the products of mindless natural forces that care nothing about us, we will have to reexamine a great deal else" (p. 159). Isn't that statement the credo of a creationist? Remember the ancient saying: "Persians are recognized by their tall hats"?

Johnson's assault on science has nary a chance of influencing the development of science. The noise produced by his books, lectures, and articles is just a nuisance. Those serious scientists who replied to the writing of that arrogant and prejudiced dilettante did him a favor, displaying a lot of politeness. As to his predictions regarding the imminent collapse of Darwinism and neo-Darwinism, I can repeat the words of the Russian poet Vladimir Mayakovski. Once, when he was publicly reading his poems, a listener asked, how long, in the poet's opinion, his poems would survive. Mayakovski answered: "Well, drop by in a thousand years to talk about it."

NOTES

1. For example, see Phillip E. Johnson, *Darwin on Trial* (Downers Grove, Ill.: InterVarsity Press, 1993); *Reason in Balance* (Downers Grove, Ill.: InterVarsity Press, 1995); *Defeating Darwinism by Opening Minds* (Downers Grove, Ill.: InterVarsity Press, 1997); *The Wedge of Truth* (Downers Grove, Ill.: InterVarsity Press, 2000); "How to Sink a Battleship," in *Mere Creation*, ed. William A. Dembski (Downers Grove, Ill.: InterVarsity Press, 1998), pp. 446–53.

2. Johnson, *Darwin on Trial*, p. 14.

3. Ibid.

4. Brian Spitzer, "The Truth, the Whole Truth, and Nothing but the Truth? Why Phillip Johnson's *Darwin on Trial* and the 'Intelligent Design' Movement are neither Science—nor Christian," Talk Reason [online], www.talkreason.org/articles/honesty.cfm [August 6, 2003].

5. Johnson, *The Wedge of Truth*, p. 14.

6. Nancy R. Pearcey, "You Guys Lost," in *Mere Creation*, p. 73.

7. Johnson, *Defeating Darwinism by Opening Minds*, p. 10.

8. Johnson, *Darwin on Trial*, p. 160.

9. Johnson, "How to Sink a Battleship," p. 448.

10. Johnson, *Darwin on Trial*, p. 13.

11. Ibid.

12. For example, among the sources where nonbiologists can find a clear presentation of the essence of the theory of evolution that, in my view, convincingly shows the fallacy of Johnson's position are "Biology and Evolutionary Theory," The Talk.Origins Archive: Exploring the Creation/Evolution Controversy [online], www.talkorigins.org/origins/faqs-evolution.html; Talkdesign.org: Critically Examining the "Intelligent Design Movement" [online], www.talkdesign.org; and Talk Reason [online], www.talkreason.org.

13. Johnson, "How to Sink a Battleship," p. 448.

14. Ibid., p. 449.

15. Johnson, *Defeating Darwinism by Opening Minds*, p. 12.

16. Ibid., p. 74n.

17. Johnson, *The Wedge of Truth*, p. 135.

18. Johnson, *Defeating Darwinism by Opening Minds*, p. 72.

19. Johnson, *The Wedge of Truth*, p. 42.

20. George R. Cooper, "Information Theory," *Van Nostrand's Scientific Encyclopedia* (New York: Van Nostrand Reinhold, 1976), p. 1360.

21. Johnson, *The Wedge of Truth*, p. 54.

22. Kenneth R. Miller, "Life's Grand Design," *Technology Review* 97, no. 2 (1994):24.

23. Johnson, *Darwin on Trial*, p. 12.

24. Del Ratzsch, "Design, Chance, and Theistic Evolution," in *Mere Creation*, p. 289.

25. Del Ratzsch, *The Battle of Beginnings* (Downers Grove, Ill.: InterVarsity Press, 1996).

26. Robert T. Pennock, *Tower of Babel* (Cambridge, Mass.: MIT Press, 2000).

PART 2

MENTAL ACROBATICS

How Religious Writers Prove the Compatibility of the Bible with Science

4

VERIFYING THE ETERNAL BY MEANS OF THE TEMPORAL?

An Introduction to the Reviews of Books Asserting Harmony Between the Bible and Science

The problem of the relationship between biblical stories and scientific data has often occupied the minds of many people, believers and skeptics alike. Many believers would like to reconcile the biblical account with science, whereas many skeptics would like to find proofs in favor either of the biblical account or of science. This interest can be easily understood, since science provides strong rational evidence supporting its conclusions, whereas religion satisfies a deep emotional desire to believe in miracles, which lifts humans above their animal nature and defy the tragedy of unavoidable death.

There are three alternative worldviews one may choose from. One choice is to accept religious dogma as the ultimate truth and hence to deny scientific data and theories as being contrary to the divine revelation of truth. Another choice is the opposite, to adhere to the scientific view and to reject the biblical story, viewing it at best as mere literature, and at worst as a baseless concoction.

There is, though, one more choice, an attempt to reconcile the scientific data and the biblical account.

The advantages of the third choice are obvious, because if such an attempt is successful, it enables one to accept the intellectually rigorous claims of science while saving one's emotionally satisfying religious beliefs.

In response to the demand, a multitude of books, papers, Web sites, etc., have appeared recently, all of them offering various arguments aimed at proving the compatibility of the Bible and science.

Surveying the literature which discusses the relationship between the

Bible and science, one cannot fail to notice several features common to the overwhelming majority of these publications.

One such feature is that the books and papers in question do not belong to scientific literature.

In scientific publications, their authors usually ask a certain question and then proceed to find the answer to it, supposedly without a preconceived idea as to what the answer may be. In an alternative form of scientific publications, their authors suggest a certain hypothesis, or often several competing hypotheses, and then try to make a rational choice between those hypotheses, based on factual evidence. The conclusions suggested in a scientific publication usually are not offered as the ultimate truth, but rather as the most likely one out of several choices, within the framework of the available data. It usually implies that in case additional data are discovered, the conclusion may be reconsidered. (The above description is not necessarily true for mathematical publications, where often a proven theorem may be the closest thing to the absolute truth.)

One more typical feature of scientific publications is that their authors are supposed to try themselves to consider possible arguments against their conclusions, weighing them as impartially as possible against their own views.

Another type of literature, which is in a certain sense opposite to the scientific one, is polemic literature. Recall that the word *polemic* itself stems from the Greek word *polemos*, meaning "hostility" or "war." Polemic literature attempts to convince the reader of the validity of the authors' views or beliefs. In this type of literature, authors commonly emphasize arguments in favor of their position and downplay counterarguments.

One more difference between scientific and polemic literature is that the scientific literature is usually addressed to everybody equally, provided the readers have a sufficient educational background to understand the subject. In contrast, polemic literature is usually addressed to a specific audience, so that the argumentation is adjusted to the expected readership.

Publications dealing with the relationship between the Bible and science (which will be reviewed in part 2), are overwhelmingly of the polemic type. Contrary to their authors' usual claims of objectivity (i.e., comparing the biblical story with scientific data in a nonprejudiced way), more often than not the actual goal of these writings is to prove the supremacy of the religious revelation. For example, in publications discussing the Torah versus science, the Torah is viewed as the indisputable depository of the ultimate truth, whereas only the scientific data are subjected to a critical discussion.

A good illustration of the above observation can be provided by a quotation from *In the Beginning* by professor of physics Nathan Aviezer, who writes in a section titled "Credo," "If I were to find that traditional Judaism appeared to be inconsistent with certain aspects of modern science, this would in no way weaken my commitment."[1] (The commitment mentioned in this quotation is to Aviezer's religious faith). In *The Science of God* by Gerald L. Schroeder, we find the following frank statement by the author: "I have an agenda, to demonstrate the harmony between science and the Bible."[2] Pursuing an agenda is not what is usually considered an unbiased approach to subject. Similar statements can be found in publications by authors who are Christians. For example, mathematician William A. Dembski, who is one of the most prominent promoters of the intelligent design theory, writes in his introduction to *Mere Creation*, "As Christians we know naturalism is false."[3] Dembski makes this statement before having discussed any arguments either in favor or against his antinaturalism views. Similar statements, often made in a rather uncompromisingly categorical manner, can be found in many other publications by defenders of the Bible. For example, one of the most prolific adversaries of the evolution theory, professor of law Phillip E. Johnson, more than once expressed in his books (for example, *Darwin on Trial*) a bluntly contemptuous attitude toward those scientists who do not share his rejection of the evolution theory (see chapter 3).

It is not hard to figure out to whom the literature in question is addressed. True believers usually are not concerned with the compatibility of the biblical story with science. Therefore, these true believers are not much interested in books and papers aimed at proving in rational terms something they deeply believe to be true anyway. There are, though, many believers who adhere to their faith simply because they grew up in a religious family, or are accustomed to certain rituals and modes of behavior, or have a strong emotional need in faith. It could be proper to say that such people actually wish to believe, and are searching for rational arguments supporting the religious dogma. The literature in question is addressed mainly to this category of readers, and therefore uses arguments which are expected to be sufficiently convincing to these "doubting believers."

There is one more category of potential readers of the literature in question—skeptics. It is this category of readers the writers of the books and papers in question view as their opponents. The writers of polemic literature usually realize that arguments which are sufficiently convincing for "doubting believers" may not be convincing to skeptics. Therefore, most of the literature in question is implicitly addressed also to skeptics, utilizing more or less clever and ingenious arguments designed to convince the latter of the writer's views and beliefs.

While arguments which are sufficiently convincing for "doubting believers" may not necessarily be distinguished by strict logic and scientific rigor, the arguments which are supposed to work for skeptics must meet much more stringent requirements. One requirement, which is necessary for the argument to work, is that they are impeccably logical and consistent. Besides that requirement, one can imagine several types of argumentation, to wit:

a. Arguments which are both relevant and correct
b. Arguments which are scientifically correct, but irrelevant
c. Arguments which are relevant, but scientifically wrong
d. Arguments which are both irrelevant and wrong

One can also imagine all kinds of arguments which do not belong to any of the above four "pure" types, but rather being of an in-between type, displaying partial features of more than one of the above types (for example, being partially correct, etc.).

Obviously, only arguments of type (a) can be convincing to skeptics. Nevertheless, the books and papers aimed at proving the compatibility of the Bible with science often make use of arguments belonging to types (b) through (d). While such arguments may be sufficiently convincing to "doubting believers," not to mention the true believers, they can hardly have any effect on skeptics.

From another angle, the literature in question has the following features:

1. The levels of requirements are different for the Bible and science. The biblical story is usually not subjected to a critical evaluation in accordance with any criteria. It is assumed to be true, and if it seems illogical on the face of it, this is assumed to simply signify an insufficient understanding of its real meaning. Scientific data, though, are required to be proven in the most unambiguous and uncontroversial way.

   If the biblical account is or seems to be compatible with scientific data, these data are viewed as a rational proof of the biblical account. If, though, the scientific data contradict the biblical account, this is viewed simply as a misunderstanding (if the scientific data are firmly proven) or as the falsity of scientific data (if the latter have not been proven beyond doubt).

2. Consequently, there are four situations, each treated in its own way, to wit:
   a. If a certain biblical assertion is (or seems to be) in agreement with proven scientific data, then such scientific data are referred to as a scientific confirmation of the biblical story.
   b. If a certain biblical assertion seems to be in agreement with some scientific theory, which has not been firmly proven, this scientific theory is often referred to as if it is proven (for example, the big bang theory).
   c. If a certain biblical assertion seems to contradict a proven scientific theory, attempts are made to interpret the biblical story in a way compatible with scientific data (for example, the six days of creation). As it will become evident when we discuss particular books and papers, such attempts often put the writers in a rather awkward position, forcing them to perform some mental acrobatics.
   d. If a certain biblical story seems to contradict a scientific theory which has not been unequivocally proven (or seems to be not unequivocally proven), various arguments are forwarded aimed at disproving the scientific theory (for example: the theory of evolution).

There are, among the multitude of books dealing with the relationship between the Bible and science, some which cover all four situations listed above. The books in this category can be referred to as being of "Bible and science" type. Examples of this type of publications are the already mentioned books by Schroeder and Aviezer. In such books one can find examples of all four types of argumentation listed above.

Other books cover only three, two, or even one of the above four situations. In particular, there are books in this category dedicated solely to rebuttals of Darwin's or post-Darwinian theories of evolution.[4] Such books usually explicitly deny the scientific theory in favor of the biblical story. Sometimes it is done in a rather brazen way, rejecting the scientific view in a categorical and even disdainful manner. Others do the same in a subtler manner, trying to profess respect for science and somehow creating the impression that the author is ashamed of being considered antiscience.

In another respect, one can imagine two types of arguments in favor of the biblical story. One type is rational argumentation, and the other an extrarational one. The extrarational arguments are obviously excessive for any true believer, since in a believer's view the Bible itself supplies more than enough of the strong, even undeniable, evidence which itself is of extrarational type and makes unnecessary any additional extrarational arguments. As to the skeptics, they cannot be swayed by only extrarational arguments. Therefore, the books and papers aimed at supporting the biblical story use some allegedly rational argumentation, thus appealing to the skeptics' minds rather than to their emotions. In view of that, the argumentation forwarded in the books in question is a legitimate object of discussion in rational terms, and revealing flaws in those arguments is what those publications call for.

This chapter is meant as an introduction to the detailed discussion of various books and papers devoted either to proving the harmony between the Bible and science or to rebuttals of some scientific theories that contradict the biblical story. This discussion is the subject of the following chapters.

There are too many books about the Bible versus science controversy to make it possible to review all or even most of them in one place. I have chosen for review several books which seem to be representative of that genre and have gained substantial popularity. In chapters 5 through 7 the books by those writers who approach their topic from a Christian perspective are discussed, while in chapters 8 through 11 the books by those writers who approach it from the standpoint of Judaism are discussed.

NOTES

1. Nathan Aviezer, *In the Beginning: Biblical Creation and Science* (Hoboken, N.J.: KTAV, 1990), p. 124.

2. Gerald L. Schroeder, *The Science of God: The Convergence of Scientific and Biblical Wisdom* (New York: Free Press, 1997), p. 51.

3. William A. Dembski, introduction to *Mere Creation*, ed. William A. Dembski (Downers Grove, Ill.: InterVarsity Press, 1998), p. 13.

4. See, for example, Michael J. Behe, *Darwin's Black Box: The Biochemical Challenge to Evolution* (New York: Simon and Schuster, 1996) (see chapter 2 in this volume); Lee M. Spetner, *Not By Chance: Shattering the Modern Theory of Evolution* (New York: Judaica Press, 1997) (see chapter 11 in this volume); Phillip E. Johnson, *Darwin on Trial* (Downers Grove, Ill.: InterVarsity Press, 1993) (see chapter 3 in this volume).

5
A CRUSADE OF ARROGANCE

Hugh Ross is a prolific author of books, Web postings, etc., all dedicated to but one goal—to prove the complete compatibility of modern science with the word of the Bible, and, more specifically, with his particular brand of the Christian faith. In this chapter I will discuss some features of Ross's books to see if he has accomplished his goal in a way sufficiently convincing for skeptics.

ROSS AS A SCIENTIST

We learn from the information provided in Ross's books that he is the head of an organization named Reasons to Believe. This organization adheres to the "doctrinal statements of the National Association of Evangelicals and of the International Council on Biblical Inerrancy."[1] Ross has also served as a minister of evangelism. Hence Ross's credentials as a proponent and defender of the Christian faith (in its non-Catholic and non-Mormonic version) are obvious.

On the other hand, the readers of Ross's books are advised that he has a Ph.D. in astronomy from the University of Toronto and that he had been a postdoctoral fellow at Caltech, studying quasars and galaxies. Indeed, in his curriculum vitae we find a list of five published papers on certain astrophysical matters in which Ross is either the author (in two of them) or coauthor (the last such publication dated 1977).[2] However, in order to judge Ross's competence as a scientist we do not need to study these five old short papers with their rather narrow topics. We can gain a rather good understanding of Ross the scientist by relying on what we can read in his books on the Bible versus science subject.

Let us look, for example, at Ross's book titled *The Fingerprint of God.*[3] In this book Ross reviews various cosmological theories, thus showing his familiarity with the relevant literature. It is hard, though, to figure out who are meant to be the readers of that book. On the one hand, Ross seems to explain the cosmological theories briefly and in a simplified manner, which would be natural if the book were written for laypersons. On the other hand, Ross's presentation is full of references to rather sophisticated scientific concepts which can only confuse a layperson. For example, on pages 42 and especially 44, Ross provides mathematical equations of the general theory of relativity, which obviously would be incomprehensible for laypersons, more so since they are not explained in the text and do not actually add any useful information.

The overall style of Ross's presentation of scientific theories, in which he, on the one hand, mentions dozens of scientific terms hardly familiar to nonexperts, but which, on the other hand, is often lacking in clarity, makes one wonder how well he himself understands those theories. It is hard to avoid the feeling that the above equations and terms are presented for the sole reason of impressing the readers with Ross's scientific qualifications and knowledge.

In his book *Creation and Time* there is a box titled "Decay and Work," where Ross offers a simplified explanation of the laws of thermodynamics.[4] While some imprecision inherent in that explanation can be forgiven if we assume that it is addressed to laypersons, there is one item in it which stands out as a sign of Ross's own possible insufficient understanding of the subject. Let us quote Ross: "Because of the principle of pervasive decay, heat energy can be transformed into mechanical energy (or work). . . . The maximum amount of heat energy that can be transformed into work is proportional to the difference between the temperature of the hot body and the temperature of the cold body divided by the temperature of the hot body."

We see in that quotation the assertion, repeated twice, that heat energy can be transformed into work. Such a statement makes a physicist pause because, if accepted literally, it is contrary to basic concepts of thermodynamics. One such basic concept holds that energy can be transformed from any specific form into any other form. For example, chemical energy can be transformed into electric energy, mechanical energy can be transformed into thermal energy, etc. However, the quantity named *work* is not considered in thermodynamics as a form of energy, which a body can possess, but rather as a quantity which measures the amount of energy transformed from one form into another form or transferred from one body to another body. While a body can possess a certain amount of energy in any specific form, it cannot "possess work." Work can be performed in a process, wherein a transformation (or transfer) of energy occurs, but as soon as the process is completed the work is nowhere to be found. If some energy were "transformed into work," it would mean this amount of energy has disappeared. Actually, the fact that some work was indeed performed can only be established by comparing the total amounts of energy in its various forms before and after the process. Therefore the statement by Ross about "heat energy transforming into work" in its literal interpretation is meaningless.

Still, we might consider the above statement by Ross as a display of a sloppy style rather than of misunderstanding. However, such a charitable interpretation falls apart when we continue reading the book in question.

In *Creation and Time* we also read: "As the universe expands from the creation event, it cools, like any other system obeying the laws of thermodynamics. When the heat energy of a system fills a greater volume, there is less energy per unit volume to go around" (p. 135).

The quoted statement by Ross confirms the suspicion that he has an insufficient understanding of some fundamentals of physics.

In order to clarify our statement, we have to talk about some concepts of thermodynamics. Necessarily, this discussion must be somewhat technical, even though still omitting, for the sake of simplification, some subtle points of thermodynamics.

The fundamentals of thermodynamics are expressed as the four laws of thermodynamics, which are postulates based on the generalization of the observed behavior of macroscopic bodies. They describe the behavior of systems which contain a very large number of constituent elements (molecules, atoms, ions, etc.). The real meaning of the four laws is revealed in another part of physics known as statistical physics, which shows that the laws of thermodynamics are not absolute but only express the most probable way the systems behave.

The law of thermodynamics that is relevant to the above quotation from Ross is the first law. The first law of thermodynamics is actually the law of energy conservation for macroscopic systems. It deals with three quantities, named internal energy, heat, and work. It states that in a thermodynamic process the change of a system's internal energy equals the algebraic sum of work and heat. The "heat" in that law, like work, is not considered a form of energy per se, which a body can possess. It is a measure of such energy transfer to or from the system which occurs via random interactions of the system's constituent particles with particles in the system's surrounding.

If a system expands (for example, a piston in a cylinder moves up, lifting some load), the system (in this example the gas in the cylinder) performs certain work. If a work is done, the internal energy of the gas decreases (it transforms into the increased gravitational potential energy of the lifted load). The decrease of the internal energy may be, either partially or fully, compensated for by a heat influx from the surrounding medium into the system. If the compensation is only partial, or is completely absent, the internal energy of the system decreases.

The part of the internal energy which is relevant to the problem in question is the thermal energy. Essentially it is the kinetic energy of the randomly moving constituent particles of the system. The measure of that thermal energy is called temperature. Hence, if the decrease of the thermal energy *which was caused by the work done by the system* is not fully compensated for by the heat influx, temperature of the system drops.

The important point here is that it is not the expansion itself which is the reason for the temperature drop. There are at least two situations when expansion is not accompanied by a temperature drop. One is the situation when the heat influx from the surrounding medium completely compensates for the decrease of internal energy caused by performing work (such as lifting a load). Moreover, if the heat influx from the surrounding medium exceeds the amount of work, the temperature of the system increases, despite the system's expansion. One more situation where expansion does not result in a temperature drop is an expansion in which no work is done. A well-known example of the latter is the expansion into vacuum. Imagine a box divided by a partition into two compartments, one compartment containing gas at some pressure and temperature and the other being empty. If we remove the partition, the gas will rapidly expand into the empty compartment. Since gas expands into vacuum, i.e., without resistance, no work is done in this expansion.

If we believed Ross, the temperature of the expanding gas would drop because, as he said, "when the heat energy of the system fills the larger volume, there is less energy per unit volume to go around" (p. 135). This expression is meaningless. In reality, as has been established, both theoretically and experimentally, in thermodynamics, the described expansion into vacuum is an *isothermal* process in which the temperature of the gas remains constant.

The concept of heat as a substance that is diluted as volume increases has been abandoned by physics as incorrect since Anglo-American scientist Benjamin Thompson, Count Rumford, authored the concept of heat as a form of motion (1796). Ross seems not to know that. His explanation assumes that temperature drop is caused by the expansion itself, which is contrary to thermodynamics.

In another one of Ross's books, *The Fingerprint of God*, we find again a similar explanation of the universe's cooling at expansion. We read, "In the standard big bang model, the universe expands smoothly and adiabatically from the beginning onward"; in a footnote to this, Ross explains: "Under adiabatic expansion the temperature will drop due to the expansion alone without loss of heat from the system."[5]

The term *adiabatic* denotes a process in which there is no exchange of thermal energy between the system and its surrounding. The situation we discussed earlier, in which a gas expanded into an empty compartment, was both *adiabatic* and *isothermal*, so there was no temperature drop, contrary to Ross's assertion. Ross, again, seems not to know that the temperature drop occurs only if the expansion is accompanied by work done by the system against external forces, while "expansion alone" causes no temperature drop.

In regard to the reason for the universe's cooling in the process of its expansion, thermodynamics, contrary to Ross's assertion, can't provide a direct explanation. The main problem is the uncertainty as to how thermodynamics can be applied to the behavior of the universe as a whole, as the expansion of the universe occurs into "nothing" (i.e., the universe is not surrounded by another medium as the adiabatic systems studied in thermodynamics are). If we accept the big bang theory, we have to realize that thermodynamics is not applicable to the big bang itself (actually to anything that occurred within the so-called Planck time, which is 10^{-43} sec). The laws of physics may, though, be legitimately applied to anything that occurred after the Planck time. This includes the four laws of thermodynamics, in particular the first and the second laws.

According to the first law the total energy of the universe remains constant regardless of any processes taking place in the course of the universe's expansion. (There are good reasons to believe that the net energy of the universe is zero.) According to the second law, the entropy of the universe increases along with the latter's expansion. The rate of expansion exceeds the rate of the creation of order (i.e., of stars and galaxies, etc.), so despite the emergence of gradually increasing order, which is accompanied by the local decrease of entropy, the overall entropy of the universe increases in agreement with the second law of thermodynamics.[6]

The temperature drop occurs because the initial thermal energy of the universe gradually converts, first, into the *rest mass* of the emerging stars, and second, into the kinetic energy of the motion of those clumps of emerging matter moving away from the initial seed of the universe. Judging from Ross's preposterous statement about the temperature drop being the result of energy diluted in ever-increasing volume, he has a rather nebulous understanding of elementary concepts of physics and cosmology.

Having thus established Ross's actual level of scientific competence, we can judge all of his ideas in regard to the compatibility of the Bible with science with that knowledge in mind.

ROSS AND NON-CHRISTIANS

As mentioned before, Ross's discussion is always from the viewpoint of a believing Christian. There are many books treating the same problem as Ross does, offering arguments in favor of compatibility of the Bible with science, but written from different standpoints. Some of these books have been written from the vantage point of Judaism by authors such as Gerald L. Schroeder and Nathan Aviezer (whose books are reviewed in chapter 9 and 10, respectively).[7] It is interesting to note that Ross and authors like Schroeder and Aviezer often use identical arguments and sometimes share the same errors.

For example, Schroeder and Ross discuss the nature of pre-Adam "hominids" in very similar terms. Both assert that the humanlike creatures who roamed the earth for many thousand years before the date when, according to the Bible, Adam was created by God from dust, were really not human since they did not possess "soul."

Both Ross and Schroeder show a very similar misunderstanding of some basic concepts of thermodynamics. Both suggest the same incorrect explanation of the universe's cooling after the big bang as being due to "heat dilution in ever increasing volumes" (according to Schroeder), or "when the heat energy of a system fills a greater volume, there is less energy per unit volume to go around" (according to Ross).

Since the books by Ross and Schroeder appeared during the same period of time, it is hard to judge whether Ross borrowed his incorrect statement from Schroeder, or Schroeder borrowed his absurd explanation from Ross, or whether they both came up with the same naive idea independently from each other by sheer chance. What is of interest, though, is that neither Schroeder nor Aviezer make any reference to Ross, and Ross never mentions Schroeder or Aviezer. Of course this can be understood in view of different agendas of Ross on the one hand, and Schroeder or Aviezer on the other. As we will see, though, Ross goes much further than Schroeder or Aviezer in his obvious aversion to acknowledging the contribution of writers who do not share his religious beliefs.

While Schroeder and Aviezer ignore the books by Christian writers even though they share the same attitude to the question of compatibility of the Bible and science, they, however, do not denigrate non-Jewish writers of such books. As to Ross's attitude, it can be determined if we look at some quotations from his writings.

In Ross's *Creation and Time* we read, "The beginning occurred only a few billion years ago and places the cause of the universe outside, that is, independent of, matter, energy, space, and time. Theologically this means that the Cause of the universe is independent of and transcendent to the universe. The Christian faith is the only one religion among the belief systems of humankind that teaches such a doctrine about the Creator. (Several religions like Judaism, Islam, and Mormonism accept as valid at least portions of the Old and New Testaments but every one of them, outside of Christianity, denies, at least in part, God's transcendence and extra-dimensional attributes.)"[8]

It is easy to imagine the possible reaction of, say, adherents of Judaism to the above statement which alleges that faithful Jews accept only parts of their own Scriptures.

There is little doubt that Moslems, Jews, or Mormons, listed by Ross as inadequately interpreting the notion of the creator, would just shrug off his remarks about their beliefs as the ranting of a religious bigot. I offer the readers a choice between two interpretations of the above statement. One is that Ross's knowledge of religions other than Christianity is not much better than his understanding of thermodynamics. The other is that Ross's religious arrogance blinds him to anything beyond his extremely narrow field of vision.

Ross's contempt of beliefs other than of his own narrowly interpreted Christianity is also evident from more passages from *Creation and Time*. For example, Ross describes how he read the Bible for the first time and how he was impressed by it. He compares the biblical story with stories accepted in other religions in the following words: "In my study of the history of astronomy, I had read dozens of creation stories from the world religions, and I wondered if this one would be like the rest. The others were good for a few laughs, with their ludicrous descriptions and inventive disordering of events" (p. 144).

Of course, each religion claims the monopoly on the ultimate truth. It is easy to imagine an adherent, say, of Buddhism or Hinduism who finds the biblical story being good for a few laughs. However gaily they or Ross laugh at each other, somebody else who will as gaily laugh at both can always be found.

Ross's narrow-mindedness makes one suspect that all his arguments are so subordinated to his agenda that they should be viewed with the utmost caution in regard to their veracity.

WAR ON TWO FRONTS

It becomes immediately evident when reading Ross's books that he is in the unenviable position of fighting on two fronts. Ross adheres to the position of the Bible's inerrancy, according to which everything in the Bible is consistent and absolutely true. In promoting and defending this position, Ross fights against skeptics who see many contradictions between various parts of the Bible, as well as between the Bible and science. In particular, Ross and his fellow "old-earth creationists" assert that the biblical story as told in the Book of Genesis is absolutely true, and its apparent incompatibility with scientific data is only due to misinterpretation of the Bible's text. On the other front, Ross also fights against the so-called young-earth creationists who reject any attempt to interpret the Bible in any manner other than literally.

The "young-earth creationists" maintain that every word of the Bible must be accepted in its simple literal meaning, and if it contradicts science, that is just too bad for science. For example, modern astronomical data, together with the facts established by physics (such as the value of the speed of light), indicate the enormous dimensions of the universe and correspondingly give its age as about twelve to fifteen billion years. This does not deter the young-earth creationists who shrug off the scientific data and claim that the large size of the universe is just an illusion. Likewise, according to the young-earth creationists, the fossils found in various strata of the earth's crust have been deposited by God to create an illusion of billions of years of the earth's history. Of course, the young-earth creationists are not in the least baffled by the question as to why God would indulge in the deception of scientists. The ways of God are unfathomable, and the word of the Bible is not to be doubted in any manner—that is the unshakable conviction of the young-earth creationists.

Of course, the denial of the scientific data by the young-earth creationists means the extreme obscurantism, but their position, however absurd, is fully consistent and hence logically unassailable.

Ross is of the opinion that sticking to the position of the young-earth creationists can only discredit the Bible and make it susceptible to ridicule. Therefore he fights against young-earth creationists in order to redeem creationism in general by proving that this viewpoint can be reconciled with the scientific data.

In doing so, Ross resorts to semantic acrobatics, attempting to interpret this or that expression in the Bible in a way that allegedly proves the compatibility between the Bible and science. In doing so, he steps out from the pit of the extreme (but logically consistent) obscurantism only to fall into a pit of arbitrary assertions that run contrary to common sense; he promotes merely another version of obscurantism, one which is, moreover, utterly inconsistent and void of reason.

In particular, *Creation and Time* contains many sections devoted to debunking young-earth creationism. Paradoxically, in his rejection of young-earth creationism Ross assumes the position of a defender of science, borrowing his arguments from scientific theories.

In chapters 9 and 10 of *Creation and Time* Ross discusses arguments forwarded by those young-earth creationists who suggest the so-called creation science which purports to substantiate the young-earth thesis via scientific discourse. He correctly concludes that the young-earth creationists failed miserably to debunk the data on the universe's age. In particular, Ross indicates that the argumentation of the young-earth creationists involves the following fallacics: (1) faulty assumptions; (2) faulty data; (3) misapplication of principles, laws, and equations, and (4) failure to consider opposing evidence (p. 103).

Very good, Dr. Ross.

In this dispute, facts are on Ross's side. He should have, however, likewise adhered strictly to facts when interpreting the text of the Bible. Unfortunately, in that latter endeavor, based only on his religious beliefs rather than on an objective consideration of facts, Ross resorts to methods also involving the above four fallacies.

ROSS INTERPRETS HEBREW

A considerable part of Ross's argumentation is based on his interpretation of various phrases in the Hebrew text of the Bible. For example, in *Creation and Time* Ross discusses the Hebrew words for "day," "evening," and "morning" in order to prove that in the biblical story about the creation of the world in six days the above three words had been used metaphorically rather than literally (46). The Hebrew word for "day" is *yom*, which Ross for unexplained reasons transliterates with a caret (^) over the *o*—*yôm*. Ross transliterates the Hebrew word for "evening" (which is *erev*) as *ereb*, also for unknown reasons. In his earlier book, *The Fingerprint of God*, Ross for unfathomable reasons used another equally odd transliteration of the Hebrew words. Ross converted the word *yom* (in Hebrew three letters, *yud-vav-mem*) into the four-letter word *yowm* with a letter *w* mysteriously appearing out of thin air. The word *adam* (in Hebrew three letters, *alef-dalet-mem*) meaning "man," in Ross's creative transliteration becomes, oddly, *adham*.

Of course, these odd transliterations are not very important in themselves, but they give rise to a suspicion that Ross's cognizance of the Hebrew language is not much better than his understanding of thermodynamics. This suspicion becomes even stronger when the reader encounters a statement Ross repeats more than once, asserting that the Hebrew language has only three grammatical forms corresponding to English tenses. This assertion is made, for example, in *The Fingerprint of God* (p. 165n) and repeated elsewhere. Ross seems not to know about the seven forms of the Hebrew verbs, such as *Paal*, *Piel*, *Hitpael*, etc., each having variations for expressing past, present, and future, serving for completed as well as for incomplete actions, in the transitive as well as in the intransitive form, in the active as well as in the passive form, etc. While readers unfamiliar with Hebrew grammar and vocabulary may be impressed by Ross's use of several Hebrew words, any average Hebrew speaker would immediately recognize that Ross possesses, at best, a very rudimentary knowledge of that language.

With such dubious expertise in Hebrew, Ross does not seem to be equipped for those convoluted interpretations of Hebrew words to which he resorts to convince readers in the validity of his thesis about the agreement between the Bible and science.

One example of Ross's less than convincing attempt to reconcile the story about the six days of creation with modern scientific estimates of the age of the universe is found in *Creation and Time*. He writes, "I was beginning to discern that the original language of the Old Testament (Hebrew) had fewer nouns than English" (pp. 145–46). Ross's opinion, expressed on that page, is that the word *yom* in the Book of Genesis was used to mean different time intervals in different verses because the Hebrew language, unlike English, lacks words describing such different time intervals.

Any Israeli second grader would be able to tell Ross that the Hebrew language has enough words to denote any time interval, and even sometimes more than one word to express subtle differences in regard to the same time interval. In particular, the word "epoch" has more than one synonym in Hebrew. The most common Hebrew word for "epoch" is *tkufa*. The less common synonyms for this word are *sfira* and *idan* (the latter is of Aramaic origin but was used in biblical Hebrew as well). There are also other expressions in Hebrew which can be used for "epoch." Since the author of the Book of Genesis had all those and many other words at his disposal, the notion of the shortage of Hebrew words for denoting various time intervals is simply ludicrous. (Note that according to Christian beliefs the real author of the Book of Genesis was God Himself. Does Ross's explanation mean that God knew fewer Hebrew words than an average Hebrew-speaking kid does?)

In his desire to prove his point regardless of the means, Ross goes to such length as to bluntly distort the meaning of some Hebrew words. For example, in *Creation and Time* Ross attempts to explain away one of the inconsistencies in the Book of Genesis according to which invertebrates were created by God after mammals and just before the creation of man. To do so, Ross provides his own translation of the word *remes* (in Hebrew three letters, *resh-mem-sin*). He writes, "The Hebrew word in question is *remes*, and its broad definition encompasses rapidly moving vertebrates, such as rodents, hares, and lizards" (p. 152).

That whole quotation is pure fantasy. The Hebrew word *remes* is a noun derived from the root meaning "to crawl." Hence it has nothing to do with "rapidly moving" animals. Moreover, the correct translation of *remes* is "invertebrate," as can easily be verified by looking this word up in any good Hebrew dictionary (for example, in the authoritative *Hamilon Haivri Hamerukaz* by A. Even-Shoshan).[9]

Equally unsubstantiated are Ross's attempts to suggest his own interpretation of some other Hebrew words such as *zera* (*zayin-resh-ayin*), meaning "seed"; *etz peri* (*ayin-tzade, pey-resh-yud*), meaning "fruit tree"; and so on.

Here is what Ross writes in *Creation and Time*: "The Hebrew phrase translated as 'seeds, trees, and fruit' (Genesis 1:11–12) has been taken by some as a reference to deciduous plants. However, the respective Hebrew nouns, *zera*, *ets*, and *periy* are generic terms that easily can be applied to plant species as primitive as those that appeared at the beginning of the Cambrian era" (p. 152).

To start with, there is no phrase "seeds, trees, and fruit" in Gen. 1:11–12. The actual text of Gen. 1:11 is in Hebrew as follows: *Veyomar elohim tadshe haaretz deshe eshev mazriah zera etz peri ose pri lemino asher zaro bo*. There are several translations of that verse into English. For example, the King James Version translates the above passage as follows, "And God said, Let the earth bring forth grass, the herb yielding seed, and the fruit tree yielding fruit after his kind, whose seed is in itself." A slightly different phrasing is found in a translation of the Masoretic text: "And God said, 'Let the earth put forth grass, herb yielding seed, and tree bearing fruit, wherein is the seed thereof, after its kind."*

Despite the slight difference between the two quoted translations, both are reasonably close to the Hebrew original which, as can be seen, provides quite specific definitions of the plants in question and leaves no room for any arbitrary interpretation like that by Ross.

In continuation of his alleged explanation of the meaning of some Hebrew words, Ross asserts in *Creation and Time* that if any person "with a scientific or analytical bent . . . gets through the first chapter of Genesis, . . . that person will become a believer" (p. 147).

While Ross obviously wants the reader to believe that he himself is a person of a "scientific and analytical bent," the impression gained from his writings is actually of somebody who is either a very gullible person lacking analytical faculty, and prone to interpreting facts in any way which would supposedly confirm his preconceived notions, or just a person driven by something other than the pursuit of truth. There are scores of people of "scientific and analytical bent" whose skepticism emerged unscathed from reading and re-reading the Book of Genesis and who continue to view it as just a beautiful piece of literature that reflects a naive effort by our distant ancestors who, at the dawn of history, tried to answer the questions which have puzzled and will continue to puzzle humans and which will probably never be answered with certainty.

Ross's book titled *The Genesis Question*, his more recent publication,

*The Masoretic text is the authoritative version of the Hebrew Bible. All subsequent Bible verses in this work translated from the Masoretic text will be indicated by "MT" following the chapter and verse numbers. This translation appears in full in *The Holy Scriptures* (Chicago: Menorah Press, 1973).

contains many detailed discussions of the controversy between the Book of Genesis and scientific data. Unlike the two books we have discussed so far in which Ross usually tries to prove the compatibility of this or that statement in the Bible with the facts established by science, in this new book his presentation assumes that the compatibility in question has already been firmly established; the Bible story is accepted as unerringly true and his task is only to clarify how to properly understand the biblical text.

It is in vain that a reader would look for any proofs of Ross's categorical statements in this book. In fact, all those statements are unsubstantiated and arbitrary.

In particular, Ross continues in *The Genesis Question* his misinterpretations of Hebrew words. Ross repeats the same assertion that Hebrew has a very limited vocabulary: "Hebrew, however, with its tiny vocabulary (more than thousand times smaller than English) must manifest more flexibility" (p. 65).

The full, unabridged edition of the *Random House Dictionary of the English Language* (1987) comprises 315,000 entries. The multivolume *Oxford English Dictionary* comprises about 700,000 entries. Hence, if we believe Ross, the vocabulary of Hebrew must comprise not more than about 700 words. In fact, even the abridged, one-volume version of the Hebrew dictionary *Hamilon Haivri Hamerukaz* by Even-Shoshan contains over 70,000 entries. Its full seven-volume version is of course much larger. As to the vocabulary of the Hebrew Bible, the vocabulary of the Torah (Pentateuch) alone (which is 111,970 words long) comprises 14,691 different Hebrew words, among them 7,361 nouns (not counting various grammatical forms of words with the same roots).[10]

Also in *The Genesis Question*, Ross asserts that Hebrew, unlike English, has no words for "epoch," "era," and "age." He maintains that in Hebrew "no other word besides *yom* carries the meaning of a long period of time" (65). As we discussed earlier, such statements demonstrate Ross's ignorance of Hebrew coupled with his arrogance in counting on the equal or even larger ignorance of his readers.

In that book we see again odd transliterations of Hebrew words, testifying to Ross's very limited knowledge of that language. For example, the Hebrew word *gavoah* (meaning "high") in Ross's version becomes *gaboah,* and the expression *hagvohim* (meaning "the high" in plural) becomes the almost-unrecognizable *hugebohim* (p. 145).

One telltale example of Ross's ignorance of Hebrew (and of some general facts of grammar as well) occurs when he tells the readers that the word *behema* is a plural form for the word *behemoth* (p. 48). This statement would cause a fit of laughs in the first grade of any school in Israel. Apparently, Ross does not know the difference between the plural form of a noun and collective nouns. *Behema* (in Hebrew four letters, *bet-hey-mem-hey*, which also can be transliterated as *b'hemah*) means "cattle," and like in English, it is a collective noun in the singular form. *Behema* is not a plural form of *behemoth* (as Ross suggests); on the contrary, the Hebrew word *behemot* (*bet-hey-mem-vav-tav*) is the intensive plural of *behemah*. This word also serves as a noun in the singular form denoting "hippopotamus." Hebrew nouns in the plural form end either with *im* (if they are masculine) or with *ot* (feminine). In particular, the correct plural form for *behemot* is *b'hemotim*, and for *behemah* (or *b'hemah*) it is *b'hemot* (in the contemporary colloquial Hebrew *b'hemot* is often used as a generic word to denote animals).

Likewise, Ross pretends to clarify the meaning of a number of other Hebrew words, such as *sheres* (a cumulative name for many small animals whose correct transliteration should be *sheretz*), *op* (meaning "fowl"; the correct transliteration should be either *of* or *ohf*), *nefesh*, and others, in his habitually arbitrary manner, twisting the alleged translation of those words to fit his thesis.

Of course, Ross is not under obligation to be fluent in Hebrew. However, pretending to be an expert in that language while hardly being familiar with even its simplest features is one of the most appalling peculiarities of Ross's publications.

Like many other writers of books in that genre, Ross is selective in his interpretation of the controversies in the Genesis story. For example, he discusses at length the story of the creation of living creatures according to Genesis 1, trying to show that the seemingly odd order of creation (insects created after mammals, etc.) stems simply from an incorrect translation of Hebrew words. At the same time, he seems not to notice, for example, some glaring contradictions between Genesis 1 and Genesis 2, in regard to that order of creation. For example, in Genesis 1 we are told that man (*adam*) was created after all the plants and animals. However, in Genesis 2 the creation of animals is reported to have happened after that of *adam*. Indeed, the creation of *adam* is reported in Genesis 2.7. In the next few verses the creation of the Garden of Eden is described (which appears to have possibly been created after Adam). Only after that does Genesis 2 report on the creation of animals. Here are the corresponding lines in Gen. 2.18–2.19 MT: "And the Lord God said: 'It is not good that the man should be alone; I will make him a help meet for him.' And out of the ground the Lord God formed every beast of the field, and every fowl of the air, and brought them unto the man to see what he would call them."

Apparently, even with all his creative interpretation of Hebrew words, in this case Ross was not able to come up with such an alleged translation of the Hebrew text which would make the controversy seem to disappear. Hence, he chose not to notice it.

SOME SPECIFIC INCORRECT STATEMENTS IN ROSS'S BOOKS

In this section I will discuss a few assorted examples of unsubstantiated statements in Ross's books.

In *Creation and Time* Ross writes, "Bipedal, tool-using, large-brained primates (called hominids by anthropologists) may have roamed the earth as long ago as one million years, but religious relics and altars date back only 8,000 to 24,000 years. Thus, the secular archaeological date for the first spirit creature is in complete agreement with the biblical date" (p. 141).

Indeed? What, then, about the biblical chronology, which tells us that Adam was created from dust only about 6,000 years ago? Is 6,000 the same number as 24,000 in Ross's view?

In the same book we read, "What astronomers and physicists are discovering in these new measurements is that the Being who brought the universe into existence is not only personal, creative, powerful, and intelligent to an unimaginable degree, but He is also aware of and sensitive to the needs of humanity" (p. 133).

The quoted paragraph demonstrates Ross's arrogance. Have astronomers and physicists voted to authorize Ross to speak for them? While the discoveries by astronomers and physicists do not exclude the possible existence of a creator of the universe, they also provide no direct indications whatsoever for the existence of such creator. To believe or not to believe in the supernatural being mentioned by Ross is a matter of personal choice. Both views are equally legitimate but neither of them has any direct connection to any scientific discoveries. As to Ross's assertion attributing to the possible supernatural being such qualities as caring about the needs of humanity, based on discoveries in astronomy and physics, the obvious lack of any substantiation for that notion is just another evidence of Ross's propensity to substitute the desired for the actual.

Moreover, Ross asserts that the work of secular scientists not only provides evidence in favor of his beliefs in the creator, but also specifically supports his particular version of Christian faith. For example, in *Creation and Time* he writes, ". . . even some atheists are more able to acknowledge that the big bang implies Jesus Christ than are our young-universe creationist friends" (p. 131). For atheists and believers alike, the big bang theory implies Jesus Christ no more than a sunset implies that some sorcerer has grabbed the sun and pulled it down from the sky. Ross does not provide a shred of evidence which would show how any scientific discoveries support his specific belief in Jesus the Savior. The reason for that is quite simple. There is no such evidence either for or against Ross's thesis anywhere in physics and/or astronomy.

In *The Genesis Question* Ross suggests an interpretation of those few lines in Genesis which describe the murder of Abel by Cain and the subsequent dialog between Cain and God (p. 102). In that interpretation, which is not based on any specific information given in Genesis, Ross appears to think that he can read God's mind. Of course, if Ross knows the unspoken thoughts of God, there is no wonder why he so confidently judges the ideas and motivations of regular mortals elsewhere. From a skeptic's viewpoint, however, all those discussions by Ross appear to be just *obscurum per obscuribus* (explaining the obscure by means of the more obscure), if we take the liberty of using that ancient Latin expression.

Also in *The Genesis Question* Ross allocates several sections to the discussion of the Flood and Noah's activities. In these sections, true to form, Ross provides a number of dubious considerations in order to rationalize the obvious implausibility of the story about a wooden ark accommodating pairs of all living creatures, etc. Without discussing these pseudoexplanations, let us point out just one odd contradiction in Ross's treatment of that subject. Ross tells us that, according to some astronomical data (such as the Vela supernova explosion), the Flood occurred between thirty thousand and fourteen thousand years ago (p. 173). Look, however, at the biblical chronology, according to which there were ten generations between Adam and Noah. The duration of life of every one of those generations as well as the ages when Adam and each of his descendants had their first-born sons are explicitly listed in Genesis. Accounting for the biblical assertion that Adam was created about 6,000 years ago, it is easy to calculate that the biblical date for the Flood is about 4,000 years ago. Ross does not say a word about the contradiction between the biblical data and his calculation based on the Vela supernova. What, then, about the Bible's inerrancy, which Ross claims to believe in?

CONCLUSION

As we read Ross's books, we encounter time and time again endless categorical assertions that this or that conclusion of science indicates the existence of the creator. Each time he makes such an assertion, Ross provides no logical grounds for it. Whatever this or that scientific theory concludes, it never points either to or against the hypothesis of a creator at work.

For example, the theory of the inflationary big bang, which Ross favors as allegedly leading to the inevitable conclusion that there must be a creator of the universe, actually has nothing to do with such a proposition. The theory in question may be correct or wrong, but in either case the existence of a creator may be either hypothesized or denied, the grounds for both propositions being beyond scientific consideration.

As a believer Ross is entitled to interpret this or any other theory as confirming his beliefs, as much as a skeptic is entitled to doubt it. Nothing in the scientific theory itself provides any arguments in favor of any of those competing interpretations.

If a scientific theory were to assert that the universe is eternal, it would not in any way contradict the hypothesis that there is God who is transcendental, who has established the laws of physics, and who rules the universe. Equally, it would not contradict the hypothesis that there is no transcendent entity beyond the universe, or that there is a transcendent supernatural entity which does not care at all about the universe. Science is not equipped to solve that problem and is not supposed to.

If science asserts that the universe had a beginning, it does not either prove or disprove the existence of a creator, or any views as to what kind of a creator it could be. Ross's assertions to the contrary are arbitrary and lack any logical or factual foundation.

The question of whether or not scientific theories confirm the particular image of God (for example, God as referred to in the Bible) is another story. It is rather obvious that even if we accept the assertion that there is a creator of the universe, an assertion which neither contradicts nor agrees with science, it does not mean that the creator necessarily meets the description given in the Bible. Ross, while asserting time and time again that science proves the existence of a creator, also each time asserts that this creator is exactly He who is presented in the Bible, meeting, moreover, the tenets of Ross's specific brand of faith. These assertions are forwarded without the slightest attempt to logically justify them.

Ross is entitled to his beliefs. Everybody else is entitled to either share or reject his beliefs. His persistent head-hammering with unsubstantiated conclusions has no chance to convince a skeptic or an adherent of some religion other than Ross's brand of Christianity.

Religious beliefs by and large are not based on rational arguments. It seems rather obvious that religious faith most often is due to indoctrination at an early age. Indeed, children of Moslems are overwhelmingly Moslems. Children of Buddhists are overwhelmingly Buddhists. As the experience of the former Soviet Union has shown, if the parents are atheists, children generally grow up as atheists.

Another factor affecting a person's religion is the social pressure. There are almost no so-called Jews for Jesus in Israel, but this denomination has a measure of success in the United States, where the overwhelmingly Christian surrounding creates a strong social pressure conducive to Jews' adoption of a faith which is closer to the majority.

Whatever the reasons for one's religious beliefs are, they have little to do with rational arguments and scientific proofs.

Ross's persistent assertions of science being in full agreement with his particular beliefs, without any factual evidence, reveals that his books are propaganda tools having little to do with either science or the question of the existence of God.

NOTES

1. Hugh Ross, *The Genesis Question: Scientific Advances and the Accuracy of Genesis* (Colorado Springs, Colo.: NavPress, 1998), p. 235.

2. For Ross's curriculum vitae, see "About Our Founder," *Reasons to Believe* [online], www.reasons.org/about/staff/ross.html#Curriculum [November 22, 2001].

3. Hugh Ross, *The Fingerprint of God* (Orange, Calif.: Promise, 1989).

4. Hugh Ross, *Creation and Time: A Biblical and Scientific Perspective on the Creation-Date Controversy* (Colorado Springs, Colo.: NavPress, 1994), p. 66.

5. Ross, *The Fingerprint of God*, p. 103.

6. Victor J. Stenger, *The Unconscious Quantum: Metaphysics in Modern Physics and Cosmology* (Amherst, N.Y.: Prometheus Books, 1995); Victor J. Stenger, *Not by Design: The Origin of the Universe* (Amherst, N.Y.: Prometheus Books, 1988).

7. See Gerald L. Schroeder, *Genesis and the Big Bang: The Discovery of Harmony between Modern Science and the Bible* (New York: Bantam Books, 1990); Nathan Aviezer, *In the Beginning: Biblical Creation and Science* (Hoboken, N.J.: KTAV, 1990).

8. Ross, *Creation and Time*, p. 129.

9. Avraham Even-Shoshan, "Remes," *Hamilon Haivri Hamerukaz* (The abridged Hebrew dictionary) (Jerusalem: Kriyat Sefer, 1974), p. 678.

10. Computerized counts of different Hebrew words in the vocabulary of the Torah (Pentateuch) were performed independently by professor of biblical studies and pastor James D. Price as well as by professor of computer science Brendan McKay, whose private communications on this matter are greatly appreciated.

6

THE SIGNATURE OF AN IGNORAMUS

Canadian Preacher Grant Jeffrey Proves the Bible's Inerrancy

Grant R. Jeffrey is the author of a number of best-selling books all devoted to one aim: to prove that the Bible reveals the absolute truth, is impeccably consistent, and is indeed the word of God. Each of his books, while approaching his task from slightly different angles, have many common features and repeat similar or even identical arguments in many respects. Therefore, to judge the overall level of Jeffrey's discourse and to form an opinion of the quality of his argumentation, it seems sufficient to choose only one of his books, reasonably representative of Jeffrey's writing. His book *The Signature of God* seems to meet the above condition.[1]

A reader who opens the book in question encounters several quotations from Jeffrey's admirers. As one of them wrote in regard to some previously published book by Jeffrey, "It is absolutely the most intriguing eye-opening book on end-time events I've come across in a long, long time. It's terrific!" The rest of the opinions about Jeffrey's books equally extol their virtues.

In this chapter I will discuss Jeffrey's arguments in order to see whether or not they indeed so brilliantly prove his case. I'll start my discussion with chapters 10 and 11 in Jeffrey's book, which I will briefly discuss together, followed by chapter 12, because these chapters vividly demonstrate the level of Jeffrey's discourse and the degree of his adherence to facts. After that, I will briefly discuss the rest of Jeffrey's book.

JEFFREY DISCUSSES THE BIBLE CODE

Chapter 10 in Jeffrey's book is titled "The Mysterious Hebrew Codes." In this chapter Jeffrey tells the readers about what he calls the "staggering phenomenon of hidden codes beneath the Hebrew text of the Old Testament" (p. 202).

The Bible codes controversy will be discussed at length in chapter 14; therefore, in this chapter I will refer to Jeffrey's treatment of that controversy very briefly. The alleged code consists of meaningful words which appear in the Hebrew original of the Pentateuch in the form of so-called ELSs (Equidistant Letter Sequences). For example, look at the word *chapter*: its first letter is *c*, its third letter *a*, and its fifth letter *t*. These three *equidistant* letters form an ELS which spells a meaningful word, *cat*, with a *skip* of 2.

There is nothing amazing or unexpected in the appearance of numerous ELSs in any sufficiently long text, be it the text of the Torah or a Manhattan phonebook. Each language contains an enormous number of words and phrasal combinations, and a statistical estimate shows that many of them are expected to appear as ELSs in every text by sheer chance.[2]

Jeffrey's discussion of the alleged Bible code is fraught with numerous inaccuracies and distortions of the real story. Since the Bible code is the subject of a separate chapter in this book, I will omit detailed discussion of Jeffrey's treatment of that subject and will limit myself to just one example of how Jeffrey presents the desired as if it is the actual. On page 208 of his book we read, "Another fascinating feature of this phenomenon was found in Genesis 2, which deals with the Garden of Eden. Scientists found twenty-five different Hebrew names of trees encoded within the text of this one chapter. The laws of probability indicate that the odds against this occurring are one hundred thousand to one." Apparently believing that no human, even Moses himself, could so skillfully encode the names of twenty-five trees as ELSs in a relatively short text, Jeffrey offered an open challenge to compile a text of a comparable length in English containing ELSs for the names of any twenty-five trees. Convinced of the invincibility of his position, Jeffrey offered to pay $1,000 to anybody who would meet the challenge. In a few weeks, Gidon Cohen of York, England, presented an English text of some 300 words in which the names of twenty-nine trees occurred as ELSs. Later, Cohen compiled a text in which a whole popular poem was "encoded" as a set of ELSs. This "encoded" poem can be seen on Brendan McKay's Web site.[3]

However untrustworthy chapter 10 in Jeffrey's book may be, it pales in comparison with chapter 11, titled "The Name of Jesus Encoded in the Old Testament: Yeshua Is My Name." This chapter in Jeffrey's book is largely based on publications of a messianic pastor, Yacov Rambsel.[4] A review of Jeffrey's (and Rambsel's) discussion of the alleged code related to Jesus will be given in chapter 14 of this book, where the complete lack of substantiation in Jeffrey's and Rambsel's claims will be shown.

If these writers were scientists in the pursuit of truth rather than religious zealots striving to interpret anything they could lay their hands on as confirmations of their preconceived views, they would have tested many other, nonbiblical texts to see if analogous ELSs can be located in those texts as well. They never did. If they did, they would easily find an endless number of ELSs spelling the Hebrew name of Jesus, *Yeshua*, as well as the combinations of that name with various other words cited by Rambsel, such as *Yeshua shmi* ("my name is Yeshua"), *dam Yeshua* ("blood of Yeshua"), *Yeshua yakhol* ("Yeshua can"), etc., in those sources as well, thus depriving the biblical occurrences of such ELSs of any special meaning. Many examples to that effect can be viewed at the Web site where the same type of ELSs are shown, taken from the book of a contemporary Israeli writer and in the textbook of geography of Israel (see also chapter 14).[5]

SEVEN TIMES SEVEN

Chapter 12 in Jeffrey's book is titled "The Mathematical Signature of God in the Words of Scripture." In this chapter Jeffrey tells the readers, with his usual abundance of such epithets as "fascinating," "incredible," and "staggering," about the multiple occurrences of number 7 in the text of the Bible, allegedly forming an intricate pattern which, if we believe Jeffrey, could exist only due to God's deliberate design: "This character of God is consistent with the revealed phenomenon of staggering complexity involving mathematical pattern within the text of the Scriptures" (p. 230).

In his description of the pattern involving the number 7, Jeffrey refers to one Ivan Panin as a "fascinating and famous mathematician" (ibid.). In fact Panin hardly can be referred to as a mathematician since he had no mathematical education or published any mathematical papers; his dubious claim to fame is based only on his work concerning the occurrences of the number 7 in the Bible.[6]

Let us again quote from Jeffrey: "Panin completed an astonishing study during the course of fifty years that revealed the most amazing mathematical pattern beneath the surface layer of the text of the Bible" (p. 231).

Jeffrey proceeds to explain the way numbers are written in Hebrew. This language has no separate characters for numerals, which are instead represented by letters of the alphabet. Here are the numerical values of the Hebrew letters:

alef = 1	*lamed* = 30
bet = 2	*mem* = 40
gimel = 3	*nun* = 50
dalet = 4	*samekh* = 60
hey = 5	*ayin* = 70
vav = 6	*pey* = 80
zayin = 7	*tzade* = 90
khet = 8	*qoph* = 100
tet = 9	*resh* = 200
yud = 10	*shin* = 300
kaf = 20	*tav* = 400

To express any number, combinations of the above letters are used. For example, 133 can be written as 100 + 30 + 3, that is, *qoph-lamed-gimel*, or it can be written as 90 + 40 + 3 (*tzade-mem-gimel*), or as 60 + 40 + 20 + 10 + 3 (*samekh-mem-kaf-yud-gimel*). Of course, not all possible combinations are commonly used, as there are some traditional ways to write numbers. For example, 35 is usually written as 30 + 5 (*lamed-hey*) rather than 20 + 10 + 5.

Then Jeffrey suggests imagining a similar system applied to the English alphabet. If this were the case, the numerals would be represented as follows:

A = 1	N = 50
B = 2	O = 60
C = 3	P = 70
D = 4	Q = 80
E = 5	R = 90
F = 6	S = 100
G = 7	T = 200
H = 8	U = 300
I = 9	V = 400
J = 10	W = 500
K = 20	X = 600
L = 30	Y = 700
M = 40	Z = 800

Then, for example, 133 could be represented either as SLC or as RMC, or, say, as QNBA, and the like.

Having explained the rule for representing the numerals in Hebrew, Jeffrey turns to what he calls "A Listing of the Phenomenal Features of Sevens Found in Genesis 1" (p. 233).

Jeffrey cites the very first sentence of the book of Genesis, which in Hebrew letters is *bet-resh-alef-shin-yud-tav, bet-resh-alef, alef-lamed-hey-yud-mem, alef-tav, hey-shin-mem-yud-mem, vav-alef-tav, hey-alef-resh-tzade.* (In this rendition, I have separated words by commas, and the letters within words by hyphens. Also, in the Hebrew original the text is read from right to left, while in the above rendition the letters are listed from left to right.) The traditional translation of this sentence is, "In the beginning God created the heaven and the earth."

The listing of the "Phenomenal Features" in the above sentence is given by Jeffrey on page 233 as follows:

1. The number of Hebrew words. . . . . Seven
2. The number of letters equals 28 (28/4 = 7) . . . . Seven
3. The first three Hebrew words
   contain 14 letters (14/2 = 7) . . . . Seven
4. The last four Hebrew words
   contain 14 letters (14/2 = 7) . . . . Seven
5. The fourth and the fifth words have 7 letters . . . . Seven
6. The sixth and the seventh words have 7 letters . . . . Seven
7. The three key words:
   God, heaven and earth have 14 letters (14/2 = 7). . . . . Seven
8. The number of letters in the
   four remaining words is 14 (14/2 = 7) . . . . Seven
9. The middle word is the shortest with 2 letters.
   However, in combination with the
   word to the right or left, it totals 7 letters. . . . . Seven
10. The numeric value of the first, middle and
    last letters is 133 (133/19 = 7) . . . . Seven
11. The numeric value of the first and last letters of all
    seven words is 1393 (1393/199 = 7). . . . . Seven

Having listed these eleven "phenomenal features," Jeffrey tells the readers that "when professors on the mathematical faculty at Harvard university were presented with this biblical phenomenon they naturally attempted to disprove its significance. . . . However, after valiant efforts these professors were unable to duplicate this incredible mathematical phenomenon."

Although I am not a professor of mathematics at Harvard, I will give examples illustrating the complete lack of significance in the alleged "phenomenal features of sevens" in the quoted sentence from the Book of Genesis.

Before providing those examples, let me turn first to Jeffrey's persistent attempts to exaggerate the significance of the alleged discoveries he is so fascinated with. Some of the eleven "phenomenal features" listed above are just repeating each other. For example, feature 3 is that the number of letters in the first three words of the sentence in question is 14. Simple arithmetic indicates that the number of letters in the last four words is also 14, since the total number of letters in the sentence is 28. Hence, feature 4 is actually a consequence of features 2 and 3, and listing it as an additional "phenomenal feature" is just a device to increase the number of alleged "phenomenal features" and thus impress the readers with that number. Likewise, feature 8 is just an arithmetic consequence of features 2 and 7, and its separate listing is again due to Jeffrey's zeal in trying to hammer his beliefs into his readers' brains, by whatever means.

Feature 10 is that the first, the middle, and the last letters of this sentence have the numerical value of 133, which is divisible by 7. However, as is quite often the case with Jeffrey's assertions, this statement is misleading. Since the sentence consists of an even number of letters (28), it has no middle letter. To get the quoted number, 133, it is necessary to add the numerical values of four rather than of three letters, namely the first (*bet*), the last (*tzade*), and the two letters in the middle of the sentence, *mem* and *alef.*

Finally, let us view the examples I promised above. I will list the features observed in my examples in exactly the same manner as Jeffrey does. As my first example, I will write a short statement regarding the first sentence in the Book of Genesis: *In this sentence I see no miracles.* Looking at this statement, it is obvious that it was not contrived using any special tricks, as its gist is germane to the ongoing discussion. What we see in that sentence is as follows:

1. The number of words is 7 . . . . Seven
2. The number of letters equals 28 (28/4 = 7) . . . . Seven
3. The first three words contain 14 letters (14/2 = 7) . . . . Seven
4. The last four words have 14 letters (14/2 = 7) . . . . Seven
5. The word in the middle (*I*) is the shortest, but together with the word to the left and the two words to the right it totals 14 letters (14/2 = 7). . . . . Seven
6. The numerical value of the last letters of all seven words is 329 (329/47 = 7). . . . . Seven

7. The number of letters in the words occupying odd positions is 21 (21/3 = 7) .... Seven
8. The number of letters in the words occupying even-numbered positions is 7 .... Seven
9. The number of letters in words ending with vowels is 14 (14/2 = 7) .... Seven
10. The numerical value of the first letters of words ending with vowels is 259 (259/37 = 7) .... Seven
11. The numerical value of the first and last letters of all words ending with vowels is 329 (329/47 = 7) .... Seven
12. The number of vowels in all the words ending with vowels is 7. .... Seven
13. The numerical value of all letters in the words whose length is an odd number is 119 (119/17 = 7) .... Seven
14. The numerical value of the last letters of the words whose length is an odd number is 14 (14/2 = 7) .... Seven
15. The numerical value of the first letters of the words whose length is an even number is 329 (329/47 = 7)...... Seven
16. The numerical value of last letters of the words whose length is an even number is 315 (315/45 = 7) .... Seven
17. The numerical value of the first letters of nouns is 140 (140/20 = 7) .... Seven
18. The numerical value of the last letters of nouns is is 105 (105/15 = 7) .... Seven
19. The numerical value of the first and the last letters of nouns is 245 (245/35 = 7). .... Seven
20. The numerical value of all letters of the verb is 105 (105/15 = 7) .... Seven
21. The sum of numerical values of every *seventh* letter (i.e., of *e*, *i*, *i*, and *s*) is 245 (245/35 = 7) .... Seven
22. (where also 35/5 = 7) .... Seven

I stopped listing the "phenomenal" seven-related patterns in my sentence when I completed twenty-two items, thus making it exactly twice as many as in Jeffrey's list for the first sentence of Genesis. Many of the above features are exactly like those listed by Jeffrey and Panin for the first sentence in Genesis. Some of the features listed by Jeffrey and Panin for the beginning sentence of Genesis (features 5, 6, and 7) are absent from the list of features in my example. On the other hand, in my example there are features which are absent from Jeffrey and Panin's list but seem to be the even more interesting coincidences. It is easy to imagine Jeffrey's delight if he could include such "phenomenal" features in his list regarding the first sentence of the Bible.

Could it be that I came up with my sentence by "incredible" (one of Jeffrey's favored epithets) luck and that such coincidences are extremely rare? Let us look at another example of a sentence, which is again relevant to our discussion: *The sevens occur in every old text.*

While I would add that "sevens" happen in new texts as often as in the old ones, what we see in the above sentence is as follows:

1. It consists of 7 words. . . . . Seven
2. It contains 28 letters (28/4 = 7). . . . . Seven
3. The first three words contain 14 letters (14/2 = 7) . . . . Seven
4. The last four words also contain 14 letters (14/2 = 7) . Seven
5. The middle word ("in") is the shortest with 2 letters. However, in combination with the word to the right or left it totals 7 letters. . . . . Seven

6, 7, 8. The numerical value of all letters which are in odd positions in an odd-numbered word and also in odd positions in the sentence totals 1715 ($7 \times 7 \times 7 \times 5 = 1715$) . . . . Seven, Seven, Seven

There are more seven-related coincidences in this sentence, and readers are welcome to look for them.

Maybe it was again my "incredible" luck to come across a second sentence with all those seven-related coincidences? Here is one more: *No wonder you see seven in Genesis*. Again, this sentence is not about some arbitrary topic. It is relevant to the subject of discussion.

1. How many words in that sentence? Surprise, surprise . . . Seven!. . . . . Seven
2. How many letters? Of course, the reader has already successfully guessed—28 letters (28/4 = 7) . . . . Seven
3. How many letters in the first four words? Is it 14? Yes, 14 (14/2 = 7) . . . . Seven
4. How many letters in the last three words? Yes, Mr. Jeffrey, 14 letters (14/2 = 7). What a miracle!. . . . . Seven
5. The numeric value of the first letters of all seven words is 875 (875/125 = 7). Isn't that nice? . . . . Seven
6. The numeric value of the last letters of all seven words is 1316 (1316/188 = 7). Incredible! . . . . Seven
7. What is the sum of numerical values of the first and last letters of all seven words? Check it—it is 2121 (2121/303 = 7). Isn't this "incredible," "surprising," and "staggering," Mr. Jeffrey, using your favorite expressions?. . Seven

8. Also, one more "miraculous" coincidence in the sentence in question: the sum of numeric values of every *fourteenth* letter (14/2 = 7) . . . . Seven
9. is 105 (105/15 = 7) . . . . Seven

Maybe such coincidences were easier to locate because the sentence has twenty-eight letters? Let us see. Here is a sentence, expressing Jeffrey's view of the Harvard mathematicians: *Harvard mathematicians were confounded by Panin's puzzle.* Here are some of the patterns in that sentence.

1. It contains 7 words.
2. It consists of 49 letters (49/7 = 7).
3. The numerical value of all letters is 5390 (5390/770 = 7).
4. The verbs in that sentence have 14 letters (14/2 = 7).
5. The first word has 7 letters.
6. The numerical value of vowels in the nouns is 343 (343/49 = 7).
7. Many more "sevens" can be identified in that sentence.

When one of my friends read the draft of this chapter, he sent me his opinion of it. Here is an excerpt from his message:

> Dear Mark, after reading your article on Jeffrey/Panin, I decided that PROFESSOR MARK PERAKH'S NUMERICAL SENTENCES ARE WONDROUS!
>
> Imagine my surprise to discover that this sentence has
>
> - Number of words = 7
> - Number of letters = 7 × 7
> - Total numerical value = 7 × 7 × 68 (Not to mention many other patterns of 7)

There is little doubt that many readers can easily compile many more examples of sentences with a host of allegedly miraculous seven-related patterns.

Maybe creating sentences with a pattern of sevens is easier in English than it is in Hebrew? Let us then try some Hebrew examples.

The initial words of Genesis in the Hebrew original are *B'reshit bara Elohim et hashamaim veet haarets*, which is translated in the King James Version as "In the beginning God created the heaven and the earth." While that translation is reasonably close to the original, the expression *b'reshit* could be more accurately translated as "first" ("First God created . . ."). The expression "in the beginning" is more precisely matched by the Hebrew expression *b'hatkhalah*. I will use the latter expression in my example. Furthermore, while in Genesis 1 we read the word *bara* translated in the King James Version as "created," in Genesis 2 the word *yatsar* is mostly used instead, translated as "formed." I will use *yatsar* in my example.

Here is my example: "*B'hatkhalah yatsar Elohim et haadamah veet hamaim*." The translation of this Hebrew sentence is: "In the beginning God formed the land and the waters."

Rendered in Hebrew letters, this sentence is as follows:

> *bet-hey-tav-chet-lamed-hey, yud-tzade-resh, alef-lamed-hey-yud-mem, alef-tav-hey-alef-dalet-mem-hey, vav-alef-tav, hey-mem-yud-men.*

It is easy to see that my example differs from the first sentence of Genesis in its choice of words (and hence of letters) except for the name *Elohim* ("God") and the preposition *alef-tav* (*et*), which has no equivalent in English but is used in Hebrew to indicate the proper relationship between a verb and a noun.

What we see in the above Hebrew sentence I composed is as follows:

1. The number of words is 7 .... Seven
2. The number of letters is 28 (28/4 = 7)................... Seven
3. The first three words have 14 letters (14/2 = 7) .... Seven
4. The last four words have also 14 letters (14/2 = 7)......... Seven
5. The middle word *(alef-tav)* is the shortest with 2 letters. However, combined with a word to the left or to the right, it totals 7 letters .... Seven
6. The three "key" words—"God," "land," and "waters"— have 14 letters (14/2 = 7)............................. Seven
7. The remaining four words also have 14 letters .... Seven
8. The sum of numeric values of first and last letters of all seven words is 1120 (1120/160 = 7) .... Seven

Listing the above features, I followed exactly what Jeffrey did for the first sentence in Genesis.

Maybe the last example was easy to compose because the gist of that sentence was somehow similar to that of the first sentence in Genesis? Well, let us try to compose some other Hebrew sentence, relevant to the subject of our discussion. Here is such a sentence: *Bedugmaot haele ein shum nesim o plaiim*. It translates as follows: "In these examples there are no wonders or miracles."

In Hebrew letters, this sentence is as follows: *bet-dalet-gimel-mem-*

alef-vav-tav, hey-alef-lamed-hey, alef-yud-nun, shin-vav-mem, nun-samekh-yud-mem, alef-vav, pey-lamed-alef-yud-mem.

We see in that sentence the following features:

1. It contains 7 words.................................Seven
2. It has 28 letters (28/4 = 7)...............................Seven
3. The first three words have 14 letters (14/2 = 7) ....Seven
4. The last four words also have 14 letters (14/2 = 7).........Seven
5. The sum of numeric values of every seventh letter, if read in the Hebrew original from left to right, is 308 (308/44 = 7)................................Seven
6. The sum of numeric values of the last letters of all seven words is 581 (581/83 = 7).......................Seven
7. The sum of numeric values of the first, last, and two middle letters is 392 (392/56 = 7).....................Seven

Since we have used three examples of 28-letter sentences in English, let us make our exercise symmetrical by adding one more example in Hebrew. Here is one more Hebrew sentence: *Behekhlet kal limtzo sheva betokh girsot shonot.* Its translation is "It is definitely easy to find seven in various versions of texts." Obviously, this sentence is again relevant to the subject under discussion.

In Hebrew letters it is as follows: *bet-hey-khet-lamed-tet, qoph-lamed, lamed-mem-tzade-alef, shin-bet-ayin, bet-tav-vav-khet, gimel-resh-samekh-vav-tav, shin-vav-nun-vav-tav.*

The features we see in that sentence are as follows:

1. It contains 7 Hebrew words.........................Seven
2. It consists of 28 letters (28/4 = 7)....................Seven
3. The first four words have 14 letters (14/2 = 7)...........Seven
4. The last three words also have 14 letters (14/2 = 7).......Seven
5. The middle word (*shin-bet-ayin*) has only three letters, but when combined with a word to the left or to the right, it totals 7 letters.......................Seven
6. The first and the second words have 7 letters............Seven
7. The third and the fourth words have 7 letters............Seven
8. The numeric value of the first letters of all seven words is 357 (357/51 = 7)...........................Seven
9. The sum of numeric values of every *seventh* letter........Seven
10. is 560 (560/80 = 7)...............................Seven

I believe the above examples provide ample evidence that seven-related patterns are common, not only in the above examples, but in any texts, both in English and in Hebrew. I leave it to the readers to try identifying more of the "sevens" in my examples.

Hence, we see that many seven-related coincidences occur by chance in any text, biblical and nonbiblical. Therefore Panin's and Jeffrey's fascination with this alleged miracle is just an indication that they lacked the willingness to properly verify their "incredible" observations before claiming to have discovered a miracle.

The readers can now decide whether they should believe Jeffrey's assertion in *The Signature of God* that professors of mathematics from Harvard University were unable to "construct a sentence about any topic they [chose]" in English which would incorporate features similar to those Panin discovered in Genesis. Using Jeffrey's own expressions we can justifiably assert that Jeffrey has displayed "unbelievable arrogance" when he was "sitting in judgment" (p. 82) of professors of mathematics.

On page 234 Jeffrey tells us that he tried to compile a text containing patterns of sevens himself but failed and therefore came to the conclusion that no human could have done it unless directly guided by God. Then Jeffrey continues by asserting that even a supercomputer could not produce a passage with "sevens" like those found in the Bible.

First of all, even if that statement were true, it would be irrelevant because no one claims that the occurrence of the patterns of sevens in the Bible is a result of a deliberate effort by human writers. The most reasonable explanation of the patterns in question is that they occur *by chance* in any text of sufficient length.

Second, regardless of the origin of the patterns of sevens, Jeffrey's assertion that no human mind—and not even a supercomputer—would be capable of creating such intricate mathematical webs in a text calls for a reminder that Jeffrey was equally confident that no human could create a text containing ELSs for twenty-five names of trees. This confidence cost Jeffrey $1,000. It would be wise for him to be a little more cautious when making such predictions.

Just as with the "sevens," Jeffrey tried to create a text with ELSs for twenty-five names of trees and failed. This failure led him to the conclusion that this task was also beyond the capabilities of anybody but God. The obvious flaw of this argument is that it was based only on Jeffrey's personal experience. To explain what I mean, I will give an example.

I would first challenge Mr. Jeffrey to write a sonnet, a short poem with a strictly prescribed form. A sonnet consists of fourteen lines, usually divided into four stanzas, the first and the second of which are each four lines long, and the third and the fourth each three lines long. The lines must rhyme following a certain pattern. Of course, the sonnet, besides following the strictly prescribed form, must also express certain feelings and/or ideas in a poetic form. I don't think we can expect a well-constructed sonnet from Jeffrey any time soon.

However, if, contrary to expectation, he does create a reasonably good sonnet, my second challenge to him would be to write a wreath of sonnets. This is one of the most difficult forms of poetic creation, consisting of fifteen sonnets that all must relate to a single common idea, but from various angles, subtly exploring it in depth. The last line of sonnet 1 must become the first line of sonnet 2, the last line of sonnet 2 must be the first line of sonnet 3, and so forth. The first lines of each of the first fourteen sonnets must constitute the fifteenth and final sonnet. This structure of interlocking rhymes, lines, and phrasal expressions contains an intricate mathematical pattern, partly involving "sevens" and no less complex than that of the sevens in the Bible at which Jeffrey is so amazed.

Imagine that Jeffrey tried to create a meaningful wreath of sonnets. There is a good chance he would fail. However, even if Jeffrey, as well as many of his readers, cannot write wreaths of sonnets, many poets can. Some have written more than one such wreath, without a computer, and many of these creations have been acclaimed as great poetry.

Would Jeffrey claim that, since he cannot write a wreath of sonnets, nobody can? Then why does he think that nobody can compose a text with the patterns of seven just because he tried and failed?

Conclusion: the allegedly "phenomenal" features involving the number 7 discovered by Panin in the Bible and reported in Jeffrey's book have no significance and appear in the biblical texts by chance, as they also do in any other text. Jeffrey's fascination with his "phenomenal" features is just one more proof of the unreliability of his statements and of his lack of sufficient knowledge and understanding of the subject he so brazenly chooses to discuss.

Finally, one more comment on the last two sections. The suggestion of an alleged "code" and of alleged patterns of sevens could make sense only if the text of the Bible were a precise replica of the original text, letter for letter. Obviously realizing that, Jeffrey repeats several times in *The Signature of God* a claim that every letter of the Bible has been preserved precisely from when Moses wrote the Torah, as it was dictated, letter by letter, by God: "Jesus Christ, Himself, affirmed that the actual letters composing the Scriptures were directly inspired by God and were preserved in their precise order throughout eternity" (p. 209).

While eternity is too long a period of time to experimentally verify Jeffrey's statement, we can do so for a much shorter period, namely for about sixteen centuries.

There exists a monumental compendium of commentaries and interpretations of the Torah written over the course of several centuries by Jewish sages. It is called the Talmud and consists of many books. Jeffrey is obviously not familiar with that source of information. If he were, he would know, for example, of a tractate named *Kiddushin,* which is found in the Talmud. Here is a quotation from that tractate written many centuries ago: "The sages of the previous generation were called *soferim* [meaning 'those who counted'] for they counted all the letters in the Torah. Thus they said that *vav* in the word *Gachon* [Leviticus 11:42] is the middle letter of the Torah; the words *darosh darash* [Leviticus 11.16] are the middle of the words; the verse *Vaitgalach* [Leviticus 13:33] is the middle of the verses" (*Kiddushin* 30a).

If, though, we look at the Koren edition of the Torah we possess today, we find the following: (1) The middle letter of the Torah is *aleph*, in the word *hu* (Leviticus 8:78), which is located at a distance of 4,829 letters from *vav* in the word *Gachon*, which appears only once in the Torah; (2) the middle verse is *Vaiten alav et hachoshan* (Leviticus 8:8), which is at a distance of 164 verses from *Vaitgalach*; and (3) the middle word of the Torah in the Koren edition is *achat* (Leviticus 8:26), which is at a distance of 743 words from *darash darosh*. Hence, the text of the Torah as we know it today differs from that known to the *soferim* centuries ago, by hundreds of words and verses and by thousands of letters. Of course, the fact that the text of the Bible underwent many changes in the course of its long existence has been well known to experts in that matter. This fact alone makes all the suggestions about the ELS code or the pattern of sevens allegedly woven by God into the Bible devoid of any meaning.

ARROGANCE OF A RELIGIOUS ZEALOT

To discuss each and every faulty argument in Jeffrey's book would require writing yet another book, and I don't think Jeffrey's opus deserves the time and effort necessary to complete such a detailed refutation of his pseudo-proofs. Therefore, in this and in the following sections I will discuss only some selected points in Jeffrey's book, starting with the rude references by Jeffrey to those who do not share his beliefs.

Fools or Liars?

On page 16 of *The Signature of God* Jeffrey writes: "Any person who honestly believes that all the marvelous complexity of this universe simply happened by chance is a fool. If he is not a fool, yet still claims to believe that this incredibly complex universe is a result of random chance, then I must conclude that he is not being honest." Let us recall names of some people who, according to Jeffrey, were therefore either fools or liars.

One has only to look at Albert Einstein's "Autobiographical Notes," where he unequivocally calls himself a skeptic.[7] He also made similar assertions elsewhere (for example in his letter of September 28, 1945, to Guy H. Raner Jr., where he said he was an atheist).[8] Others espousing similar beliefs include Niels Bohr, Lev Landau, Bertrand Russell, and Stephen Hawking. Many more Nobel Prize winners, outstanding physicists, mathematicians, chemists, writers, and philosophers either did not share Jeffrey's belief that the universe emerged as the result of an action by God or doubted that view and did not adhere to a definite belief in that matter. All these people of great intelligence and integrity, in Jeffrey's view, are either fools or liars. Accounting for the simple fact that Jeffrey himself has no credentials in any field of science, isn't his claim a display of this arrogance?

Many years ago, I knew an illiterate man, a peasant who had a penchant for discussing physics of which he has, to put it mildly, only an approximate knowledge. One of his unshakably held views was that an electric bulb lights up when "a plus and a minus meet." He pestered me with demands to discuss this and other questions and was very confident that he succeeded in muzzling me with his wise arguments. Of course, since his ken in physics was next to zero, I had no way to disabuse him of his ridiculous views. Now, discussing Jeffrey's book, I feel the same frustration facing the impossible task of arguing with somebody who has none of the background necessary for a reasonable discourse, be it the question of probability calculation or any of Jeffrey's alleged proofs which I am going to review.

Jeffrey Discusses Thermodynamics

Chapter 6 in Jeffrey's book is titled "Scientific Proof that the Bible is Accurate." In that chapter Jeffrey very convincingly proves that he has no proper understanding of certain facts of science. Let us look, for example, at Jeffrey's attempt in *The Signature of God* to enlist thermodynamics as a tool to support his beliefs: "The second Law of Thermodynamics describes the fact that all systems and elements of the universe tend to disintegrate to a lower order of available energy or organization. . . . In fact, the second Law of Thermodynamics, the Law of Entropy, absolutely proves that the theory of evolution is nonsense" (p. 112). Jeffrey bases his assertion on the proposition that evolution means an increase in the complexity of organisms, i.e., the increase in the level of organization, i.e., the decrease of entropy. In Jeffrey's view, this would be contrary to the second law of thermodynamics and that is why, in his view, the evolution theory is nonsense.

Of course, we already know that, in Jeffrey's view, everybody who disagrees with his beliefs is a fool, and that professors of mathematics at Harvard are half-wits incapable of "constructing a sentence" with a certain mathematical structure. From the quote above we see that Jeffrey's opinion of physicists and biologists is also not very high. Indeed, if we believe Jeffrey's assertions about the laws of thermodynamics, physicists do not understand these laws since they interpret them differently from Jeffrey. Likewise, the many biologists who support some version of the evolution theory must be dunderheads as well, as they do not realize that their view is nonsense.

Let us see how Jeffrey interprets the second law of thermodynamics. When explaining that law, Jeffrey shows that he has not grasped a very important element thereof. This law maintains that spontaneous processes in *closed macroscopic systems* are accompanied by the increase of entropy, i.e., by an increase in the degree of disorder. While the word *macroscopic* is important, it has little bearing on Jeffrey's misinterpretation of the law in question. However, the word *closed* is quite germane for this discussion. Contrary to Jeffrey's assertion, the second law of thermodynamics does not say that *all* systems and elements of the universe tend to move toward larger entropy (i.e., greater disorder) but only *closed* systems. By a closed system thermodynamics means a system which does not exchange matter and energy with its surroundings. Living organisms are by no means closed systems. They are homeostatic systems—i.e., they automatically tend to maintain certain ranges of temperature and composition—but they continually exchange matter (food and excrement) and energy (heat and work) with their surroundings. Actually, each healthy living organism, be it a plant, animal, or human being, continually and largely successfully resists the increase of its entropy. If Jeffrey's interpretation of the second law were true, not just evolution, but the very existence of life would be impossible. Each time a child is conceived, a process starts that culminates in the child's birth during which the entropy of the fetus, and with it of the mother's body as a whole, continually decreases. Does this process contradict the second law of thermodynamics? If we believed Jeffrey, the answer would be yes; the correct answer, though, is no. The fetus as well as the mother's body are not closed systems. The entropy of a system which is not closed can very well decrease or stay constant without contradicting any laws of thermodynamics in the least. Life, among other things, is a process in which the entropy of a system fluctuates around a certain more or less permanent level, accompanied by the increase of the entropy of its surroundings, and with it of the universe as a whole.

Without discussing here the merits or shortcomings of the evolution theory, we can state that it is fully compatible with the second law of thermodynamics, while Jeffrey's interpretation is just a display of his illiteracy in physics. (I will not discuss here Jeffrey's attempts to use the first law of thermodynamics to support his views, not because it is any better than his treatment of the second law, but just to avoid an excessively detailed discussion.)

The examples of Jeffrey's misinterpretations of scientific facts, erroneous statements, and unsubstantiated attempts to interpret this or that quotation from the Bible as allegedly confirming his beliefs can be multiplied, but those few examples discussed above provide enough material to realize how dismayingly ill founded his efforts are to convince readers of his beliefs.

Jeffrey Calculates Probabilities

I have mentioned in the preceding sections that Jeffrey's overall ignorance of science is, among other things, evident from his handling of probabilities. The proper calculation and interpretation of probabilities will be discussed in chapter 13. From that chapter, it must become clear that Jeffrey's understanding of probability theory does not extend beyond some popular explanations. With the typical self-confidence of a dilettante, Jeffrey does not shy away from trying to *teach* probability to the readers whom he apparently expected to be even less familiar with the subject than he himself is. His explanation of the multiplication of probabilities and of the odds encountered when estimating a probability of an event that combines several consecutive tests (pp. 171–72) does not go beyond the most rudimentary understanding of that complex quantity. Since, however, this question will be elucidated in detail in chapter 13, in this chapter I intend not to discuss how Jeffrey calculates probability but rather to show how awkwardly he tries to use the probability concept to support his assertion that the Bible contains numerous predictions of future events and that these predictions have been precisely fulfilled.

In *The Signature of God*, Jeffrey writes that "the Old Testament contains over three hundred passages that refer to the coming of the Messiah." He goes on to assert that forty-eight of these passages allegedly provide "specific details about the life, death and resurrection of Jesus" (p. 170). He then quotes some of the alleged predictions, insists that they all came true, and calculates the probability that those predicted events could have happened by chance. Of course, the probabilities he calculates turn out to be very small. Hence, Jeffrey concludes triumphantly, the Bible has precisely predicted future events and must therefore be God's word.

Let us look at these "predictions" related to Jesus. The simple fact is that none of them mention the name of Jesus of Nazareth. Interpreting the quotations in question as related to Jesus is an arbitrary choice by Jeffrey and other defenders of the concept in question. Jeffrey seems to be impressed by the *wording* of those passages in the Old Testament that in his view relate to Jesus because it is quite similar to the *wording* of some verses describing the life and death of Jesus in the New Testament. Is that surprising? Obviously, the authors of the New Testament were well versed in the Old Testament and were used to quoting it. They borrowed many expressions from it in their writing, even when they were not using it to prove the truthfulness of their contention that the coming of Jesus was predicted in the Old Testament. Since, though, they actually had such an agenda, it was even more natural for them to utilize the expressions from the Old Testament to show that this or that event was a fulfillment of a prediction delivered through a prophet.

Most of the Old Testament's passages in question can be interpreted in various ways, and there is no evidence that they imply Jesus of Nazareth. Let us review a few examples.

On page 172, Jeffrey quotes from Genesis 49:10: "The scepter shall not depart from Judah, nor a lawgiver from between his feet, until Shiloah comes; and to him shall be the obedience of the people."

Then Jeffrey quotes from the New Testament (Matt. 2:1): "Now after

Jesus was born in Bethlehem of Judea in the days of the Herod the king, behold, wise men from the East came to Jerusalem." To view the latter quotation as the fulfillment of the prediction in Genesis 49:10 requires a lot of imagination. Jeffrey estimated that the fulfillment of this "prediction" had the probability of 1 in 2,400. Even if this estimate were mathematically correct (which it is not by any stretch of the imagination—see chapter 13), it is meaningless because there is no evident connection between the quotation from Genesis and that from Matthew. What has the quotation about a lawgiver from Judah to do with the quotation about wise men coming to Jerusalem after Jesus was born in Bethlehem? There is no reason to assume that the lawgiver from Judah was meant to be Jesus of Nazareth.

On page 173 Jeffrey quotes another passage from the Old Testament (Isa. 40:3): "The voice of one crying in the wilderness: 'Prepare the way of the Lord; make straight in the desert the highway for our God.'" The quotation from Matthew 3:1, 2, which, in Jeffrey's view, signifies the fulfillment of the prediction in Isaiah 40:3, is, "In those days John the Baptist came preaching in the wilderness of Judea, and saying, 'Repent, for the kingdom of heaven is at hand.'" Jeffrey calculates the probability that the appearance of John the Baptist could be predicted by chance as 1/20. Again, even if Jeffrey's calculation itself had meaning (which it does not), there is no reason to connect the above passage from Isaiah with that from Matthew. The only common point in the two passages is the use of the word "wilderness." Is that sufficient reason to assert that Isaiah meant John the Baptist?

In his next example, Jeffrey calculates the probability (in his estimate, 1/50) that Jesus' coming to Jerusalem *on a donkey* could be predicted by chance. His quotation from Zechariah 9:9 reads: "[The] King is coming to you; he is just and having salvation, lowly and riding on a donkey, a colt, the foal of a donkey." Then Jeffrey provides a quotation from Luke 19:35 which describes how Jesus came to Jerusalem *riding on a colt*. Is this really a proof of a fulfilled prediction? Zechariah spoke about some unnamed king. There is no reason to assume he meant Jesus. In another place in his book, Jeffrey devotes a whole section to the discussion why the ancient Israelis used no horses (pp. 250–53). What then is unusual in that particular prediction by Zechariah that whichever king he had in mind would ride into Jerusalem on a donkey?

I rest my case now, without reviewing the rest of Jeffrey's claims of allegedly fulfilled predictions, none of which is better substantiated than the three quoted above. All of them are actually arbitrary interpretations of quotations without any factual basis, accompanied by a meaningless calculation of probabilities.

Jeffrey Counts the Population of the Earth

Let us quote from a section of Jeffrey's book titled "The Population of the Earth": "Obviously, a huge discrepancy exists between the evolutionists' suggestion of man's origin approximately one million years ago compared to the Bible's declaration of man's creation by God approximately six thousand years ago" (p. 125). Indeed, the discrepancy is huge, and that statement is one of the very few correct statements in Jeffrey's book. However, even this generally correct assertion requires amendments. First, the conclusion that man appeared on the earth roughly one million years ago is based on archaeological data; therefore it is not merely a suggestion by evolutionists, but a scientific theory supported by a large body of evidence. Of course, the biblical story about the creation of man about six thousand years ago is indeed a "suggestion" lacking any evidence besides the assertion in the Bible itself.

To support his belief in the inerrancy of the biblical story, Jeffrey offers demographic calculations. He starts his calculations by assuming that the entire present population of the earth consists of descendants of just one couple of survivors of the Flood. This cataclysm, according to Jeffrey's suggestion, took place some forty-three hundred years ago. Since, according to the Bible, there were eight survivors, i.e., four couples after the Flood, Jeffrey presents his account for only one rather than for four couples as a manifestation of his impartiality. His next assumption is that every family throughout history produced on the average 2.5 children. Finally, Jeffrey suggests that the average lifespan of a generation was forty-three years. Then the number of generations between the time of the Flood and today is estimated as one hundred. Based on these three estimates—one original couple, 2.5 children per couple, and one hundred generations—Jeffrey calculated that the population of the earth today must have reached about five billion. In Jeffrey's opinion, "it is fascinating that the earth's population today is almost identical with what we would expect if mankind began repopulating the earth after the Flood forty-three hundred years ago" (p. 126). What a triumph for the believers in the Bible's story!

On the other hand, if man existed for about one million years, says Jeffrey, then we would have to account for 23,256 generations. Repeating the calculation based on the same assumptions as used for the biblical account, we would arrive at an enormous number of people that must live on the earth today, namely 10 to the power of 2091.

Jeffrey does not tell the readers how exactly he calculated the number

—about five billion—of people living today. However, his assumptions—only one couple starting the proliferation after the Flood, one hundred generations after the Flood, the lifespan of forty-three years per generation, and 2.5 children per couple—enable us to figure it out. If every forty-three years every 2 people were replaced, on the average, by 2.5 people, then every forty-three years the population of the earth increased by 2.5/2 = 1.25 times (i.e., by 25 percent). Then the total of one hundred generations must result in the increase of the population by 1.25^{100}, which equals about 4.9 billion. Apparently, that is how Jeffrey came up with the quoted estimate of five billion.

The above calculation shows why Jeffrey arbitrarily chose to account for only one couple of Flood survivors rather than for all the four couples of the biblical account, and why he arbitrarily chose the lifespan of a generation to be forty-three years. It was contrived a posteriori to get the desired number—about five billion people living today. In doing that, Jeffrey conveniently forgot to multiply his number by 2, the initial number of the Flood survivors he chose to account for. If he did multiply, his number would not be about five but about ten billion, i.e., almost twice the actual population of our planet today. If, though, he had accounted for Noah, Shem, Ham, and Japheth with their wives, rather than for only one surviving couple, he would have arrived at about forty billion instead of five billion.

However, the actual absurdity of Jeffrey's calculation is much worse, as it can be seen if we try to apply his method of calculation to some other situations. Jeffrey professed his unshakable belief in the biblical story. According to that story, for example, when the Jews escaped from slavery in Egypt, the Jewish nation consisted of six hundred thousand men plus an unknown number of women and children. If we assume that the number of women was roughly equal to that of men, and that each couple, by a very conservative estimate, had at least one child, the total number of Jews fleeing Egypt must have been at least about 1.5 million (some rabbinical sources estimate that number to be closer to 2.5 million, but let us choose a more conservative estimate). If that number is correct, a natural assumption is that the total population of Egypt was at least several million. Indeed, archaeological data indicate that about 3,300 years ago, when the Exodus of the Jews supposedly occurred, the population of Egypt was about 2.5 million. Of course, besides Egypt, there were other countries, so the total population of the earth must have been, at that time, many millions.

Let us therefore see if any of the two numbers—either 1.5 million Jews or 2.5 million Egyptians living at the time of the Exodus—can be reconciled with Jeffrey's calculations. The Exodus, according to the Torah's chronology, took place about 3,300 years ago. Then the time interval between the Flood and the Exodus, according to the biblical story, was about 1,000 years. If the lifespan of a generation, according to Jeffrey's assumption, was 43 years, then the number of generations that lived between the Flood and the Exodus must have been 1000/43, which rounds up to 23. If the proliferation after the Flood started with only three couples (Shem, Ham, and Japheth with their wives) the population of the earth at the time of the Exodus, according to Jeffrey's calculations, must have been 6×1.25^{23}, which is only about 1,016 people, rather than millions! Yes, Mr. Jeffrey, your calculation is hopelessly at odds with the very biblical data you so firmly believe in.

Now look at the time that elapsed after the Exodus. Let us estimate only the population of Egypt, which, as we have mentioned before, was, some 3,300 years ago, about 2.5 million people. During that time the number of generations—each living, as per Jeffrey's assumptions, for 43 years—must have been 3300/43, which rounds up to 77. Then the population of Egypt in our time, if we apply Jeffrey's calculation, must have become $2.5 \times 10^6 \times 1.25^{77}$, which is over 72 trillion (72,000,000,000,000) people. Even if we account for possible migration of people from and to Egypt, the absurdity of that number is too obvious. Hence, whatever way we analyze Jeffrey's estimation of the population growth, we find that his calculation yields either unreasonably low numbers or absurdly large numbers.

The archaeological data unequivocally show that man existed long before that day about 6,000 years ago when, according to the Bible, the first man was created out of dust. Even on the remote islands of the Pacific, populated later than the continents, a human society already existed more than thirty thousand years ago.

Of course, Jeffrey offers no arguments which would refute the archaeological evidence.

Jeffrey Confirms the New Testament's Account

Chapter 5 in *The Signature of God* is titled "The Historical Evidence about Jesus Christ." Rather than to discuss at length all alleged proofs of the authenticity of the Gospels offered by Jeffrey, which would require many pages of discussion, let us concentrate only on a few examples of Jeffrey's discourse which are typical of his treatment of facts.

First, on page 91 we see a section titled "Evidence about Jesus from

Flavius Josephus." Flavius Josephus, whose original name was Yoseph Ben Mattatiahu, was at one time a commander of Jewish warriors who defended the fortress of Yothapatha in Galilee against the onslaught of Romans under the command of the future emperor of Rome, Vespasian. When the fortress fell to the Romans, Yoseph surrendered to the Romans and later became a Roman citizen of high stature. He wrote several books, some of which, including *Antiquities of the Jews*, survived until our time. He wrote that book some sixty years after the supposed death of Jesus of Nazareth. Flavius Josephus was born in 37 C.E., i.e., several years after the date (about 33 C.E.) assumed as that of Jesus' crucifixion. He could not have witnessed the events described in the four gospels if these events indeed occurred in the first three decades of the first century. All Josephus could do was to repeat a story he heard from someone else. Possibly, he had access to the sacred texts of what he called the "Christian tribe," since in his time a Christian community existed in Rome. Therefore, the passage in Josephus's book telling the story about Jesus has none of the evidentiary value that witness testimony would. Moreover, some details of that story have caused suspicion that the passage in question is a later insertion by some Christian scribe in order to forge testimony from a supposedly unbiased historian. Josephus himself was a Jewish priest, an observant Jew who would not view Jesus as the Son of God or the Messiah. This is evident, in particular, from another book by Josephus, *Against Apion*, which was probably the first book ever to denounce anti-Semitism and defend Judaism. Therefore the passage in question sounds quite out of line with Josephus's style and attitude. In the passage in question, Jesus is referred to as "Christ." It is hard to imagine that, characterizing Jesus, Flavius Josephus, who was fluent in Greek, would use the word *Christ*, which stems from the Greek word meaning "the Messiah." He also knew very well that it was unthinkable for Jewish elders and priests to suggest to a Roman governor to condemn a Jew to the cross. Being well aware of Jewish customs and religion, which, among other things, forbade transferring a Jew for trial to Gentiles (especially during the holiday of Passover!) and for whom crucifixion was an abomination anyway, Flavius Josephus would hardly accept the story about Jesus at face value and repeat it without comments. These considerations give weight to the suggestion that the passage in question is a later insert by some Christian forger. However, even if Josephus had indeed written the passage in question, it has no evidentiary significance being just hearsay, a story he obtained from some other, unknown source.

Other references by Jeffrey to various non-Christian sources (Suetonius, Pliny the Younger, Lucian of Samosata, and Cornelius Tacitus) are no more convincing than that with Flavius Josephus. They boil down to repeating some limited information about the activities of early Christians, with vague references to the founder of their "cult," of whom none of the quoted authors had any firsthand knowledge. All of them simply repeated the story they heard about the early Christians and the alleged founder of their creed.

All other references by Jeffrey are of the same quality: vague stories which the cited authors heard from someone else and of which they had no personal knowledge.

One of the most preposterous (and actually mendacious) claims by Jeffrey is that the Dead Sea scrolls contain proofs of the Gospel's authenticity; however, this claim is unsubstantiated. The Dead Sea scrolls comprise hundreds of fragments of various manuscripts, many of which originated at the time of Jesus' supposed life and death. There are among the scrolls parts of all of the Old Testament books except for the Book of Esther—but not a single fragment which could be proven to be a segment of the New Testament—and multiple civic and religious documents of that time. The fact, quite damaging to Jeffrey's beliefs, is that there is not a single word about Jesus of Nazareth, or Jesus born in Bethlehem, or Jesus son of Joseph and Mary, or any Jesus (*Yeshua*), in any of the Dead Sea scrolls. All of Jeffrey's unfounded attempts to connect this or that fragment of the Dead Sea scrolls with Jesus, the alleged founder of Christianity, are without merit and border on a deliberate distortion of facts. Of course, Jeffrey could say that the absence of proof is not a proof of absence, and I would agree with such a statement. Unfortunately, Jeffrey chose another way, to twist the meaning of certain quotations from the scrolls in question in order to support his beliefs. Let us review a couple of examples.

On page 99 Jeffrey writes, "In 1991 the world was astonished to hear that one of the unpublished scrolls included incredible references to a 'Messiah' who suffered crucifixion for the sins of men." According to Jeffrey, the "scroll also describes the Messiah as a 'leader of the community' who was 'put to death.'" The fragment of a scroll in question, the total of only five lines, is believed to have been written by an Essene scribe. It actually refers to Isaiah but makes no mention of Jesus or anybody from Nazareth, or anybody born in Bethlehem or anybody crucified in Jerusalem about 33 C.E., and, contrary to Jeffrey's assertion, has no features which would be common with any part of the New Testament except that the Messiah was a "shoot of Jesse." The latter reference is of no special meaning in regard to Jesus, because it was the established tradition in Judaism that the Messiah would come from the descendants of King David (whose father's name was, in its English rendition, Jesse). Likewise, the reference in that scroll to the Messiah's hands and feet being pierced does not mean anything, because such treatment of condemned men was usual for the Romans at that time. Thousands of Jews were crucified by the Romans, and the hands and feet of the victims were routinely broken and pierced. Jeffrey's statement that "this scroll confirms the historical truthfulness of the New Testament record about Jesus and His crucifixion" is just an attempt to put a favorable spin on something which actually has no bearing on his thesis.

Later Jeffrey tells about another "fascinating discovery" of a scroll mentioning the Son of God (p. 101). Again, this scroll, reflecting the beliefs of an unknown scribe, makes no reference whatsoever to the Jesus of the gospels. Jeffrey finds similarity between the wording in that scroll and some phrases in Luke 1:32 and 1:35. Such similarity is not surprising at all: Luke, according to the Christian tradition (which, however, has been disputed) was a physician. If this is true, then he must have been an educated man quite familiar with the ways of expression in his contemporary writings, sermons, and discussions of religious questions. It was only natural that some phrases in his writing were similar in style and vocabulary to those of other writers of that time. To conclude from the above similarity of style that the scroll in question confirms the gospel's tale seems to be a rather far-fetched proposition.

The story told in the four gospels supposedly refers to events of pivotal importance in the history of civilization. Hundreds of manuscripts, originated at the time of the events described in the Gospels, both known before the discovery of the Dead Sea scrolls and those found starting in 1947 in the caves at the Dead Sea, never mention the name of the supposed central figure of those events and make no obvious reference to him or to his life and death. This fact, while not necessarily proving that the Gospel is fiction, nevertheless provides an argument to the opponents of the Gospels' reliability. Jeffrey's assertions to the contrary are unfounded.

The rest of Jeffrey's examples are of the same type. He quotes some words from a scroll which have no signs of relating to Jesus, and without any factual basis for it, asserts that the writer of the scroll meant something confirming Jeffrey's beliefs. Reviewing that part of Jeffrey's book amply illustrates the lack of real proofs at his disposal and his desperate determination to create alleged proofs from whatever material is at hand.

Influence of Jesus as Proof of Jeffrey's Beliefs

At one point, Jeffrey quotes approvingly from a book by Philip Schaff, *The Person of Christ*: "This Jesus of Nazareth, without money and arms, conquered more millions than Alexander, Caesar, Mohammed and Napoleon. . . . Without writing a single line, He set more pens in motion, and furnished themes for more sermons, orations, discussions, learned volumes, works of art, and songs of praise than the whole army of great men of ancient and modern times."[9]

Is the above statement true? Except for the mention of Mohammed, it is, as long as the immense influence of Christianity on the culture of many nations is meant. It is not true, though, when it implies that such influence is unique.

Yes, an enormous body of literature, beautiful sculptures, great paintings and poems, dissertations, and historical studies have been inspired by the story of the life and death of a poor and unarmed traveling rabbi from Nazareth, regardless of whether it is history or legend. However, a similar statement equally applies to some other figures, including Mohammed and Buddha. The number of people who accept the teachings of either of these two historical figures is about as large as that of Christians; the scope of literature and art inspired by them is as impressive as that inspired by Christianity. The beautiful architecture of temples in India, Thailand, and Cambodia, or of some famous mosques, was inspired by religions having *little* or *nothing* in common with Jesus or Christianity. Such examples can be easily multiplied.

There are also many examples of evil personalities who nevertheless had very large followings and became the subjects of art, literature, and scholarly study. Of course, the maniacal Adolf Hitler comes to mind as an example. This man had neither money nor arms when he started his crusade aimed at conquering the world and annihilating Jews, Gypsies, and mentally ill people, while killing in the process also thousands of Poles and Russians, and even his fellow Germans. For twelve years, he succeeded in converting tens of millions of educated, talented, and industrious people into a nation of murderers, and only the resounding defeat of his valiantly fighting hordes and the catastrophic destruction of the country caused the awakening of his brainwashed followers.

Stalin organized a massacre of millions of his fellow citizens, but during World War II, Russian soldiers fought and died shouting his name.

There are many other examples of the absolute power of certain personalities over the life and death of their followers. Pastor Jim Jones led eight hundred of his followers to a senseless suicide. There are other similar occurrences as well; in our time, Usama bin Laden succeeded in making nineteen of his followers commit suicide in a single murderous act on an unprecedented scale.

Despite the difference in scale and motivations among these phenomena, they have important features in common. A charismatic personality captures the minds and imagination of scores of people who become faithful followers of whatever ideas, conceptions, and ways of life the originator of that particular faith or political viewpoint expounds. While the mass psychology underlying such phenomena has not been sufficiently understood, one thing is obvious: Islam, Christianity, Buddhism, Hinduism, Judaism, and other religions contradict each other in many respects. They cannot all be equally true. However, each of those religions has millions of followers. The number of followers of this or that concept or religion, and the degree of their devotion to that concept or religion, by no means signifies that the tenets of this or that religion or teaching are true. Therefore, Jeffrey's and Schaff's references to the immense influence of Christianity on the culture of the Western world, however true and impressive, cannot be viewed as proof of the doctrine of that religion or of the story told in the Gospels.

CONCLUSION

As I have mentioned, reviewing all chapters and paragraphs in Jeffrey's book would require a book of about the same size; therefore I chose to limit my review to selected sections only. It may raise a question of whether my selection of items for review was governed by an agenda. In other words, the question may be whether the sections omitted in my review deserve praise or at least are to be accepted as reasonable discourse—my answer to that question is a firm no. While my choice of items subject to review was to some extent influenced by my own areas of interest, I can assert that I have read the rest of Jeffrey's book and did not find there a single item I would accept as reasonable. Whatever topic Jeffrey discusses, be it archaeological data allegedly confirming the biblical story (pp. 69–80), or his explanation of the origin of the circumcision custom (pp. 155–56), or the many other subjects Jeffrey chooses to discuss, it is always replete with exaggerations, imprecision, inaccuracies, direct errors, and distortions, all aimed at "proving" Jeffrey's own professed beliefs.

NOTES

1. Grant R. Jeffrey, *The Signature of God* (Toronto: Frontier Research, 1996).

2. Brendan McKay et al., "Solving the Bible Code Puzzle," *Statistical Science* 14, no. 2 (1999):150; Mark Perakh, "The Rise and Fall of the Bible Code" (in Russian), *Kontinent*, no. 103 (2000):240–70.

3. Brendan McKay, "Encoding Messages is EASY" [online], cs.anu.edu/au/~bdm/dilugim/longels.html [March 31, 2002].

4. Yacov Rambsel, *Yeshua—the Hebrew Factor* (San Antonio, Tex.: Messianic Ministries, 1996).

5. Mark Perakh, "B-Codes Page" [online], www.bigfoot.com/~perakh/fcodes/ [March 31, 2002].

6. Ivan Panin, *The Writings of Ivan Panin* (Agincourt, Ont.: Book Society of Canada, 1972).

7. Albert Einstein, "Autobiographical Notes," in *Albert Einstein: Philosopher-Scientist* (New York: Harper and Bros., 1949), pp. 3–5.

8. Albert Einstein, letter to Guy H. Raner Jr., September 28, 1945; published in *Skeptic* 5, no. 2 (1997):64.

9. Philip Schaff, *The Person of Christ* (n.p.: American Tract Society, 1913), quoted in Jeffrey, *The Signature of God*, p. 96.

7

SHOW ME PROOF

A Preacher in a Skeptic's Disguise

The text of Fred Heeren's book *Show Me God: What the Message from Space Is Telling Us about God*[1] is interspersed with various extraneous matters, including imaginary conversations with the fictional editor of the book, named Carl. Most of these "interludes" have little connection to the book's main theme. There is, though, a brief fictional story (p. xx) in the very first of those "interludes" which is worth mentioning. It deals with a marketing expert who specializes in "packaging" religious books. Among other things, this "seasoned marketer" tells Heeren that the most important thing for a book is not its contents but rather the title and cover design. The title we already know, and it certainly meets the marketer's requirement. Who wouldn't be interested in seeing God? The striking title of Heeren's book is reinforced by the cover design. On the cover we see a beautiful photograph of a remote nebula, taken by the Hubble telescope. In front of that nebula, we see Einstein pointing his finger toward it, under large letters spelling out the title *Show Me God*. The implications of that combination of title and picture are obvious: The greatest scientist of the twentieth century points to space, showing us God; this means that space is supposed to tell us about God, and science (personified by Einstein) is supposed to reveal God's existence. However, any reader who wishes to look a little deeper into the cover art of Heeren's book can discover that the picture in question is a fake. The artist combined two photographs that had nothing in common. One was a genuine photo of a nebula, which did not require any alterations. The other was a photo of Einstein, taken in 1931, when the famous scientist was giving a lecture at Mount Wilson Library in California. In the original unadulterated photo, Einstein is seen looking at the audience while his hand is extended toward a blackboard as he is writing a formula. His fingers are bent, and he seems to be holding a piece of chalk. In order to make the picture fit Heeren's goal, the artist first had to remove from it the blackboard, replacing it with a view of the nebula, and, second, had to attach someone else's hand, with a finger pointing toward the sky, to Einstein's sleeve.

This photographic trick can serve as a symbol of the entire book in question. Several times, Heeren repeats that he is a skeptic in search of truth. Don't believe that! It is as true as the picture of Einstein pointing to God. Heeren's book contains numerous references to science, and lengthy explanations of scientific theories and data. To his credit, Heeren's explanations are accurate and well written. He has mostly faithfully conveyed to his readers information about contemporary cosmological science as he received it from scientists, including some of the best-known names in the field. However, the more one reads of Heeren's book, the more evident it becomes that all this scientific information is used by Heeren to disguise his actual thesis. If Heeren's book contained only the mentioned explanations of scientific theories and his interviews with prominent scientists, it would provide interesting reading. Unfortunately, using the interesting scientific material, Heeren never forgets to squeeze into it his beliefs, which have nothing to do with science but are those of an uncritical believer in the Bible's inerrancy.

Heeren is actually a preacher disguised as a "skeptic in search of truth." He tries to tell the readers that scientific data not only point to God as the creator of the universe (a view that cannot be either proved or disproved scientifically) but that these data lead directly to the belief in Jesus being God. Hence, Heeren is not just a believer suggesting arguments in favor of God's existence, but rather a propagandist for his specific religious beliefs, and in that he is on a shaky ground: He provides no substantiation for his propagandistic effort, thus devaluing his book to a religious rant feebly disguised as an objective exploration of his thesis.

HEEREN PROVES GOD BY SHEER LOGIC—OR DOES HE?

Heeren presents his book as the result of a skeptic's journey over a long path of gradual discoveries of truth, wherein he interviewed many outstanding scientists and thus arrived at the conviction that the Bible speaks the truth. However, a reader should not be deceived by that claim. There is nothing skeptical in Heeren's approach to the question of God; he is the epitome of a fervent believer and as hard-boiled a creationist as they come. This is particularly obvious from the way Heeren uses quotations attributed to various prominent physicists, such as Weinberg, Hawking, Guth, and others. To deflect possible accusations of misuse of these quotations, Heeren writes: "Though I have made every effort to quote them in context, I should mention that the intent of their side of our conversations was certainly not to offer support for any particular metaphysical belief. . . . I hope it will further serve to show that their more supportive statements were made apart from (and sometimes in spite of) their personal beliefs, not because of them" (p. xi). The plain meaning of the above quotation is that Heeren selectively used those parts of the scientists' statements which fit his goal. This is exactly what is called quoting out of context.

Heeren's actual credo is clearly delineated on pages 88 through 102, where he endeavors to prove the existence of an omnipotent, invisible God by means of sheer *logic*. Logic is certainly a respectable tool for analyzing any situation, however, logic by itself, although necessary, is not sufficient to prove a thesis. When relying on logic, two conditions have to be met. The first condition is the credibility of the premise. The second condition is a strict adherence to the chain of consecutive conclusions, each following from the preceding step of the discourse.

Heeren's allegedly logical proof of God fails on both accounts. He starts his discussion with the following section title: "Logic Demands a Cause for Every Effect" (p. 88). However, this statement has nothing to do with logic—it is a premise for the subsequent discourse, and in itself it is neither logical nor illogical. Logic is the path from a premise to a conclusion and so cannot serve to choose a premise. The latter is chosen either based on experimental evidence or as a philosophical principle. In mathematics—formalized logic in its most powerful form—the premise (usually referred to as either an *axiom* or a *postulate*) is a statement chosen on the basis of considerations alien to the subsequent logical steps. (Sometimes an axiom is defined as something which is obvious and thus requires no proof. However, the modern view of an axiom has erased the previous distinction between an axiom and a postulate.) No conclusion reached through logic can be anything more than a statement which is true *provided that we accept the premise*.

Heeren's premise is that "every effect has a cause" (which he also expressed as "From nothing, nothing comes").[2] Of course he is entitled to offer any premise he wants, but there is no reason to attribute to his premise the status of an absolute truth. Indeed, how do we know that every effect in fact has a cause? This statement is a postulate based on experience, and not, contrary to Heeren's assertion, on logic. The experience serving as a basis for the above statement is certainly very extensive. The necessary requirement for any scientific data to be accepted as part of the scientific arsenal is *reproducibility* of experimental results. Reproducibility means that all experiments conducted under identical conditions must yield identical outcomes. As discussed in chapter 12, while reproducibility is expected, it is actually only *assumed* to have been achieved because, as a practical matter, it is beyond the reach of a scientist to ensure the absolute reproduction of the same conditions when repeating experiments. Nevertheless, the assumption of reproducibility works well in science and the principle of causality is a widely used premise in science, and most probably will remain in the arsenal of science. However, the assumption of causality, contrary to Heeren's view, does not follow from logic.

But let us accept Heeren's premise and see whether his discourse meets the second requirement of strict logic. The discourse, among other things, must include no statements which contradict the preceding steps, and hence no statements which contradict the premise. Since Heeren's premise was that "nothing can come out of nothing," his conclusion must conform to that premise. Here is Heeren's ultimate conclusion from his allegedly strictly logical discourse: "A series of causes cannot be infinite. There must have been a first cause which is uncaused" (page 88).

Using Heeren's own words, "This is not rocket science" (ibid.). If there must have been an uncaused "first cause," then, by plain logic, his premise was wrong. You can't have it both ways, Mr. Heeren. If everything must have a cause (as the premise stated) then there could not be an "uncaused" first cause. If, though, there must have been an uncaused first cause, then not everything must have a cause.

Hence, Heeren's allegedly strictly logical proof of God's existence collapses on itself. Of course, Heeren's "proof" is not new at all. The exact same allegedly logical set of arguments has been suggested many times before and, while seeming very convincing to believers, failed to change the views of skeptics. The reason for such a failure of the allegedly logical argument is its contradiction of plain logic.

Does the collapse of Heeren's allegedly logical argument prove the absence of God? Certainly not. It shows though that the existence (or nonexistence) of God requires a very different type of argument and cannot be proven by logic alone.

Continuing on, Heeren tries to imagine counterarguments by atheists and offers refutations. His first point is that science cannot offer an explanation of "how matter and energy could have emerged from nothing before that." Therefore, says Heeren, "humankind is limited to the same explanation it has had from the beginning: a supernatural explanation" (p. 89). Science indeed has no universally accepted explanation for the appearance of matter and energy "out of nothing." This proves the limitations of science. However, to say that something has a "supernatural explanation" means offering no explanation. Referring to a "supernatural explanation" is admitting a failure to have any explanation, because we have no way of knowing what that supernatural explanation really is. The origin of matter and energy is unknown. All explanations, by Heeren or by anybody else, are speculations. Trying to explain it by referring to the biblical concept of God is as plausible as any other of an endless variety of ways to invent a supposed explanation.

Of course, to predict the future achievements of science is a very risky endeavor; however, my suspicion is that the question of the origin of matter and energy will never be answered. The concepts of the human mind stem from experience. I am afraid that the human mind is not capable of solving the dilemma of deciding whether the universe came out of nothing or existed always. Both concepts—"nothing" and "forever"—seem to be beyond human comprehension. We cannot comprehend the meaning of the notion that the universe always existed and will continue to exist forever, rather than evincing that idea in an abstract way, not really understanding its meaning. Likewise, we cannot comprehend true nothingness or the concept of "no time" before the big bang.

Heeren continues, asserting that matter and energy must have been created out of nothing by some "limitless being outside of time and space." Maybe so. But to make skeptics believe that assertion, it is not enough to just say so. How do you know that, Mr. Heeren? Where did that uncanny knowledge come from, besides your emotional need to believe? When, at the beginning of the book, you said you were a skeptic, you were not telling the truth.

If the big bang theory is true, and the universe as we see it appeared "out of nothing" some fifteen billion years ago, it does not answer the question of what that "nothing" was which preceded the big bang. There is no proof whatsoever that the big bang had a supernatural cause, and to assert otherwise is a display of arrogance by religious zealots who have nothing to prove their assertion besides their blind faith in some ancient legends.

Science cannot offer a firm explanation for the source of energy that exploded at the big bang, although a number of hypotheses have been suggested. Therefore, it is often said that the energy in question appeared out of "nothing." In fact, though, science now accepts as reasonable the concept that the net energy of the universe is zero and therefore no energy creation out of nothing in the big bang needs to be assumed. Likewise, since modern science maintains that time is just an attribute of matter, it is said that time started at the big bang. However, science offers no universally accepted explanation of the meaning of that assertion. If one says there was nothing *before* the big bang, not even time, the very word *before* implies that the concept of time is meant specifically, as defined in the theory of relativity. When we say, according to the special theory of relativity, that the time interval between two events is different in two frames of reference, what we mean is that two clocks, one placed in one of the frames in question, and another placed in the other, will show different time intervals between the events. The construction of the clocks does not matter, nor does the physical process utilized for time measurement, as all physical processes change their rate in the same way depending on the frame of reference to which they are "attached." What is essential, though, is that time is just a certain characteristic of a physical process whose rate depends on the frame of reference (and on the intensity of gravitation at a given location). There is no way to define time separately from any physical processes. When we say that time "started" at the big bang, what we are actually saying is that before the big bang there were no events and hence no processes with which we could associate time. However, there is no way to assert anything about whatever might have existed and occurred before the big bang. If *something* existed before the big bang, all signs of that something must have been obliterated in the big bang, if not in some event preceding the big bang. This something could very well be the source of the energy of the big bang (if a source of energy was needed at all). Hence, when science says that there was nothing before the big bang, it means that science has no way of knowing what, if anything, existed and occurred before the big bang.

Grasping at the word *nothing*, Heeren and his colleagues in the creationist enterprise suggest that, since the universe allegedly appeared out of nothing, it must have been created for a purpose by a supernatural First Cause. Of course they do not offer any substantiation for that explanation, which does not explain anything anyway.

Rather than making the arbitrary assumption of a supernatural entity being the conscious creator of our universe at the moment of the big bang, it is not less plausible to assume that another universe might have existed before the big bang. This universe could have had properties and natural laws in some respect different from our universe and these natural laws could have led to its annihilation thus providing the source of energy for the big bang (if such source of energy was indeed needed). For that supposed preceding universe to have ended up in a "big crunch," its properties would have had to differ only slightly from our universe. Does the described hypothesis have any proof? No, it does not and cannot, because such a preceding universe, whatever its laws and properties could have been, would necessarily have been annihilated without a trace in the big bang. However, the described hypothesis shows that there is no need to assume a supernatural creator for whose existence there is not a single proof except for the words of some ancient writers, who could not and did not know even a small fraction of what science has subsequently established.

Note that the hypothesis of the preceding obliterated universe is not equivalent to the theory of a universe oscillating between big bangs and big crunches (although it does not contradict that theory, either). That theory relates to *our* universe, which, however, does not seem to meet the conditions for oscillating between big bangs and big crunches (although the verdict on this problem is not yet final). The hypothesis of a possible preceding universe assumes that the latter could be in some respects different from our universe, in particular that it ended up in a big crunch, while our universe may continue expanding forever, whatever "forever" may mean.

As Heeren continues, he offers certain considerations claiming to prove what God is like. Like his preceding pseudo-proofs of God's existence, these notions are far from new. Very similar "proofs" have been suggested uncounted times before. While they may sound reasonable to a believer, they have no chance of convincing true skeptics because all of these supposed proofs are arbitrary.

Heeren's first assertion about God, presented as a section title, is that "The First Cause Must Be Independent of Its Effect." Why? Because, Heeren tells us, "logic demands" it. He continues by saying, "God must be transcendent; that is above and beyond the boundaries of its creation" (p. 90). Maybe so. However, to base a conclusion only on the assertion that it "must be so" is not quite convincing for those who have not yet been converted to Heeren's faith. There are many possible ways to attribute various qualities to the supposed First Cause, and Heeren's assertion remains just a display of his personal views not substantiated by anything but his opinion.

On the same page, Heeren further asserts that the First Cause must be omnipotent. Again, it may be so. Equally plausible may be an assumption that the supposed First Cause is extremely powerful but still not omnipotent, i.e., still not capable of doing certain things. This question has been the subject of numerous theological disputes where such dilemmas were discussed as whether or not God can create such a stone which he himself would be unable to lift, and the like. Concluding this section, Heeren says: "Once we accept the idea that the universe had a First Cause, we must also accept the fact that all the miracles of the Bible (from the parting of the Red Sea to the resurrection of Jesus Christ) are quite plausible and easily explained" (ibid.).

For skeptics, there is a very big gap between the notion of a First Cause and the plausibility of particular biblical stories. If, as Heeren urges us to do, we wish to adhere to strict logic, there is no logical path from the general concept of the First Cause to the particular biblical stories, which can be either true or false regardless of the existence of a First Cause. Even accepting the idea of a God by no means leads automatically to the belief in the divine origin of the Bible. For the latter, Heeren has so far not offered any proofs. Heeren is entitled to believe that the Bible is the word of God—it may be so. But to make skeptics accept that belief, it is not enough to simply say that it is so.

It seems worthwhile to note that Heeren makes no distinction between miracles described in the Torah (such as the parting of the Red Sea) and in the New Testament (the resurrection of Jesus). Of course, everybody knows that while both Jews and Christians believe in God and in the divine origin of the Torah, Jews do not believe in the resurrection of Jesus. This shows the absence of a direct logical connection between the acceptance of belief in the divine First Cause and beliefs in the divine origin of particular parts of the Bible, including the alleged miracles.

In the title of the next section, Heeren tells us that "The First Cause Must Be Eternal (Transcending Time)" (p. 91). Heeren does not provide any substantiation for his assertion except that it "must be" so. The Creator, says Heeren, "is without beginning or end." Maybe this is so. But the simple assertion that this is so is not at all convincing to skeptics. Science indeed provides no answer to the question, "What was there before the big bang?" Neither does the assertion that a supernatural being has always been, is, and always will be. How does Heeren know this? Did the supernatural Creator give him this information? Like in all other cases discussed by Heeren, a reference to a supernatural First Cause is not an explanation but rather an admission of the absence of an explanation.

To support his thesis, Heeren also offers an interpretation of the word often used in the Hebrew Bible for God, namely the tetragrammaton, YAHVE (in Hebrew, the four letters *yud-hey-vav-hey*). He suggests that these four letters are "apparently derived from the Hebrew for I AM" (ibid.). This interpretation is rather strained, since the four letters in question, although allowing for several interpretations, do not actually constitute any unequivocally recognizable Hebrew word or phrase.

Furthermore, Heeren refers to the theory of relativity, which has established the relativity of time; this, in his opinion, somehow justifies the concept of God being beyond time. However, the concept of the relativity of time in Einstein's theory is transparent. The special theory of relativity tells us that the time intervals between any two events are different in two frames of reference that are moving relative to each other with a certain constant velocity. The general theory of relativity tells us that time flows differently for two bodies if they are subjected to different gravitational forces. Both conclusions have nothing to do with the concept of a deity beyond time. Heeren's attempt to enlist the theory of relativity to support his thesis is without merit.

Over the next three sections of *Show Me Proof*, Heeren attributes to God three more characteristics: The First Cause must be spiritual (transcending space), must be all-knowing (omniscient), and must have personhood (pp. 92–93). I can only repeat what I have said about the preceding alleged attributes of the First Cause, namely that Heeren simply states what he wants to believe without offering even a hint of proof of the validity of his assertions, except for saying that "this must be so." If the word "must" were replaced with "may," the above assertion could be viewed as one of the possible interpretations of the facts. However, to prove that it "must" be so, rather than "may," would require a much more convincing set of arguments than those employed by Heeren. For example, the only reason to assume that the First Cause has personhood, is, according to Heeren, the amazing balance between all the forces and conditions in the universe, which, he believes, would be impossible without the creating will of a supernatural person. This is just the same familiar argument from design, to which Heeren does not add anything that has not been said many times before and which is just an arbitrary assumption.

At the end of this chapter, Heeren quotes prominent physicist and Nobel laureate Steven Weinberg. The quotation in question is supposed to support Heeren's religious beliefs. However, Heeren is cautious enough to say that "Weinberg hovers somewhere between agnosticism and atheism" (p. 94). If that is so, obviously Weinberg's views cannot support Heeren's unadulterated creationism. This is just one example of Heeren's method of using quotations from various prominent scientists to prove his own views. He uses the same tactics quoting the renowned astronomer Robert Jastrow, who, Heeren tells us, "is a self-proclaimed agnostic" (p. 149).

Quotations from scientists like Weinberg and Jastrow, while technically verbatim are essentially taken out of context, and hence misrepresent the intent of these scientists.

Heeren's treatment of such quotations raises a simple question. These scientists have at their disposal the same information as Heeren does. Moreover, Heeren actually has acquired this information from the scientists whom he interviewed. Why then do they remain "between agnosticism and atheism" or "self-proclaimed agnostics" rather than seeing the truth as Heeren does? Are they fools incapable of evaluating evidence, and is Heeren far superior to them in that respect?

In some other places of his book Heeren resorts to the direct distortion of facts. For example, he asserts that Einstein "tried to find an explanation for his general relativity equations that would not require a beginning and a Beginner for the universe, and he came away a believer in both" (p. 110). This statement, which Heeren uses to enlist Einstein as a supporter of his beliefs, is a distortion of Einstein's views. As mentioned earlier, Einstein unequivocally claimed to be a skeptic and later asserted being an atheist (see chapter 6). Whenever Einstein used the word *God*, it was never meant to refer to the biblical God and never meant that Einstein was a believer.

Of course, the personal beliefs of this or that scientist do not prove anything, but more important for this review is that Heeren could not offer a single valid argument in support of his religious beliefs.

Concluding the chapter in question, Heeren asserts that his supposedly logical discourse leads to the conclusion that God loves us. According to Heeren, his so-called logic not only proves the existence of God, but also points to all the attributes of God that he discussed, and also proves that the story told in the Gospels is true. However, for an unbiased reader, there is no logical connection between Heeren's original assumption and his final conclusion, in which he espouses blind faith in the story told in the Gospels.

HEEREN DISPROVES ALTERNATIVE THEORIES

Having dealt with his supposedly strictly logical discussion of God's attributes, Heeren devotes the whole of chapter 4 to a rebuttal of several theories suggested as alternatives to the big bang theory. Elsewhere Heeren stated that some Christians believe in the big bang, while some others do not. Heeren himself is a staunch adherent of the big bang theory, which in his view supports the biblical story. However, the mere fact that some of his coreligionists do not believe in the big bang is a good indication that the big bang theory in itself does not imply the veracity of the biblical story.

Heeren constantly asserts the alleged irrefutable logic of his position. Under the heading "A Skeptic's Question" he writes: "Even if a common sense tells you that the universe had to have a great First Cause, sometimes common sense is wrong" (p. 100). Then he attempts to demolish all explanations other than his belief in the great First Cause.

Before discussing Heeren's refutation of the alternative explanations, note that the above quoted statement incorporates a not very subtle subterfuge, attributing to a skeptic a view no genuine skeptic would ever share. No skeptic would agree that "common sense" demands the hypothesis of a great First Cause. Common sense, which elsewhere Heeren also refers to as logic, in no way requires the admission that there must have been a great First Cause. The hypothesis of the First Cause is just that—a hypothesis, which has nothing to do with common sense or logic. One may believe in the First Cause, or disbelieve, or leave the question open; any of the three choices is equally compatible with common sense.

Let us see how Heeren disproves the alternative explanations of the universe.

He starts with the brief discussion of some concepts of Eastern religions. While his view of those religions meets no objections, as they indeed provide no reasonable picture of the universe, Heeren simplifies the tenets of those religions, reducing them to a primitive one-dimensional outlook. In his rendition, all that the Eastern religions claim is, "The cause and effect are the same" (p. 100). Of course, there is much more to those religions than Heeren's description implies. Referring to his definition of the essence of the Eastern religions, Heeren says that "this contradicts the obvious principle from logic that every cause demands an effect." Here, again, Heeren demonstrates that he is quite in the dark about what constitutes logic and what the concept of a postulate entails.

Then Heeren reviews several theories (but not all of them) which have been suggested as alternatives to the presently prevalent theory of the inflationary big bang. This discussion is immaterial for the real question that interests Heeren. Whether the theory of the inflationary big bang is correct is irrelevant to the question of whether the universe has been created by a supernatural First Cause or is self-sustained. Any of the alternative theories can be reconciled both with the assumption of the supernatural First Cause and with the assumption of a self-sustained universe. Scientific theories neither prove nor disprove the existence of God. Therefore Heeren's effort to show that the theory of the inflationary big bang is the only one acceptable has no bearing on the main thesis of his book.

Viewed separately from Heeren's underlying idea of a supernatural creator of the universe, his discussion of alternative theories is a moderately entertaining tale about several of those competing scientific theories. Heeren partially acquired his information about these theories from interviewing prominent scientists.

He devotes several pages to the theory of a steady-state universe, whose main proponent has been Sir Fred Hoyle.[3] In his interpretation of Hoyle's effort to substantiate the theory of a steady-state universe, Heeren seems to claim that he can read Hoyle's mind. In Heeren's opinion, the reason that Hoyle continues to search for arguments in favor of his theory is that "a universe with a beginning requires an outside agency that cannot be explored by science" (p. 104). Of course, this is just Heeren's own favorite interpretation of the alternative theories. Neither the big bang theory nor Hoyle's steady-state theory requires any assumptions about the existence or nonexistence of an "outside agency." As of today, Hoyle's theory seems to have little chance of success (although it cannot yet be irreversibly ruled out), while the inflationary big bang theory seems to be supported by a host of observational evidence. However, the rise and fall of scientific theories occurs without relation to religious concepts. If Hoyle's theory is ultimately abandoned, it will in no way support any religious conclusions. The big bang theory in itself has no religious implications and does not require the assumption of a supernatural First Cause (but also does not contradict it).

Similar comments can be made in regard to Heeren's discussion of other alternative theories, such as "plasma cosmology" (p. 106), etc. For unknown reasons, Heeren did not discuss certain alternative theories, for example, that of Evgenii Lifshitz and Isaak Khalatnikov of the Russian Academy of Sciences, who offer an ingenious explanation of the universe's expansion requiring no initial big bang.[4] Heeren concludes his discussion of alternative theories with a section titled "There Simply Are No Natural Explanations for the Ultimate Origin." Quoting several scientists, he asserts the limitations of science. Of course, no scientist would disagree with the fact that science has its limitations. However, while every scientist would readily agree that science cannot overstep certain barriers in its quest for knowledge, Heeren, unlike most scientists, performs an acrobatic jump over a logical pit, claiming in the title of his conclusion, "Science Ends Where the Bible Begins" (p. 120). What is the basis for that assertion? One would search in vain for a basis in logic or in any evidence; the only basis for that statement is blind faith.

Whereas science indeed stops at the big bang (if the latter has indeed occurred) and cannot (and possibly never will) unequivocally answer the question of the big bang's origin, this fact provides no reason to look for an explanation in the motley collection of ancient poems, tales, historical chronicles, and fables named the Bible. It is common knowledge that the Bible contains an abundance of inconsistencies, and no casuistry, however ingenious, could reasonably explain them away. Moreover, those parts of the Bible that can be verified by comparison with scientific data often contradict the latter, be it the biblical cosmogony or the history of ancient times. These inconsistencies and contradictions are equally abundant in the Old and New Testaments. Heeren's jumps from the discussion of scientific theories to the assertion of the truthfulness of the biblical stories have no substantiation. This makes his book a mixture of a relatively entertaining popular rendition of modern scientific cosmology with the unsubstantiated ranting of a religious zealot.

HEEREN REVIEWS THE BIG BANG FROM THE BIBLE'S VIEWPOINT

Reviewing Heeren's book is by no means a rewarding job. This thick volume of over four hundred pages contains numerous repetitions of similar arguments, mostly of the type "it must be so." He returns time and time again to the same questions, chewing them from various slightly differing angles. Each time, having told the readers about some facts of science, he jumps to a conclusion which by no means follows from the preceding narrative and which boils down to an unsubstantiated assertion of the truthfulness of his personal version of religious beliefs.

The following is a typical quotation from chapter 7 (titled "The Bible and the Big Bang"): "I won't deny that there is a conflict between science and traditional beliefs among many Christians, but as I can show, there is no conflict with the Bible itself" (p. 182). This assertion can be viewed as the succinct expression of Heeren's main thesis and of the main goal of his book.

Heeren starts with the following statement: "For Jews, Christians and Muslims who wish to pay attention to science, science has brought them good news and bad news. The bad news, for some, is that the universe appears to have been created billions, not thousands, of years ago. The *good* news is that it was *created*" (p. 182).

An unbiased reader would immediately notice Heeren's not very subtle trick—an ostensible bow to science covering a little less ostensible distortion of science. The first part of his statement is correct: science indeed leads to the conclusion that, contrary to the Bible, the universe has been in existence not for a few thousand but for billions of years. This is certainly bad news for those whose emotional adherence to their faith makes it hard to swallow the findings of science. The second part of Heeren's statement, though, is false. There is no good news for believers in the conclusions of science, which in no way asserts that the universe was "created." This is an interpretation of the facts of science by Heeren and his cohorts among believers. This interpretation may be true, but it is not enough to just say so and pretend it is an unavoidable conclusion from scientific theories. If the theory of the big bang is true (as most scientists believe) it only asserts that about fifteen billion years ago our universe *emerged* and started expanding. This prevalent scientific theory asserts no less and no more than that. Replacing the term *emerged* with the word *created*, Heeren makes an unsubstantiated assumption. Of course Heeren and his cobelievers are entitled to such an interpretation of the prevalent scientific view, and their interpretation may be correct, but it may just as well be wrong, and to attribute their interpretation to science itself is a subterfuge.

As was mentioned earlier, there are many alternative interpretations of the origin of the big bang, and so far science has no means to determine which interpretation is correct. In particular, rather than assume a supernatural source of creation, it seems easier to assume that our universe was preceded by another universe, all traces of which were obliterated in the big bang (or in some event preceding the big bang) and which was the source of energy of that immense explosion we refer to as the big bang (although actually the constant net energy of the universe may well be zero).

There is a telltale admission by Heeren on page 183: "My own theological biases long included the idea that God created, not just transcendently, but recently." In view of this admission, the readers may judge whether or not to trust Heeren's claim that he is a skeptic.

Heeren's chapter 7 contains rather detailed arguments against the views of young-earth creationists, who insist on a literal interpretation of every word in the Bible. Since such a literal reading of the Bible reveals its irreconcilable contradictions with science, while Heeren wants to prove that the Bible and science are fully in harmony, he offers well-known arguments—for example, the interpretation of the word *day* in Genesis as actually meaning a much longer period of time. Of course, all these arguments have been suggested many times before, and failed to convince the young-earth creationists, who prefer to deny the validity of science rather than to subject the Bible to a nonliteral interpretation.

Some of the passages in that chapter sound rather funny. For example, Heeren says in regard to the "days" of creation as they are described in Genesis: "Charting the days reveals the symmetric beauty of days that correspond to one another both horizontally and vertically" (p. 189).

While beauty is in the eyes of the beholder, a great innovation by Heeren seems to be in discovering that days can be arranged either horizontally or vertically. This is an important contribution to physics. Regrettably, Heeren forgot to explain the exact method for arranging days horizontally and/or vertically. Are there other ways of arranging days, for example at an angle of 37 degrees to the horizon, or along the sides of a triangle, or along a circular path? Then, if days are arranged vertically, couldn't the upper day fall down and smash the lower days? And which arrangement is better, horizontal or vertical? If days are arranged vertically, do we need a stepladder to move from a lower to a higher day? (Heeren's imaginary editor Carl failed to do his job by not deleting such preposterous paragraphs like the one about "vertical" days.)

Toward the end of this chapter, Heeren offers a discussion of the Hebrew word *rakiya* (*resh-kof-yud-ayin*), which he transliterates as "raquia." Heeren repeats the interpretation of that word given several times by various writers, who all intended to find in this word a hint at the Bible's alleged indication of the universe's expansion. To achieve such an interpretation, usually reference is made to the verb *raka* (*resh-kof-ayin*), one of whose meanings in Hebrew is "to stretch." The definitive Hebrew dictionary by Even-Shoshan provides a number of meanings for the word *rakiya*, but none of them is "expanse."[5] The closest translation of "rakiya" (which in the King James Version of the Bible is translated as "firmament") would be "canopy" or "tent" (a thing which, of course, must be stretched to serve its purpose). For the ancient writers who wrote the book of Genesis, the description of the sky as a canopy or tent was a natural manifestation of their view of the apparent blue cupola above their heads. To derive from that word an indication of the expanding universe requires a very inventive imagination indeed. Some Hebrew-English dictionaries (for example, the dictionary edited by David Shumaker)[6] translate *rakiya* as "vault." Whichever of the possible translations of that word one chooses, "expanse" is one of the least reasonable.

HEEREN PROVES DESIGN—OR DOES HE?

Chapters 8, 9, and 10 in *Show Me God* are titled "Evidence for Design," "Alternative Explanations for Design," and "Implications of Design." As could be expected, Heeren tries in these three chapters to prove that the origin of the universe in general, and of life in particular, must be attributed to what has been referred to as "intelligent design."

The reference to supposedly intelligent design has recently achieved a considerable popularity, being vigorously promoted by a number of writers, who unlike their predecessors have largely abandoned the most primitive, patently wrong notions (like the assertion that evolution contradicts the second law of thermodynamics), have accumulated scientific degrees, and have elevated the discussion to a seemingly more sophisticated level. One odd feature of Heeren's book is the complete absence of references to the proponents of the intelligent design "theory." He never mentions writers who are commonly viewed as the most prominent "intelligent design theorists," such as Phillip Johnson, William Dembski, and Michael Behe. While the reasons for such omission could only be guessed, it is worth mentioning that some of the design theorists defend and promote the idea of intelligent design by using argumentation much more sophisticated than that used by Heeren. Ironically, he refers instead to the books by Hugh Ross, whom he calls an "astronomer." Ross's literary output is discussed in chapter 5 of this book, where it is shown to contain many errors testifying to Ross's inadequate understanding of his subject. The fact that Heeren refers to Ross but ignores the more sophisticated books and papers by prominent promoters of intelligent design is telltale in itself.

The three chapters in question are characterized by the same approach we have discussed regarding the preceding chapters. Heeren tells about some facts of science, interspersing his narrative with assertions that the facts in question confirm his religious beliefs. The manner in which these assertions are supported is always the same and can be summarized as "this must be so" statements.

We find in Heeren's book the worn-out assertion that the second law of thermodynamics somehow supports the Bible's story (p. 228). This assertion, which had been common in the writings of creationists of the past, has been largely (although not completely) discarded as wrong by the new crop of creationists.[7]

IS THE GOSPEL LOGICAL?

Chapter 11 in Heeren's book is titled "Is the Gospel Logical?" There we read the following categorical assertion: "There is no gospel without the historical events. Of all the religions, no other has left such evidence of God's involvement with humanity. . . . If ever God reached into our world, it was through a human named Jesus" (p. 274).

It is easy to see the arbitrariness of that quotation. For example, in numerous books and articles by defenders of Judaism we find very similar assertions, wherein, though, God's involvement with humanity is viewed in very different terms. Jewish sources tell us that God revealed himself to the entire nation of ancient Israelites on Mount Sinai. Since Heeren, as a Christian, must believe in the divine origin of the books of the Pentateuch, he must believe that God indeed appeared before the Israelites and spoke directly to Moses. If one believes in that story, was that event not the direct involvement of God with humanity? I am far from trying to assert that the story told in the Torah is true. Moreover, the recent archaeological discoveries seem to provide arguments against the veracity of the Torah's account. As Israeli archaeologists seem to have found, the religion of the early Israelites was not even universally monotheistic.[8] Ancient artifacts show that at the time of the alleged revelation in Sinai, many (if not all) of the Israelites worshipped two gods—one male, named Yahve, and the other female (the wife of Yahve), named Asherta (or Astarta).

Whatever the veracity of the Torah's story, the inconsistency in Heeren's position maintaining the uniqueness of the foundations of the Christian faith is apparent.

Regardless of whether the Gospels' story is unique (as Heeren seems to insist) or is just one of many different stories told in sacred books of various religions (as is obviously the case), let us discuss the alleged historical evidence which, according to Heeren, distinguishes the Christian Gospels.

The four Gospels that are the foundation of the Christian religion describe various miraculous events beyond the everyday experience of regular people. Since they are explicitly presented as miracles, their plausibility is not based on the common experiences of regular people and therefore they cannot be refuted simply because regular people have never observed miracles. They are supposed to be accepted on faith, without proof; hence, we cannot discuss the plausibility of the miraculous events described in the Gospels using the same criteria we use when discussing nonmiraculous events known from historical evidence. Believing in the miracles such as, for example, the Immaculate Conception or the resurrection of Jesus requires a leap of faith and can be neither confirmed nor rejected based on rational arguments.

However, if we want to judge the veracity of the Gospel's story, we can look at the nonmiraculous parts of it to see first if it contains any contradictions, and second if it fits the historical evidence.

What we find in the Gospels is a number of inconsistencies. For example, Matthew and Luke provide two very different versions of Jesus' genealogy. According to Matthew, Jesus's step-grandfather (i.e., the father of Joseph) was Yacov, but according to Luke, his name was Heli. The number of generations between David and Jesus, according to Matthew, was twenty-eight, but according to Luke, it was forty-two. The lists of Jesus' ancestors given by Matthew and Luke are very different. Moreover, if, as the Gospels tell us, Joseph was not actually Jesus' biological father, in what way could Jesus be considered a descendant of David? Jesus' mother was not connected to the line of David's descendants.

While in some of the Gospels we are told that, after his resurrection, Jesus appeared before his disciples in Jerusalem; in others we are told that this event occurred in Galilee. In Luke we are told that three *local shepherds* came to the manger in Bethlehem to greet the newborn Jesus; in Matthew, the same three men are referred to as three *wise men from Eastern lands*, who came following a star in the sky above Bethlehem.

Without discussing the many other discrepancies found in the four Gospels, let us now briefly summarize the storyline which is more or less common to all four of them. According to this story, a wandering preacher named Yeshua (whose name has been transliterated in English as Jesus, apparently stemming from the Greek version *Yisus*), who was born in the city of Bethlehem but grew up in Nazareth, was arrested in Jerusalem at the age of about thirty-three, and was crucified by the Roman rulers of Palestine at the behest of the Sanhedrin, the Jewish supreme court. Some details of the story include, for example, a crowd of Jews demanding to crucify Yeshua but to release a criminal named Bar Abbas (whose deeds included murder), allegedly following a custom of releasing a convicted criminal at Passover time. Of course, the story in question is very well known, so there is no need to repeat it here in all of its details. There is hardly a doubt regarding the great literary quality of that touching story which rarely fails to invoke strong feelings in a reader, as it combines poetic beauty with subtle nuances of psychological and philosophical insights.

All that readily admitted, a completely different question is whether the Gospels tell the story that actually occurred or a fiction, possibly having some basis in reality but including multiple fictional elements. Of course, believing Christians would shrug off such a question. Their unshakable belief in the veracity of the Gospels is not based on evidence, but rather on the indoctrination received in their childhood.

In order to discuss the credibility of the gospel's story, let us consider the following points:

1. If the man named Yeshua preached his version of faith in the temple, this would hardly cause such a harsh reaction from other Jews. At the time of the events described in the Gospels, scores of preachers traveled the land of Israel preaching various versions of Judaic religion, and there is no historical evidence that any of them was ever severely punished. If a man was proclaimed to be the son of God, but his parents, brothers, and sisters all obviously quite human, he would most likely be considered not quite sane, rather than a dangerous criminal. The tenets of Judaism prescribed a punishment of false prophets by stoning them to death. While there is no evidence that at the time of Yeshua such punishment was actually ever implemented, even if it were, it would be a very different occurrence from the story told in the Gospels.

At the time of the events in question, the most powerful element of Jewish society were the *perushim* (in the traditional English rendition, the Pharisees) whose influence would hardly be threatened by a traveling preacher who had but a small band of followers. Another group, the so-called *tzeddokim* (in English, the Sadducees), while differing from the Pharisees in some interpretations of the law, exercised some control over the temple but had very little influence on the Sanhedrin. Zealots and Essenes had even less influence (the latter actually sought refuge in remote corners of the land, such as the Qumran). The Pharisees who controlled the Sanhedrin had no reason to take Jesus' activities very seriously.

2. An offender would hardly be taken to the Sanhedrin. The case against Jesus was not serious enough to merit a hearing at the supreme court of the land. The main task of the Sanhedrin was to interpret religious law.

3. If, contrary to historical evidence, Yeshua were taken to the Sanhedrin, it would be found empty, since, as the Gospels tell us, the Passover

holiday had started. It is hard to believe that the elders of the Sanhedrin would conduct any business on the evening of that day, since, according to Jewish law, the Sabbath begins on Friday evening.

4. If, contrary to historical evidence, the Sanhedrin conducted a hearing of Yeshua's case, it is implausible that they would transfer Yeshua, a Jew, to the Roman authorities for punishment. Jewish religious law (*Halakha*) quite explicitly forbids extraditing a Jew to Gentiles for trial and punishment. Even if Yeshua were found guilty by the Sanhedrin, they would do everything possible to shield him from the Romans, and if they chose to punish him, it would be done by their own means. These means would never include death on a cross. Crucifixion was completely out of the question within the framework of the Judaic law.

5. There is no historical evidence to support such a custom as releasing a prisoner at Passover time. By contrast, it is well known, for example, that the Romans used to "decimate" a legion that did not show sufficient courage in battle. In that frightening procedure the soldiers of the legion would form a line, and every tenth soldier would be executed on the spot. If they were so unmerciful to their own people, their attitude to foreigners, especially to a defeated people, was much crueler. After a victorious war, the Romans would lead the imprisoned king or warlord of the defeated country in chains through the streets of Rome, and then immediately strangle him in the basement of the Mamertine jail. From what is known about the Romans, they ruled by fear, and a gesture of lenience such as releasing a prisoner on a local holiday would be alien to their mentality and practice.

6. It is implausible that a crowd of Jews would demand a crucifixion of one of their own and, moreover, would cry, "let his blood be on us and our children." For Jews, the Roman custom of crucifying people was an abomination. It was especially true if a fellow Jew was condemned to die on a cross. The fact of history is that the Romans crucified thousands of rebelling Jews, who therefore especially hated that cruel punishment. Of course, it is possible that the crowd mentioned in the Gospels consisted of only five, ten, or twenty people, and that such a small crowd of bloodthirsty scoundrels, including some friends of Bar Abbas, could possibly have gathered, trying to get the release of their cohort. It is also possible that the crowd demanding to crucify Yeshua consisted not of Jews but rather of Greeks, who at that time constituted a substantial fraction of the local population. However, the Gospels maintain the implausible version, blaming the Jews for the cruelty of the Romans.

In other words, many elements of the Gospels' story are implausible from a historical viewpoint, even if we do not discuss the miraculous parts, such as the miracles performed by Yeshua/Jesus and his resurrection. If the nonmiraculous parts of the Gospels seem to be inconsistent with the historical evidence and self-contradictory, why should we believe their miraculous parts?

It can be added that, besides the Gospels, there is not a word in any sources contemporary with Jesus' alleged ministry that mention Jesus and his story as told in the Gospels (including the absence of such information in the Dead Sea scrolls). The Roman writers who are often alleged to have confirmed the story in question, all lived much later than the events described in the Gospels. None of them had any firsthand knowledge of these events, and all of them (Tacitus, Suetonius, Flavius Josephus) only reported the stories told by Christians. The most reasonable conclusion is that one cannot rely on the Gospels as a true account of actual events. The Gospels seem to be a mix of a beautiful legend (told in four different versions) with political polemics between the adherents of the fledgling Christian religion and its original source—the traditional Judaism.

CONCLUSION

A question which seems to arise naturally after having read Heeren's lengthy opus is: why was this book published? Actually, it seems to consist of several books, each broken into many segments, intermittently following each other and having little connection to each other. One of these books is a popular tale about contemporary cosmological science. This is the best part of Heeren's story; it more or less accurately conveys the information Heeren acquired from his interviews of prominent scientists and from scientific literature. If Heeren's book contained only this part, its publication would be justified as that of a moderately good popular book about science.

Another "book" within Heeren's book is the part in which he attempts to convince readers of his religious beliefs. This part contains nothing new, as every argument by Heeren has been suggested many times before and has failed to convince anybody who was not already among Heeren's fellow believers. If *Show Me God* contained only this part, it would be just one more example of ranting by a religious zealot, propagandizing his narrow sort of beliefs. There is no connection, logical or otherwise, between the two mentioned elements of Heeren's book, artificially combined in a hodgepodge of unrelated ingredients, vastly differing in style and substance.

Then there is one more book-within-a-book consisting of imaginary conversations with the imaginary editor, Carl. This part would neither stand on its own as a separate book, nor does it add anything of interest to the overall picture painted by Heeren.

Moreover, in the farrago of heterogeneous components of the book in question, there is even a piece of fiction, telling an absurd story about the full text of the Bible being transmitted in Morse code from the entire universe. Finally, Heeren's book includes a historical review, listing names and short biographical sketches of fifty renowned scientists who all were devout Christians.

It is worth saying a few words about that list. First, a reader cannot fail to notice the uneven character of Heeren's choice of names. Along with some titans of science, like Newton or Faraday, the list includes a number of names whose place in science, while respectable, falls short of being on the level of Descartes or Kelvin. It appears that, while his goal was to demonstrate that science owes its most important achievements mainly to Christian believers, Heeren could not find enough candidates of high scientific regard for his list. To make up a list of fifty names, he was compelled to include into it some scientists of lesser importance, and, in doing so, to overemphasize their role in science. The composition of his list is more telltale with respect to the names he did not include than with respect to those he did. We do not find in Heeren's list of the most important contributors to science physicists such as Einstein, Bohr, Schroedinger, Dirac, or Heisenberg; chemists such as Arrhenius, Mendeleyev, Haber, or Kekule; mathematicians such as Gauss, Cauchy, Hilbert, or Lobachevsky; or biologists such as Darwin, Crick, Watson, or Morgan. Instead, he listed much less renowned names, like Buckland, Carver, Kidd, and Michell.

It may also be noted that many of the scientists listed by Heeren did not adhere to his narrow version of Christian faith, not to mention that their views often contradicted not only those of Heeren, but also of each other; for example, Descartes and Pascal had two quite different sets of beliefs, neither being anywhere close to Heeren's unquestionable belief in the Bible's inerrancy.

By including in his list only devout Christians, Heeren apparently wished to demonstrate that the Christian faith created a fertile ground for the development of science, and even to create an impression that the achievements of science are owed mainly to the contribution of Christian believers. Of course, such a thesis is contrary to facts. Even if we ignore the well-documented history of the persecution of scientists and suppression of the scientific discoveries by various Christian denominations, it would be easy to compile a list of great scientists who happened to be agnostics, atheists, believing Jews, Muslims, or Hindus. The fact is that the personal religious persuasions of individual scientists have little relation to their scientific achievements. Therefore Heeren's list of Christian believers among prominent scientists, even if we ignore its arbitrariness, is meaningless and irrelevant to the theme of his book.

The inevitable conclusion from the preceding sections of this chapter is that Fred Heeren failed to prove in *Show Me God* either the harmony between science and the Bible or even the veracity of the latter.

NOTES

1. Fred Heeren, *Show Me God: What the Message from Space Is Telling Us about God* (Wheeling, Ill.: Day Star, 2000).
2. Ibid., p. 88.
3. Fred Hoyle, *The Intelligent Universe* (New York: Holt, Rinehart, and Winston, 1983).
4. For a discussion of Lifshitz and Khalatnikov's theory, see Stephen J. Hawking, *A Brief History of Time* (New York: Bantam Books, 1996), p. 51.
5. Avraham Even-Shoshan, *Hamilon haivri hamerukaz* (The abridged Hebrew dictionary) (Jerusalem: Kriyat Sefer, 1974), p. 686.
6. David Shumaker, ed., "The Hebrew-English Dictionary," in *Seven Language Dictionary* (New York: Avenel Books, 1978), p. 283.
7. Del Ratzsch, *The Battle of Beginnings* (Downers Grove, Ill.: InterVarsity Press, 1996).
8. See, for example, Zeev Hertzog, "The Bible: No Findings on the Locations" (in Russian), *Vremia Iskat*, no. 3 (2000):121.

8

CHALLENGING THE CHALLENGE

Did a Book Published over Twenty Years Ago Indeed Prove the Harmony between the Torah and Science?

The collection of articles titled *Challenge: Torah Views on Science and Its Problems,*[1] edited by Aryeh Carmell and Cyril Domb, was published in 1976 and again in 1978. Moreover, many of the articles in this collection originally appeared even earlier, in the fifties and sixties.

A natural question at this point would be, Why review a collection more than twenty years after its publication? The answer is that the articles of that collection offer many arguments in favor of the compatibility of the Torah with science that have been repeated time and time again in numerous subsequent publications that often add little new to the ideas and views already expressed more than twenty, thirty, even forty years ago. If one wishes to trace the origin and development of the modern popular trend aimed at proving the invincibility of the religious view of the world as it faces its obvious contradictions to scientific theories, the collection in question is a good source of pertinent information.

There is no mention in the book in question of such concepts as "intelligent design" in its modern version (see chapter 2) or "irreducible complexity" of biochemical systems (see chapter 1). These concepts gained popularity many years after the publication of the *Challenge* collection. However, by and large, the authors of the twenty-three articles gathered in that collection have impressive credentials. Most of them have advanced degrees in physics, chemistry, biology, or engineering. Many of them are also rabbis. The overall level of discourse is characterized by ingenuity of arguments.

Of course, not all the papers in that book are on the same level. There are a few articles characterized by elementary misunderstanding of science (like the paper by Menachem Schneerson) or by an improper use of mathematics to supposedly prove God's existence the same way the laws of physics have been "proven" (the paper by G. Schlesinger). A detailed review of these two articles follows. Some other papers are just short notes lacking elaborate argumentation. Still, most of the material can be viewed as being among the most powerful pieces of argumentation in favor of the compatibility of the Torah with science. However, despite that sophistication, the collection in question has failed to prove its point as far as skeptics are concerned.

Indeed, all of the authors in this collection are firm believers in the Torah's inerrancy. There is no question for them of whether the Torah tells the truth—they take it for granted that the Torah is the repository of the ultimate truth. Of course, from such a position it follows automatically that whatever science has established or may establish in the future must necessarily be compatible with the Torah. If such a notion is adopted as the premise for all the discourse that follows it, there remains nothing to prove, as the proof is already contained in the premise.

Of course, the described attitude is unsatisfactory for skeptics who are looking for arguments in favor of either science or the Torah in those cases where there is an obvious contradiction between them. *Challenge* provides no arguments which would satisfy skeptics, as it caters to believers only and is designed to explain to them why they should not be puzzled by the seeming contradictions between their beliefs and the facts of science.

The collection in question consists of four parts, to wit: "Areas of Interaction"; "Creation and Evolution"; "The Secular Bias"; and "Ethical Problems." Each section is examined separately below.

SECTION 1, "AREAS OF INTERACTION"

The authors of the papers in the first section assert, often without offering supporting arguments, that whatever science says is automatically in tune with the word of the Torah. The style of these articles is restrained, but the essence is uncompromising and categorical. They conduct their discourse mostly in general terms, avoiding the discussion of any particular points of controversy between the Torah and science. However, unless one blindly believes in the Torah's inerrancy, all the sophistication and ingenuity displayed in the arguments of these papers remain unconvincing.

Some quotations from the collection should serve to illustrate the position of its authors. In his paper titled "Religious Meaning of Science,"

William Etkin writes, "When we learn to comprehend a new geometry, a new chemical concept of gene structure, a new statistical analysis of the evolutionary process, a new theory of instinct, or any other of the great theoretic triumphs of contemporary science, we recognize that somehow we are in tune with the Creator and His creation" (p. 38). In a paper titled "Genesis and Geology," Aaron Vecht writes, "Science is but the description of nature, and since both nature and Torah are the work of the one God there can be no basic incompatibility between them" (p. 177). In "Evolution—Theory or Faith?" Harry Marcell writes, "Whatever scientific theories are eventually held to account for the way things came about in the world, they will always only suggest how God created: they can never supplant the recognition of creation itself" (p. 195). Similar views have also been expressed by many other authors in the collection.

Unfortunately, statements such as those quoted above are not supported by factual evidence and remain just personal views of the authors.

This section also contains an article by Cyril Domb, one of the collection's editors, titled "The Orthodox Jewish Scientist." In its second section, titled "Philosophic Outlook," Domb claims, "The fundamental assumption of all science that there is a regular pattern in nature so that experiments performed under identical conditions will lead to identical results is very much in accord with religious tradition." However, a few lines later, the author asserts that "God, who is responsible for these 'natural' laws, can revoke them on any particular occasion" (p. 21).

There is an obvious contradiction between the two statements. If we accept Domb's second assertion, a scientist can never be sure that identical conditions will necessarily lead to identical results (as Domb's first statement claims). If God can intervene and revoke 'natural' law at any moment, there is no way to be sure that identical conditions will result in identical outcomes of an experiment. Professor Domb, you can't have it both ways. This is a sad example of how logic may fall victim to an overriding desire to find support for religious beliefs in science.

SECTION 2, "CREATION AND EVOLUTION"

The second section contains papers which deal more specifically with the controversy between Genesis, geology, and evolution. It opens with a paper by Rabbi Menachem Schneerson that in a certain sense stands alone, for its level of discourse is below the majority of articles in the collection. Since, however, the author of that paper was acclaimed as a great thinker of this century, let us take a closer look at his article.

Rabbi Schneerson's "Letter on Science and Judaism"

Rabbi Menachem M. Schneerson, the seventh Lubavitcher Rebbe, is characterized in the collection as "one of the outstanding Torah personalities of the present generation" (p. 142).[2] (The quoted characterization was printed in 1978; since then Rabbi Schneerson has passed away.) The authority Schneerson enjoyed in his lifetime among his followers was enormous. Many of them viewed him as a modern Moses or even as a Messiah. Here is a telltale detail: when a magnificent synagogue was built in Miami Beach, Florida, stones were reportedly brought for its foundation from two places—the *Kotel Hamaaravi*, the western wall of the destroyed Temple in Jerusalem, and the house in Brooklyn where the seventh Lubavitcher Rebbe lived.

From *Challenge* we learn that Schneerson studied at the University of Berlin and at the Sorbonne. Unfortunately, the biographical segment in that collection does not tell us either which subjects Rabbi Schneerson studied or for how long. There is no information available regarding any possible contribution by Schneerson to any specific field of science. It seems safe to assume that he never performed any scientific work in any field of science. It also seems safe to assume that he had no personal experience in conducting scientific experiments, sorting out and interpreting experimental data, developing any scientific theories, participating in discussions of any specific scientific ideas, or generally being involved in any real scientific activity, which is the only way a person acquires a real understanding of what a work of science is all about.

Reading Schneerson's article leads to the conclusion that he had no real understanding of the scientific method and of the essence of the scientific exploration of reality. His paper is an odd mix of platitudes and misrepresentations of science. One such platitude is his assertion that "at best science can only speak in terms of theories inferred from certain known facts." How true! Why, though, this situation, which is not disputed by any scientist, should be viewed as a weakness—as Rabbi Schneerson seems to imply—remains his secret. Yes, science speaks "in terms of theories inferred from known facts." How much more credible the Torah would be if it also spoke in terms of theories inferred from known facts! If, as Rabbi Schneerson indicated, using theories inferred from known facts is a limitation of science, what about the Torah, whose statements are not inferred from any known facts but are simply unsubstantiated assertions without factual basis?

The essence of scientific method will be discussed in general terms in chapter 12. For now, let us see how Rabbi Schneerson treats this subject. His misrepresentations of science include an explanation of two methods utilized by science, extrapolation and interpolation. The explanation in question reduced the scientific method to only these two possible variations, which of course is a gross oversimplification. Moreover, he distorted the essence of these two methods.

Interpolation, Schneerson taught us, is a method whereby, "knowing the reaction under two extremes, we attempt to infer what the reaction might be at any point between the two" (p. 144). If we replaced the word "reaction" with the word "value," the above definition would be an adequate one for a mathematical operation of interpolation. However, it falls short of being a proper definition of any legitimate procedure employed in science. In physics, chemistry, and biology, the simplistic inference of what the "reaction" might be at an intermediary point between two points where the reaction has been studied is not a proper way to develop a theory. Any interpolation, if it takes place at all, is never a bare guess but is always based on certain information enabling the researcher to reasonably predict the behavior of a system under study between the two known "extremes."

That such interpolations are legitimate and not at all arbitrary is seen from the great successes of science, which have led to the enormous progress of technology we all witness. The very picture presented by Schneerson, whereby there is information available at some two points, say *A* and *B*, and from that, information related to a point *C* located between *A* and *B* is inferred, is in itself a distortion of the scientific procedure. If an interpolation (which is a legitimate course of action in experimental science) is employed it is not normally based on the information related simply to two extreme points.

Let us discuss the question of a legitimate interpolation by using a specific example. Since Schneerson used the term "reaction" it seems appropriate to consider an example from chemistry. Imagine that a study is conducted whereby the dependence of the rate of an electrochemical reaction on temperature, current density, solution composition, etc., is investigated. One of the common methods of experimental study is to change one of the parameters (for example, temperature) gradually while keeping all the rest of the parameters constant (within a certain range). The researcher chooses a discrete set of values of temperature, for example, 300 K, 320 K, 340 K, 360 K, 380 K, 400 K (K stands for Kelvin, which is the unit of thermodynamic temperature; 1 K equals one Celsius degree). The researcher makes an effort to keep the variations of current density, solution composition, and all other parameters as small as possible, and measures the reaction rate at the listed six values of temperature. She repeats the measurements many times. When she is satisfied that the repetition of measurements generates values which all are within the same margin of error, she applies some mathematical treatment to her data, for example, the least square fit. The result of the described meticulous procedure is some curve reflecting the dependence of the reaction rate on temperature, corresponding to the fixed values of current density, solution composition, etc. Then the entire procedure is repeated for another value of current density, or for another value of concentration of solution components, etc. After many such measurements have been completed, the researcher has a family of curves, each showing the dependence of the reaction rate on temperature, but for various current densities or various concentrations of the solution components. This procedure is very far from the simplistic picture given by Schneerson, whereby the data for two extremes are used to infer the data for an intermediate point. The rate of reaction for, say, a temperature of 310 K, which is actually between the points measured—at 300 K and 320 K—is estimated not just from the two values at 300 K and 320 K but from the entire *consistent* combination of multiple experimental points. The next step is to form an explanation of the experimental curves in question. Such an explanation is not arbitrary, but rather is based on the enormous body of knowledge accumulated in science. Since the theory must explain a multitude of experimental data rather than just two values at some two points, as Schneerson suggested, there are usually not too many choices which would reasonably fit all the experimental points. Finally, when a theory is developed which seems to account for the entire set of experimental data, it is used to predict the outcome of other experiments. If in the course of the further studies by various researchers the predictions of the theory are reasonably confirmed, the theory becomes a part of the scientific arsenal, as a plausible interpretation of facts. It is never viewed as the absolute truth, but usually every good scientific theory contains at least a grain of truth in it. This example illustrates that Schneerson's description of interpolation falls short of being an adequate presentation of a scientific method.

After explaining interpolation, Schneerson goes on to speak about extrapolation, which, he asserted, is inferior to interpolation. He gave the following definition and an example: "The method of extrapolation, whereby inferences are made beyond a known range, on the basis of certain variables within the known range. For example, suppose we know the variables of a certain element within a temperature range of 0° and 100°, and on the basis of this we estimate what the reaction might be at 101°, 200°, or 2000°. . . . Of the two methods, the second (extrapolation) is clearly the more uncertain. Moreover, the uncertainty increases with the distance away from the known range" (p. 145).

Like in the case of interpolation, Schneerson's description is a simplification and hence a distortion of a real scientific procedure. No scientist would ever simply guess what the "reaction" would be at 101° or 2000° based solely on the data for the range between 0° and 100°. Any extrapolation, if employed in genuine scientific research, is based on a multitude of data which establish a well-documented *trend.* Besides the particular set of data at the scientist's disposal, he always bases his extrapolation also on the enormous wealth of multifaceted knowledge accumulated in science about the "reaction" in question.

Scientific theories are not built upon either simple interpolation or simple extrapolation, but rather on a combination of various mutually controlling methods and firmly established trends.

Schneerson continues: "[A] generalization inferred from a known consequent to an unknown antecedent is more speculative than an inference from an antecedent to consequent" (p. 145). To illustrate that assertion, Schneerson offers an example: "Four divided by two equals two. Here the antecedent is represented by the dividend and the divisor, and the consequent—by the quotient (2). . . . However, if we know only the end result, namely the number, 2, and we ask ourselves, how can we arrive at the number 2, the answer permits several possibilities, arrived at by means of different methods: 1) 1 plus 1 equals 2; 2) 4 – 2 equals 2; 3) 4 / 2 equals 2" (ibid.). This arithmetic platitude, contrary to Schneerson's view, is irrelevant to the question of the validity of scientific theories. It is arithmetically correct that the number 2 can be obtained by an endless number of arithmetic procedures. However, in scientific research the inference from a consequent to an antecedent is never made simply based on some number alone. If a researcher obtains, as a result of a measurement, a certain individual number, be it 2 or anything else, he never tries to draw any conclusion as to what caused this number by limiting his discussion to that number alone. Any conclusion "from consequent to antecedent" is offered on the basis of a multitude of data that show a distinctive trend, and by taking into account the large body of information accumulated by science about the reaction in question as well as other, similar reactions.

Furthermore, Schneerson's assertion that "a generalization inferred from a known consequent to an unknown antecedent is more speculative than an inference from an antecedent to consequent" is factually wrong. The procedure Schneerson refers to as an inference from a consequent to antecedent is the most common in science, and boils down to developing a theory explaining a set of known facts. On the other hand the procedure he refers to as inference from antecedent to consequent is actually using a theory to predict the outcome of experiments yet to be performed. More often than not, the former is less speculative than the latter, which is contrary to Schneerson's assertion. If a set of experimental data is sufficiently large, a theory explaining it can be reasonably substantiated. On the other hand, predicting the results of future experiments is a more speculative endeavor. Therefore the actual occurrence of events predicted by a theory is normally viewed as a more convincing argument in favor of that theory than simply an explanation by a theory of the already available data.

Of course, scientific theories can be wrong. If that is the case, they usually have a very short life. Every theory, even if it explains a certain set of data very well, is always subjected to multiple unmerciful tests, probing the limits of its applicability. The process of establishing an accepted scientific theory is very complex and quite different from the simplistic picture painted by Schneerson (see also chapter 12).

Schneerson continued his attack on the validity of science by listing a number of weaknesses plaguing scientific theories. Among those weaknesses is, for example, that scientific theories "have been advanced on the basis of observable data during a relatively short period of time." Since Schneerson's thesis is that the Torah provides more reliable information than science, a legitimate question is, What are those "observable data" which form the foundation of the Torah's story? There are no such data for either long or short periods of time. Why, then, should we prefer the Torah's story to scientific theories?

Another weakness of science is, according to Schneerson, that "on the basis of such a relatively small range of known (though by no means perfectly) data scientists venture to build theories by the weak method of extrapolation, and from the consequent to the antecedent, extending to many thousands (according to them, to millions and billions) of years!" (p. 146). This quotation shows once again Schneerson's misrepresentation of scientific theories. The age of the universe has been estimated in science through many different methods, all providing fairly consistent numbers. All these estimates are based on firmly established regularities with no indications that any such regularity could not have been at work at any time in the past. Of course, there is no way to conduct a direct experiment to test if a certain regularity (for example, the constant rate of radioactive decay for certain elements) indeed had been at work, say, a billion years ago. However, the large body of experimental evidence provides a reasonable foundation to believe that the regularity in question indeed was a feature of the world a billion years ago as it is now. If he was so inclined, Rabbi Schneerson might believe that, for example, the rate of the radioactive decay was not constant over the course of millennia. By the same token, I may believe that, say, the Torah was written by an Egyptian *chaty* (vizier) named Amenemhat in the year 1900 B.C.E., about whom is at least known that he indeed lived in that remote epoch.[3] However strongly I may hold such a belief, it has no evidentiary value and does not make true the assertion that Amenemhet was indeed the real author of the Pentateuch.

It seems appropriate at this point to discuss an example of how science deals with hypothesizing an antecedent from a consequent. Using Schneerson's classification, the theory I suggested for tensile stress origin belongs to the pure "from consequent to antecedent" type.[4] This theory plausibly explains various observed data by postulating a certain behavior of a certain type of defect in crystals called dislocations—namely, the egress of these dislocations to the surface of crystals. At that time there existed no experimental technique which would enable direct observation of the dislocations. When I came up with the idea of the egress of these dislocations, it was not based on direct evidence; my idea was based on imagination, as it consistently seemed to explain a wide variety of experimental data I had accumulated. Of course, my theory was of the "consequent to antecedent" type, according to Schneerson's classification. Its foundation was in logic and consistency, since it was based not on some single number, as in Schneerson's example, but rather on a multitude of facts and on the observed trends. In this form, the theory was published. Several years later, new advances in electron microscopy enabled scientists to see the dislocations directly. This vindicated the creators of the dislocation theory, once again demonstrating the power of scientific inference. Soon afterward, some English physicists observed directly the *egress* of dislocations to the crystals' surface, which I surmised several years earlier to be the reason for the tensile stress in films. Of course, before the direct observation of this phenomenon, people like Schneerson could argue that my theory was based on the use of a weak method "from consequent to antecedent"—that it was based on data obtained for a limited range of conditions, etc. However, the entire history of science proves the power of scientific inference as well as the high plausibility of good scientific theories.

Several years later I set out to develop a theory which would explain the anisotropy of stress I observed in certain magnetic films. Again, my tools were logic and the plethora of experimental facts I accumulated. I suggested a theory which explained the observed anisotropy through the magnetic properties of dislocations at play in the course of their egress. The theory explained in a fairly plausible way the entirety of the observed phenomena.[5] However, nobody has yet been able to verify directly the assumed behavior of the moving dislocations; hence, that theory has so far no direct experimental proof. Therefore I can't assert that the theory in question is true, as I could with the earlier theory. I tend to view the theory of anisotropy as plausible, though, because of its ability to explain a multitude of facts logically. So far this theory has not met objections from other scientists.

These two cases exemplify two types of scientific theories. To one type belong the theories which have a direct experimental confirmation. Of course, there is always a possibility that new experimental evidence may contradict an accepted theory. More often than not, though, contradictory data do not necessarily prove the theory wrong, but rather reveal the boundaries of the theory's validity. To the second type belong theories which have no direct experimental confirmation. The plausibility of such theories is based, first, on their logical consistency and the ability to reasonably account for all known facts and, second, on the fact that other scientific theories, which have been confirmed by direct experiments, were developed by the same process of scientific inference, which therefore is known to often provide significant insight into reality.

Schneerson specifically argues against the theory of evolution. One of his categorical statements is, "If you are still troubled by the theory of evolution, I can tell you without fear of contradiction that it [*sic*] is not a shred of evidence to support it" (p. 246). Rabbi Schneerson seems to have had a limited knowledge of the subject he dared to discuss. While the theory of evolution has many yet-unanswered questions, to insist that there is no evidence supporting it is a display of ignorance of the matter. There is an enormous amount of evidence supporting the theory of evolution, even though some of that evidence is incomplete.[6]

Continuing his discussion, Schneerson displays his position as an adherent of young-earth creationism. The defenders of that position maintain that the age of the universe is exactly as the Bible tells us, namely less than six thousand years, and all the evidence pointing to the much older Earth (a few billion years) or the universe (about twelve to fifteen billion years) is simply an illusion. He says, "Even assuming that the period of time the Torah allows for the age of the world is definitely too short for fossilization (although I do not see how one can be so categorical) we can still readily accept the possibility that God created ready fossils, bones or skeletons (for reasons best known to Him), just as He could create ready living organisms, a complete man, and such ready products as oil, coal, or diamonds, without any evolutionary process" (p. 147).

I can readily accept that the moon is made of green cheese, and you can readily accept that in Australia people walk with their bodies hanging upside down, and your friend can readily accept that sunset occurs when a sorcerer who dwells beyond the horizon grabs the sun and pulls it into a cave. If such suppositions were viewed as legitimate, reasonable explanations of the facts, maybe Schneerson's "readily accepted" assumption could also be considered seriously. Otherwise the idea offered by the young-earth creationists and shared by Schneerson, of God having created, for unknown reasons, ready fossils, bones, or skeletons, etc., remains in the realm of fairy tales and hardly deserves serious discussion, since it lacks any semblance of substantiation.

Other Articles

Other papers in the second section of the collection differ substantially from the one by Schneerson, most of them displaying a much higher level of sophistication. For example, the paper by Harry Marcell, "Evolution—Theory or Faith?" discusses in detail various difficulties encountered by the theory of evolution. These difficulties are real. For example, Marcell describes in detail the "design" of the apparatus enabling some snakes to produce a potent venom and inject it into the bodies of their victims. The apparatus in question is very complex and looks like the product of a very ingenious design. The theory of evolution does not offer a detailed explanation of how exactly the apparatus in question developed, step by step, via a natural, unguided process. To Marcell's credit, he does not categorically assert that the natural development of the apparatus in question was impossible; he hints, though, at the improbability of such a natural development by pointing to its high complexity, and invites readers to make their own conclusion. His obvious implication is that attributing the development of the snake venom apparatus to a natural unguided process is implausible. While the mechanism of snake venom production and use is indeed fascinatingly complex and finely tuned, evolutionary biologists have suggested quite reasonable explanations of how such biological mechanisms could have naturally developed.[7] Therefore Marcell's discourse, with all of its ingenuity and eloquence, remains just that—an interesting story lacking evidentiary significance.

In the section in Marcell's article titled "Science and Pseudo-Science," the author says that his article is "not directed against the true scientist who holds his theories tentatively and knows that they are always subject to revision as facts accumulate." It is hard not to fully agree with Marcell on that point. A few lines further he continues: "Our thrust is directed against the exponents of 'scientism' who inflate biological theories into cosmic philosophies" (p. 189).

Should we then interpret Marcell's attitude as not trying to disprove scientific theories when they are not "inflated into cosmic philosophies" but still contradict the Torah? What about the theory of evolution, which in itself is not a "cosmic philosophy" but a powerful theory in biology, supported by a wealth of empirical evidence? Marcell's critique of the difficulties of the theory of evolution, exemplified by his reference to the snake venom apparatus, is directed not against any inflation of the theory of evolution to a cosmic philosophy, but against the very essence of that biological theory itself. So much for the consistency of Marcell's discourse.

Another paper, "A Critical Review of Evolution" by Morris Goldman, constitutes a full-fledged denial of Darwinism as being incompatible with the Torah. This is how Goldman explains the essence of Darwinism: "The living things change from one form to another as a result of accidental events, and not as a result of deliberate purpose on the part of the Divine power. . . . God is irrelevant in the Darwinian evolutionary scheme, and that is what is wrong with it for a Jew" (p. 217). Of course, other interpretations of Darwinism asserting its compatibility with religious beliefs have also been suggested.[8] But if we were to accept Goldman's definition, a resulting problem would be how to reconcile this view with the statements from section 1 of *Challenge* (quoted in the preceding sections) according to which there cannot be any contradiction between the Torah and science.

Having pointed to the principal contradiction that, in his view, exists between theory of evolution and Judaism, Goldman unequivocally asserts that this contradiction in itself is sufficient reason to conclude that the theory of evolution must be wrong. He sees his task as "how to demonstrate even to the non-believer the falseness of the secularist beliefs" (p. 218). The main argument used by Goldman to refute Darwinism boils down to the assertion that "Darwinian reasoning is completely beyond testing" (p. 221). This argument, regardless of its correctness (actually, it is far from being correct) sounds very strange coming from somebody who wants us to accept his beliefs as if the Torah's story is not "completely beyond testing."

Ultimately, no article in section 2 provides any fact-based argumentation which would show the plausibility of the Genesis story in order to make it at least equally as reasonable as scientific theories, let alone its being better substantiated than the theory of evolution. Evolutionary scientists freely admit that this theory has not answered all the questions about the origin of species and related problems. Nevertheless, it has successfully and consistently explained a vast variety of empirical observations.[9] Therefore denials of its validity based only on stressing its weak points, as has been practiced by the authors of the articles in section 2, fail to be convincing to an unbiased reader.

SECTION 3, "THE SECULAR BIAS"

The article by Carl Klahr opening the third section, titled "Science versus Scientism," seems to present the program for all the papers in this section and purports to be supportive of genuine science while refuting what is referred to as "scientism." The latter term has been defined by Klahr as follows: "It is a conviction on the part of many scientists and teachers of science that the only valid answers to almost all the questions of fact or philosophy must come from extrapolation of science" (p. 289). Klahr does not provide references to any scientists or teachers of science who have ever suggested the above view. (Personally, I cannot recall a scientist saying that philosophy must come from extrapolation of science.)

Klahr then provides a long list of statements which allegedly evince the views of that mysterious, never-observed breed he calls "scientologists." This list is a funny mix of various unrelated claims, some of them a reasonable reflection of facts, and others unsubstantiated suppositions. For example, all "scientologists," according to Klahr, adhere to, among other views, the following two statements. Statement 6 says, "Growing up in Samoa and growing up in New York are essentially identical; only the artifacts are somehow different. The human animal is only superficially different in various cultures." Statement 12 says, "There are billions of planets in the universe with intelligent forms of life living in some of them" (p. 290).

It would be interesting to know, would a person qualify as a "scientologist," rather than a scientist, if she thinks that growing up in Samoa is different from growing up in New York, but on the other hand believes that there are billions of planets with life possibly existing on some of them?

Klahr seems to claim that he has a superior knowledge of the variety of topics covered by the list of statements in question, from sociology to biology and from history to physics. For example, he implies that he knows for a fact that there are no planets in the universe on which there exists life. And if somebody thinks otherwise, then, according to Klahr, this somebody cannot be viewed as a scientist but, by definition, must be referred to by the pejorative term "scientologist."

However odd Klahr's assertions sound, his article may be viewed just as a not very serious attempt to remove the halo of respectability from some forms of scientific method, those where science (legitimately) deviates from a mere collecting of facts and embarks on hypothesizing their explanation.

Rabbi Schlesinger's "Empirical Basis for Belief in God"

Another paper, by G. N. Schlesinger, is titled "The Empirical Basis for Belief in God." According to the information provided in *Challenge*, Schlesinger is a professor of philosophy of science and a rabbi. His credentials include two books on the philosophy of science. Hence, we see in him a highly qualified participant in the dispute regarding the existence of God, combining knowledge of science (at least in some general sense, as he seems not to have contributed to any particular branch of science, but only to the philosophy thereof) with training in convoluted discourse using intricate argumentation, as practiced in the Torah and talmudic studies.

In his paper, Schlesinger sets out to prove the existence of God mathematically, at least to the same extent and in the same sense as the laws of physics can be viewed as "proven."

Of course, no law of physics has been "proven." All these laws are postulates based on the interpretation and generalization of experimental and observational data (see chapter 12). They are all necessarily subject to revision as new experimental evidence emerges. However, despite all the limitations of the scientific approach, we all are witnesses to the enormous success of scientific exploration of the world. Within legitimate limits, laws of physics work very well. Because of the great successes of science, the laws accepted in physics and related disciplines are justifiably viewed as great achievements of the human mind.

Schlesinger maintains that the existence of God can be justified to the same extent of plausibility as the best scientific laws. To prove his point he indicates that every law of science is accepted only because we have first adopted some underlying presuppositions. These presuppositions, according to the author, are usually taken for granted, but are actually unsupported by evidence and, moreover, are unsupportable.

As an example, Schlesinger discusses the discovery of planet Neptune, which was predicted before its actual observation. This prediction was based on the application of Newton's laws. However, says Schlesinger, the theory developed by Newton is by no means the only one possible describing the motion of celestial bodies. There can be an endless number of laws governing the motion of celestial bodies, and, in particular, one can imagine any number of theories other than that of Newton which would predict the existence of Neptune in a particular area of the solar system equally well. To substantiate his claim, Schlesinger suggests an example of a law which, while differing from that of Newton, would provide the same prediction in regard to Neptune.

Let us see how Schlesinger constructs his example. Newton's well-known second law can be (for the particular case of a constant mass) written

$$a = \frac{F}{m}, \qquad (8.1)$$

where a stands for acceleration, F for force, and m for mass.

Let us assume, says Schlesinger, that some scientist named Whewton suggested a different law from the one above. The supposed "Whewton's law" would add one more term to the expression of Newton's: let us denote the additional term W. Then Whewton's law of mechanics would look as follows:

$$a = \frac{F}{m} + W. \qquad (8.2)$$

Schlesinger then suggested choosing the following definition of W:

$$W = \mathrm{A} \sin \frac{T!\pi}{n},$$

where A is an arbitrary constant, T is the temperature at the center of the earth rounded up to an integer, and n is the value of T at a particular time (Schlesinger chose that date to be midnight, January 1, 2001; his paper was published in 1976). He further assumed that T is greater than n at any time before January 1, 2001, but gradually decreases. With all these arbitrary assumptions, at any time before January 1, 2001, $T!/n = I$ is an integer; since sin $(I\pi)$ is zero if I is an integer, then $W = 0$. Therefore, at least until January 1, 2001, the additional term W remains equal zero, and both Newton's second law and the alternative Whewton's law give identical results. Many other forms of W could be chosen, and an endless number of values can be assigned to A, and all these alternative laws would predict the existence of Neptune equally well. However, these alternative laws are all different from Newton's second law. In the example discussed above, if Whewton's law were true, then after January 1, 2001, all celestial bodies would drastically change their behavior.

Of course, says Schlesinger, scientists take it for granted that Newton's law is much more trustworthy than Whewton's law. "This is a matter of prior judgment which is not and could not be based on evidence available to anyone before the year 2001," says Schlesinger. All this discourse demonstrates, maintained Schlesinger, is that whatever statement is accepted in science is necessarily built upon presuppositions that are taken for granted. In his further discussion Schlesinger proceeds to show that God's existence can be justified as reasonably as any of the scientific laws.

Before reviewing the next part of Schlesinger's discourse, let us summarize the above discussion of the alleged arbitrariness of scientific laws. Schlesinger's view seems to be that Newton's second law and what he calls Whewton's law—but what more properly should be called Schlesinger's law—have the same degree of validity. Since, until January 2001, no evidence could be obtained disproving Schlesinger's law, then, according to Schlesinger, his law was as good as Newton's second law or any other alternative law that would predict the existence of Neptune equally well.

It is easy to see the fallacy of Schlesinger's example. Newton's law, as with every law of physics, is a postulate. But it is a postulate that is by no means arbitrary. On the contrary, its plausibility is well founded in an enormous amount of consistent evidence whose interpretation in the form of Newton's second law is also highly logical. It enables scientists to make numerous predictions regarding the behavior of moving bodies, and so far these predictions have been fulfilled with an amazing accuracy. In the beginning of the twentieth century, certain limitations of this law were discovered which led to the establishment of its area of validity; this area, despite the newly discovered limitations, remained very large.

The additional term suggested by Schlesinger as an example was also a postulate. However, that is the only feature common to the two laws in question. Schlesinger's law, unlike that of Newton, was not based on any evidence but rather was artificially *contrived* in such a way that until January 1, 2001, no nonzero values would be added to the results following from Newton's law. Since this added term, until January 2001, was zero, then, until the date in question it not only could not be disproved, but was also unverifiable and hence not a law of science. Moreover, even if, on January 1, 2001, all laws of mechanics abruptly changed, this still would not constitute a proof of Schlesinger's concocted law, because its mathematical representation contains a quantity (the temperature at the center of the earth) which must be measured with a reasonable accuracy at the precisely predetermined moment of time. Schlesinger did not offer any method enabling one to conduct such a direct measurement. Moreover, all data serving as a basis for a scientific theory must be reproducible. However, as soon as midnight of January 1, 2001, passed, there was no way to repeat the measurement. In other words, the additional term added by Schlesinger had only the appearance of a meaningful formula, but was actually a meaningless concoction designed as an example of an imaginary situation which did not and will not actually exist. Both before and after January 2001, it has no meaning in the scientific sense, and there is no reason to discuss this unsubstantiated addition. Scientists do not take for granted Newton's second law, and professors of philosophy such as Schlesinger must know it. Newton's second law has been accepted for reasons. Nobody suggests that it is the absolute truth; but, based on an enormous amount of available evidence, it is a very reasonable approximation of reality, and we have all the reasons to base our predictions and conclusions on that law, as long as it is applied within the limits of its validity.

The question of the possible validity of the artificially contrived law of Schlesinger is essentially a different question. The actual question is whether the laws of nature—which, as we believe, are reasonably approximated by laws of physics—will remain the same in the future or will change at some moment in the future. Schlesinger may argue that at any moment in the future the laws of nature may abruptly change. However, such an assertion can be neither confirmed nor rejected by any means available to science. Therefore Schlesinger's entire discourse, employing a contrived alternative law, has no relation to science and is irrelevant in regard to the philosophy of science as well as any potential practical implications of his discourse.

Let us now review another approach by Schlesinger, in which he uses an estimation of probability for God's existence. To discuss this discourse, we have to follow the author's entire line of argument rather closely. To this end he uses the following mathematical relationship: if $p(A)$ is the probability of event A, and $p(B)$ is the probability of event B, and $p(A|B)$ is the probability of event A provided event B has actually occurred, and $p(B|A)$ is the probability of event B provided event A has actually occurred, then

$$\frac{p(A|B)}{p(A)} = \frac{p(B|A)}{p(B)}. \tag{8.3}$$

This equation is actually a simplified form of Bayes's theorem. It is known that use of Bayes's theorem is predicated on a number of conditions.[10]

Schlesinger first applies this equation to the prediction of the discovery of Neptune, based on Newtonian mechanics. His reasoning goes as follows: Let $p(N)$ be the probability that Newton's laws are true. Let $p(P)$ be the probability that in a specific part of the solar system there is a hitherto undiscovered planet to be named Neptune. Let $p(N|P)$ be the probability of Newtonian mechanics being correct provided a hitherto unknown planet named Neptune has indeed been discovered in a specific part of the solar system, as predicted by Newton's theory, and let $p(P|N)$ be the probability of discovering a new planet in a specific part of the solar system, provided Newton's laws are correct. Then, according to the previous equation, we can state that:

$$\frac{p(N|P)}{p(N)} = \frac{p(P|N)}{p(P)}. \tag{8.4}$$

The value of $p(P|N)$, says Schlesinger, is obviously much larger than $p(P)$. Indeed, $p(P)$ is the probability of discovering a new planet in a specific part of the solar system. Such discoveries have been extremely rare (in the last couple thousand years before the discovery of Neptune, only one new planet, Uranus, was ever discovered); hence, the probability of discovering one more unknown planet must be very small. On the other hand, if Newton's laws are true, then Neptune must necessarily exist in a specific part of the solar system, so $p(P|N) = 1$. Hence, says Schlesinger, the right side of equation (8.4) is a very large number, meaning that the left side must also be a very large number. This means that $p(N|P) \gg p(N)$. In other words, the discovery of Neptune increases very much the probability of Newtonian mechanics being true.

Schlesinger then proceeds to apply an identical method to religious beliefs. As an example, the author discusses the well-known Bible story about the destruction of the walls of Jericho at precisely that moment when the Israelites, led by Yehoshua Bin-Nun, completed their march around the city. His chain of notions is as follows. Let $p(D)$ be the probability that God exists, and $p(J)$ the probability that the walls of Jericho would collapse at the mentioned moment of time. Let, further, $p(D|J)$ be the probability that God exists provided the walls of Jericho indeed collapsed at the precisely defined moment of time, and $p(J|D)$ the probability that the walls of Jericho would collapse at the mentioned moment of time provided God indeed exists. According to equation (8.3), we can state, says Schlesinger, that:

$$\frac{p(D|J)}{p(D)} = \frac{p(J|D)}{p(J)}. \tag{8.5}$$

Schlesinger then maintains that $p(J|D)$ is "obviously" much larger than $p(J)$ because, in his view, if God exists, then it is "quite likely that He should listen when He is called upon by His true followers" (p. 409). In other words, Schlesinger says that the probability of Jericho's walls collapsing at a precisely defined moment, which is a very unlikely event, is quite small, but if God exists, then the same event must become much more probable. From that he concludes that, by virtue of equation (8.5), $p(D|J)$ must also be much larger than $p(D)$. The conclusion: the collapse of Jericho's walls makes the probability of God's existence immensely more probable.

If we compare this discussion with that related to the case of Neptune's discovery, we see that $p(D)$ in equation (8.5) is an analog of $p(N)$ in equation (8.4) and $p(J)$ in (8.5) is an analog of $p(P)$ in (8.4), while $p(D|J)$ in (8.5) is an analog of $p(N|P)$ in (8.4) and $p(J|D)$ in (8.5) is an analog of $p(P|N)$ in (8.4). It is easy to see, however, that there is no actual analogy between these probabilities. The main difference is between $p(N|P)$ in (8.4) and $p(D|J)$ in (8.5). In (8.4), $p(N|P)$ was the probability that Newton's theory is true, provided a new planet is discovered precisely in a specific part of the solar system, as predicted on the basis of Newton's theory. Event P (the discovery of a new planet) has actually occurred as predicted. In (8.5), though, $p(D|J)$ is the probability of God's existence provided the walls of Jericho did indeed collapse at a defined moment of time. Unlike event P in (8.4), which actually occurred, there is no evidence that event J in (8.5) ever occurred. If one believes in the veracity of the Bible's account, most commonly one also believes that God exists. However, as an alleged proof that God's existence is highly probable, equation (8.5) is illegitimate, being an example of circular reasoning. The most recent archaeological data suggest that the story about the collapse of Jericho's walls is rather doubtful.[11] Hence, Schlesinger's scheme is hardly helpful for estimating the probability of God's existence.

Schlesinger seems to realize the weakness of his example, so, for his ultimate argument he discusses another example, again using the same scheme of symmetric probabilities according to Bayes's theorem. In this case, his argumentation is almost identical with an "abduction" argument suggested twenty years later by William Dembski and Stephen C. Meyer.[12] Dembski and Meyer's argument has been shown to be based on the implicit assumption that God must have *wished* to create the universe in the big bang.[13] Schlesinger's argument is similarly based on the assumption that God wished to create life.

Unlike in the case of Newton's theory and the discovery of Neptune, there exists no evidence that God, if he exists, must necessarily have wished to provide conditions for man's existence. Schlesinger may think he knows what God wished or did not wish to do, but this is nothing more than Schlesinger's private opinion, which has no evidentiary value. Hence, Schlesinger's suggestion that God's existence can be proven with the same degree of certainty as scientific theories is not convincing.

SECTION 4, "ETHICAL PROBLEMS"

Whereas various aspects of the relationship between science and the Torah are discussed in the first three sections of *Challenge*, the articles in section 4 deal with ethical problems from the Torah's viewpoint rather than with the "science vs. Torah" controversy. Here we find a discussion of such problems as the ethical aspect of organ transplantation from the viewpoint of Judaism, the Jewish view on population control, etc. Since these topics are beyond the main theme of this book, I leave this section without further discussion.

CONCLUSION

As was said at the beginning of this chapter, the main idea evinced by the authors of *Challenge* was to assert that the Torah in no way contradicts the achievements of modern science. Unfortunately, to prove that assertion, the authors of the collection did not offer anything beyond unsubstantiated statements, often contradicting each other and avoiding the discussion of the obvious inconsistencies abundant in the Torah's story. Furthermore, some of the authors could not constrain themselves to following the stated purpose of the work, and resorted to denials of scientifically established facts. Overall, despite the impressive credentials of many of the authors in that collection, and despite the often seemingly considerable sophistication of their discourse, skeptics would not be moved to abandon their skepticism and accept the statements of these believers-with-science degrees on the basis of the discourse they presented.

NOTES

1. Aryeh Carmell and Cyril Domb, eds., *Challenge: Torah Views on Science and Its Problems* (New York: Feldheim, 1976).

2. Lubavitcher Rebbe is the title of the spiritual leader of a Jewish religious sect named Habad (its followers are referred to as Lubavitcher Hasids). This sect emerged at the end of the eighteenth century, its name stemming from the hamlet Lubavitch (in what is now Belarus), which was the center of the sect until 1916. Altogether there were seven Lubavitcher Rebbes, most belonging to the Schneerson family. More recently their center has been moved to Brooklyn, New York. After the recent death of Rabbi Menachem Schneerson, however, no heir to this title has emerged.

3. Arthur Cotterell, ed., *The Encyclopedia of Ancient Civilizations* (London: Rainbird, 1980), p. 26.

4. M. Ya. Popereka (a.k.a. Mark Perakh), "On the Origin of Internal Stresses in Electrolytically Deposited Materials" (in Russian), *Fizika metallov i metallovedeniye* 20 (1965):753–62; *Vnutrenniye napryazhenia elektrolitichski osazhdayemykh metallov* (Internal stress in electrolytically deposited films) (Novosibirsk, USSR: Zapadno-Sibirskoye Knizhnoye Izdatelstvo, 1966); English translation published for the National Bureau of Standards and the National Science Foundation by the Indian National Scientific Documentation Centre (New Delhi, 1970); "A Theory of Internal Stresses in Electrolytically Desposited Materials" (in Russian), in *Trudy Tret'ego Mezdunarodnogo Kongressa po Korrozii Metallov* (Proceedings of the Third International Congress on Metallic Corrosion), vol. 3 (Moscow: Mir, 1965), pp. 350–56.

5. Mark Perakh, "Anisotropy of Spontaneous Macrostress in Ferromagnetic Films Induced by Magnetization," parts 1 and 2, *Journal of the Electrochemical Society* 122, no. 9 (1975):1260–62, 1263–67.

6. Richard Dawkins, *The Blind Watchmaker: Why the Evidence of Evolution Reveals a Universe without Design* (New York: W. W. Norton, 1978).

7. Ibid.

8. Kenneth R. Miller, *Finding Darwin's God: A Scientist's Search for Common Ground between God and Evolution* (New York: Cliff Street Books, 1999).

9. For a series of well-written articles in support of the theory of evolution, see The Talk.Origins Archive [online], www.talkorigins.org.

10. Colin Howson and Peter Urbach, *Scientific Reasoning: The Bayesian Approach* (La Salle, Ill.: Open Court, 1993).

11. Zeev Hertzog, "The Bible: No Findings on the Locations" (in Hebrew), *Haaretz* (Tel Aviv), October 29, 1999; Russian translation published in *Vremya Iskat*, no. 3 (2000):115–22.

12. William A. Dembski and Stephen C. Meyer, "Fruitful Interchange or Political Chitchat? The Dialog between Science and Theology," in *Science and Evidence for Design in the Universe*, ed. Michael J. Behe, William A. Dembski, and Stephen C. Meyer (San Francisco: Ignatius, 2000), p. 213.

13. Mark Perakh, "The Anthropic Principles—Reasonable and Unreasonable: And the Fallacy of the Abduction-Type Inference to a Supernatural Source of the Big Bang," Talk Reason [online], www.talkreason.org/articles/anthropic.cfm [April 12, 2002].

9

THE END OF THE BEGINNING

Nathan Aviezer Explains How to Interpret the Book of Genesis

In this chapter I will discuss Nathan Aviezer's book *In the Beginning: Biblical Creation and Science* as well as his paper "The Anthropic Principle."[1] *In the Beginning* was reprinted several times and translated into several languages. It is considered by Aviezer's cobelievers as a very convincing and well-substantiated interpretation of those parts of the biblical story which contradict the data of science but which, according to Aviezer, are actually in full harmony with science if they are properly interpreted.

A professor of physics, Aviezer is a prominent member of a group of scientists, who, while constituting only a relatively small fraction of the scientific community, are quite active in a persistent effort to reconcile scientific data with the biblical story. Aviezer's writings are rather typical of the literature in question. While he avoids the most egregious misstatements and crude errors we see in some other publications (for example, in books by Gerald L. Schroeder; see chapter 10) Aviezer's books and papers contain imprecise assertions, illogical conclusions, and even direct errors (as in his treatment of probabilities).

PROBABILITIES ACCORDING TO AVIEZER

In "The Anthropic Principle" Aviezer asserted that the probability of intelligent life emerging by chance was so negligible that the role of a guiding supernatural mind must be accepted as the only possible explanation for our existence. In this regard, Aviezer discussed the question of probability calculations. Aviezer argued against the criticism suggested by Professor

Raphael Falk of the Hebrew University of Jerusalem. According to the reference in Aviezer's paper, Professor Falk's critique has been published in a Hebrew-language journal called *Alpai'im*.[2] However, I could not obtain that journal and therefore, in my discussion of the dispute between Falk and Aviezer, I must rely on the quotations in Aviezer's paper

As quoted by Aviezer (p. 19), Falk says, "According to Aviezer's logic, the probability that I am writing these lines with a dull yellow pencil, using my left hand, sitting at my kitchen table, on the third floor of a specific Jerusalem address—this probability is completely negligible. Nevertheless, all these events happened and they clearly mean nothing."

Aviezer then attempts to demolish Falk's argument. While some readers may find that Falk's example was not the best possible, Aviezer's response is inconsistent. Aviezer does not notice that, in trying to disprove Falk's argument, he actually repeats exactly the same argument, this time considering it to be a valid confirmation his own views.

In the course of his dispute with Falk, Aviezer discusses an example given by G. N. Schlesinger.[3] In that example, a dollar bill is pulled from a wallet and is found to have the serial number G65538608D. If one calculates the probability that an arbitrarily chosen bill has exactly this number, it turns out to be less than one in ten billion. Nevertheless, the extremely improbable event took place, but we find nothing surprising in that. Aviezer proceeds to analyze faults in that statement; referring to the famous physicist Richard Feynman, he asserts that the actual probability for the bill to have the quoted serial number must be accepted as 100 percent:

> Following Feynman's advice, we shall clearly define the event described above, which immediately leads to the conclusion that there is a probability of 100 percent that the *dollar bill pulled from the wallet has G65538608D for its serial number*. Why? Because this *number* was chosen by looking at *the serial number on the $1 note*. In other words, one was simply asking: "What is the probability that the *serial number on the note is the serial number on the note*?" And the answer to this question, clearly, is 100 percent. Since the event was not improbable at all but certain, there is no reason whatever to be surprised by its occurrence. (pp. 13–14, emphasis added)

This is a reasonable explanation on Aviezer's part. There is, though, one point Aviezer seems to have not noticed—his argument is exactly what Falk said.

To illustrate my point, let us replace the italicized words in the quotation from Aviezer. The event we will discuss now is not the serial number on a note, but the existence of the universe as we see it, including the existence of intelligent life. For example, "$1 note" will now be replaced with "universe as we know it," etc. We get now the following discourse, preserving Aviezer's logic:

> Following Feynman's advice, we shall clearly define the event described above, which immediately leads to the conclusion that there is a probability of 100 percent that the *universe which we see is such as we see it.* Why? Because *this universe* was chosen by looking at *this universe.* In other words, one was simply asking: "What is the probability that the *universe we see is the universe we see*?" And the answer to this question, clearly is 100 percent. Since the event was not improbable at all but certain, there is no reason whatever to be surprised by its occurrence.

There is no logical difference between the example of the dollar bill (or with Falk's writing his comments with a yellow pencil) and that of the universe. While intending to refute Falk, Aviezer has asserted, actually in agreement with Falk, that after an event has occurred, a discussion of its probability is meaningless, as its occurrence becomes a certainty.

Falk's statement is also correct in a different sense, contrary to Aviezer's interpretation of these statements. For example, in the case of a dollar bill, the question can legitimately be asked: What is the probability that a dollar bill that is still in the wallet and whose number we do not know will turn out to have the quoted serial number? The correct answer to that question is not 100 percent, but rather less than one in ten billion. Does this very small probability mean that the actual serial number of the bill cannot turn out to be the quoted serial number? No—each bill has some number on it, and to find any particular serial number on the actual bill is as probable as finding any other existing serial number. Why, then, couldn't it happen to be exactly the quoted number?

Likewise, the existence of the universe as we see it could have a very low probability if we calculated this value without knowledge of the actual universe. However, if Aviezer suggests that we calculate the probability of the dollar bill's serial number by first looking at it, which makes the probability in question 100 percent, then he has to allow us to do the same with the universe.

In light of these considerations, Aviezer's discourse in regard to the exceedingly small probability of the spontaneous development of the universe with intelligent life is unconvincing.

THE CASE OF MULTIPLE WINS IN A LOTTERY

Let us turn now to one more part of Aviezer's paper dealing with probabilities. Here Aviezer discusses an example of multiple wins in a lottery. First he considers a case where a hypothetical Chaim Cohen of Afula wins the jackpot: "Although the chances were only one in a million that the winner would be Chaim Cohen from Afula, there exist one million 'equivalent' Chaim Cohens. Therefore the substance of what I heard is that someone won the Lotto this week. And the chances for that event to happen—someone winning—are 100 percent. Hence I have no reason to be surprised" (p. 14).

This statement is correct (except for the 100 percent chance of someone winning; if we account for the possibility that more than one player chooses the same losing set of numbers, the actual probability of someone winning is about 37 percent—see the pertinent calculation in the appendix). It is essential for our further discussion to make note of that statement, and to add that the number Aviezer used—one million—has no special meaning. The above statement would be equally correct if instead of one million tickets available, there were only one hundred thousand, or one million million.

Aviezer then states that if Chaim Cohen won a second time in a row, it would be amazing. If Chaim Cohen won a third time in a row, it would indicate, says Aviezer, that there was something fishy about that lottery—most probably, its outcome was rigged in favor of Chaim Cohen: "The chances that this same person will win the Lotto once again are easily shown to be only one in a million millions. *Such events are so rare that they simply do not occur*" (emphasis added).

This last statement, however appealing it seems to common sense, is actually flawed in several respects (as will be shown in detail in chapter 13). We are indeed amazed if a particular player wins more than once in consecutive games. But we are amazed not because of the very small probability of such an event, as Aviezer thinks. The reason for our intuitive view of the multiple wins as a plausible indication of fraud is as follows.

From the analysis in chapter 13 we can conclude that when a particular player wins more than once in consecutive games, we are amazed not because the probability of winning for that particular player is very low, but because the probability of *anybody* (whoever he happens to be) winning consecutively in more than one game is much less than the probability of someone winning only once in even a much larger lottery. We intuitively estimate the difference between the two situations. However, the important point is that what impresses us is not the sheer small probability of someone winning against enormous odds. This probability may be equally small or even smaller in the case of winning only once in a big lottery, but in that case we are not amazed.

Of course, Aviezer is right when he says that a triple win by Chaim Cohen would make anyone pause before buying a ticket in such a lottery. This attitude is, though, based not on the very small probability of Cohen's winning, as Aviezer assumed, but on the difference between the probabilities of someone (whoever he may happen to be) winning more than once and someone winning in only one game, even if the probability of winning several times in a row in consecutive games of a small lottery is actually larger than that of winning only once in a large lottery.

Furthermore, remember that when, for example, in probability theory the probability of either heads or tails in the test with a coin is discussed, the assumption of an "honest coin" is implied. In the example of a triple win, though, Aviezer introduces an element that is alien to the mathematical calculation of probabilities, namely the possibility of cheating. The important point is as follows: we *know* (or, at least, suspect) that there is a distinctive possibility of cheating on the part of the lottery organizers; therefore our evaluation of the triple win by Chaim Cohen is based not just on the mathematical probability of his triple win, but also on the interfering uncalculated probability of cheating. In fact, we subconsciously compare two competing probabilities; one, that Chaim Cohen honestly won twice (or three times) in a row, and the other, that the lottery was rigged. If Chaim Cohen won three times in a row, our intuitive estimate is that the probability of the triple win for anybody (not just for Cohen) is so small that it becomes lower than the probability of cheating. If, though, we were confident that no cheating could take place, our attitude would be quite different, namely it would be safely based only on the mathematically calculated probability.

The analogy between the lottery and the spontaneous emergence of life is useful up to a certain point. However, as any analogy, it is not perfect and ceases to be legitimate if we account for the possibility of cheating; in estimating the probability of spontaneous emergence of life, there is no possibility of cheating to consider.

There are also other substantial differences between the probabilities in a lottery and in the possible processes of the spontaneous emergence of life. One such difference is that all possible events in a lottery are supposed to be equally probable. However, the postulate of equiprobability does not hold for the spontaneous development of the universe or life within it. There is little doubt that certain paths in such development could be much more probable than some other possible paths, a difference that dictates caution in discussing the universe and life in terms of a lottery—but Aviezer did not seem to account for that difference.

The most important point is, though, that the value of probability in itself, however small it may be, has no cognitive value. Aviezer's assertion that "such events are so rare that they simply do not occur" is contrary to facts. If it were correct, it would mean *no event* would occur. Indeed, each one of those "million millions" of possible events has the same (very small) probability occurring; hence, Aviezer's statement is equally applicable to all of them. However, some particular event (for example, *some* ticket, sold or unsold, being the winning ticket) whose individual probability is very small must necessarily take place, even though we don't know in advance which one. Thus, events of extremely small probability occur constantly.

When we discuss the probability of the spontaneous emergence of life, the values of that probability, however small, are irrelevant and do not contradict the hypothesis of the spontaneous emergence of life (which, it has been stated, could have occurred not through purely random events, but as a result of processes of reasonably high probability).[4]

THE SUN AND THE MOON

As another example of imprecision Aviezer allows himself in order to support his beliefs, let us look at how he tries to justify the biblical story in regard to the sun and the moon being a "large light" and a "smaller light": "The apparent sizes of the sun and of the moon are *exactly* the same. Each of these astronomical bodies has an apparent size of 0.53°" (p. 47). This statement is not exactly correct.[5] First, the visible (apparent) sizes of the moon and of the sun are not constant: they change as the moon orbits the earth and the earth orbits the sun, their orbits being not exactly circular. The *average* angular diameter of the sun (if it were watched from the center of the earth) is about 31'59", which is close to 0.533°. The *average* angular diameter of the moon is slightly less, 31'5". However, the sun's apparent size varies from the mean by about 1.7 percent, while that of the moon, by about 7 percent. The *maximum* apparent angular size of the moon is about 33'16", which is larger than even the largest apparent size of the sun. Moreover, the maxima and minima of apparent sizes of the sun and moon occur at different times. Hence, the difference between the apparent diameters of these two celestial bodies varies between two extreme situations: in one, the sun appearing larger by some 10 percent; in the other, smaller by some 8 percent. Furthermore, the distance between the earth and the moon is slowly increasing; thus the apparent diameter of the moon is gradually decreasing. While in the distant past the apparent diameter of the moon was noticeably larger than that of the sun, in some distant future it will become small enough to make a total solar eclipse impossible. Anyway, even if the a near-coincidence mentioned by Aviezer were exactly true, how such a near-coincidence can make the moon a source of light, rather than just a reflector, remains unexplained in Aviezer's book.

AVIEZER'S DISCUSSION OF THE ANTHROPIC PRINCIPLE

The term *anthropic principle* apparently started gaining popularity after 1973, when an English physicist named Brandon Carter introduced it at a gathering of scientists on the occasion of the five-hundredth anniversary of Copernicus's birth. Carter noted that the values of physical constants must be within a very narrow range in order to enable the existence of life and that the observed values of these constants happen indeed to be in that narrow range. Were any of the physical constants slightly different, life would be impossible.

Since then, the anthropic principle, as Carter's observation was labeled, has become a favorite subject of discussion by many adherents of the Bible's inerrancy, who view it as convincing proof that the universe as we see it, and intelligent life in particular, could not have emerged by chance, but must have been created for a purpose and according to a detailed design by a supernatural agent. There is vast literature devoted to the discussion and interpretation of the anthropic principle.[6] Several versions of this principle have been suggested by various authors, including the so-called weak anthropic principle (WAP), the strong anthropic principle (SAP), the participatory anthropic principle (PAP), and the final anthropic principle (FAP), each of which are defined slightly differently in various sources.[7]

Aviezer mentions neither the distinction between these variations of the anthropic principle nor their definitions. Moreover, he does not offer any explicit definition of his own for this principle, though one can infer from his article that his interpretation of this principle is very close to that of another propagandist of the Bible's inerrancy, Hugh Ross (whose literary output was discussed in chapter 5), the difference being that Ross approaches the issue from a Christian standpoint.

The interpretation of the anthropic principle by both Ross and Aviezer seems to boil down to the assertion that the universe is extremely "fine-tuned" for the existence of life, which allegedly points to its supernatural origin.

For convenience, I suggest labeling all versions of the anthropic principle that attribute the fine-tuning of the universe to a purposeful action of a supernatural agent as the supernatural anthropic principle (SNAP) and all its versions that attribute the fine-tuning to natural causes as the natural anthropic principle (NAP).

Aviezer adheres to SNAP (though without using that term). One example of Aviezer's adherence to a rather extreme form of SNAP is his contention that all the stars in the universe were created for the benefit of mankind. The contemporary scientific view maintains that atoms of the elements constituting human bodies have been created within the stars. The early universe contained no elements other than hydrogen and helium. Stars are the giant ovens where nuclear reactions result in the creation of all other elements—including carbon, oxygen, etc.—which are necessary for life. Therefore, concludes Aviezer, stars were created specifically in order to make possible the existence of humans.

There are a few questions, though, which Aviezer did not address. One is, Why are there hundreds of billions of galaxies in the universe, each containing billions of stars? One-billionth of 1 percent of that number of stars would suffice to supply all the elements necessary for the whole of humanity, billions of times over. Why, then, do so many stars exist whose existence has no effect on humans?

Moreover, it seems possible to show, without penetrating into the nuances of the anthropic principle, that SNAP is logically unsubstantiated. One of the ways to do this succinctly is by applying a simple probabilistic approach. To this end, I will use a simplified version of Bayes's theorem,[8] well known in probability theory, in the following form:

$$\frac{p(A|B)}{p(A)} = \frac{p(B|A)}{p(B)}, \qquad (9.1)$$

where $p(A)$ is the probability of the occurrence of some event A; $p(B)$ is the same for another event, B; $p(A|B)$ is the conditional probability of the occurrence of A provided B has actually occurred; and $p(B|A)$ is the conditional probability of event B provided A has actually occurred.

With this equation, it is possible to represent our two versions of the anthropic principle in probabilistic terms: For NAP, let $p(FT)$ be the probability that a certain universe is fine-tuned for life. Let $p(FT|L)$ be the conditional probability that a universe is fine-tuned for life provided life indeed exists in that universe. Let $p(L)$ be the probability that life exists in a universe, and $p(L|FT)$ the conditional probability that life exists in a universe provided that this universe is fine-tuned for life. The equation can thus be formulated

$$\frac{p(FT|L)}{p(FT)} = \frac{p(L|FT)}{p(L)}. \tag{9.2}$$

The natural anthropic principle then boils down to the statement that

$$p(FT|L) \gg p(FT). \tag{9.3}$$

The meaning of (9.3)—the probabilistic representation of NAP—is the assertion that the existence of life in a universe makes it much more probable that the universe in question must be fine-tuned for life. From (9.3), it also follows that

$$p(L|FT) \gg p(L), \tag{9.4}$$

meaning that if a universe is fine-tuned for life, this considerably enhances the probability that life will exist in that universe. This is a reasonable assumption (some readers may even view it as obvious). If (9.4) holds, (9.3) must hold as well. Hence, NAP seems to be a reasonable and logically faultless assumption.

On the other hand, the supernatural anthropic principle requires the following assumption:

$$\frac{p(S|L)}{p(S)} \gg 1, \tag{9.5}$$

where $p(S)$ is the probability that a universe was created supernaturally, and $p(S|L)$ is the conditional probability that a universe was created supernaturally, provided life exists in that universe.

Using equation (9.1), we can formulate SNAP as

$$\frac{p(S|L)}{p(S)} = \frac{p(L|S)}{p(L)}, \qquad (9.6)$$

where $p(L|S)$ is the conditional probability that life exists in a universe provided that universe was created supernaturally. To satisfy (9.5) and (9.6) we must assume that

$$p(L|S) \gg p(L), \qquad (9.7)$$

meaning that if we know that a universe was created supernaturally, this knowledge increases the probability of life's existence in such a universe, $p(L|S)$, compared to the probability of life's existence in the absence of knowledge of supernatural creation, $p(L)$. This means assuming that the supernatural creator of a universe must also have wished to create life in it. This is an arbitrary assumption, since we have no knowledge of what a supernatural creator of a universe may have wished or not wished to do. Hence, SNAP is also an arbitrary assumption and therefore logically unsubstantiated.

The conclusion that a universe which is fine-tuned for life implies a supernatural creator is an example of "circular reasoning." In order to conclude that $p(S|L) \gg p(S)$, which is a succinct representation of SNAP, we must first assume that $p(L|S) \gg p(L)$; i.e., we must assume a priori the existence of a supernatural agent who wished and planned to create life. But the latter assertion is what was to be proven by the entire discourse.

Of course, establishing the arbitrariness of an assumption does not mean that such an assumption is necessarily wrong; it does mean, though, that one may not assert that such an assumption is correct. At best, the question about the correctness of such an assumption remains open until some convincing proof of its being either correct or not is found. As the matter stands now, no such proof has been suggested, so the assertion that the values of physical constants point to the supernatural origin of the universe and life remains an unsubstantiated assumption, reflecting religious preferences rather than factual evidence.

This simple probabilistic discourse has shown that the supernatural interpretation of the anthropic principle, so popular among proponents of creationism, both of explicit and implicit kinds, is logically unsubstantiated.[9]

SIX DAYS OR SIX EPOCHS?

Another example of Aviezer's approach (which has actually been used many times before Aviezer) is his assertion that the word *day* in the beginning verses of the Book of Genesis, which describes the creation of the universe in six days, is actually meant to denote "epoch." Such a proposition may be fine in a theological discourse; it is, however, an arbitrary hypothesis not based on any factual evidence or logic when the text of the Bible is viewed from a rational standpoint.

Indeed, the Hebrew language has such words as *shniya* ("second"); *daka* ("minute"); *yom* ("day"); *shavua* ("week"); *khodesh* ("month"); *shana* ("year"); *mea* ("century"); *tkufa*, *sfira*, *minyan*, and *idan* (all synonyms for "epoch"); and many other words denoting various time intervals (as was already mentioned in chapter 5). If, as Aviezer suggests, the author of the Book of Genesis meant that the universe's creation took six *epochs*, why would that author choose of all words at his disposal the word *yom*, and not *tkufa* or, say, *sfira* or *idan*? Moreover, while mentioning each of the six alleged epochs, the author of the Book of Genesis repeats each time the phrase "and there was evening and there was morning, the second day" (or third, or fourth, etc.). Hence, it seems rather obvious that the author of Genesis, who might not be familiar with our contemporary prevailing scientific view of the age of the universe (about twelve to fifteen billion years), meant to say that indeed the universe was created literally in six days.

What is the reason for assuming that the word *yom* in Genesis really meant "epoch"? It would be in vain to search the text of the Bible for any indication that this Hebrew word was meant to denote anything but "day." There are no such direct indications anywhere in the Bible. (Sometimes the word *yamim*, the plural of *yom*, is used in the Bible metaphorically, but in the sense of "times.") This means the reasons for this interpretation must be looked for outside the Bible.

There is little doubt, though, of the actual motivation for interpreting *yom* as meaning "epoch": the contemporary, widely held *scientific* view maintains that the universe came into existence roughly fifteen billion years ago. The assumption that *yom* was actually meant to denote "epoch" is based only on Aviezer's agenda, which is to prove compatibility of the Bible with science.

It is worthwhile to recall that not long ago the scientific estimate of the age of the universe was not billions but only millions of years. If that view, now largely abandoned by science, were still accepted, all calculations designed to prove the harmony of the Bible and science would need to be altered. Isn't it somehow ironic that the adherents of the view that the Bible is in full harmony with science have to modify their interpretation of the text of the Bible each time a scientific theory is changed? Since Aviezer and other writers of the genre in question claim to be genuine believers, they are supposed to believe that the Bible supplies eternal truth, while science by definition provides only temporal truth. If the prevalent scientific view happened to be that, say, the universe was created in six seconds, it is doubtful that Aviezer would have much problem with an assertion that the word *yom* in Genesis actually meant *shniya*, i.e., "second," rather than "epoch." This is so because interpreting a word that literally means "day" as actually implying "epoch," is arbitrary, not based on any factual evidence.

THE PARADOX OF THE ORIGIN OF LIFE

It has been established in biology that the production of proteins in living organisms occurs with the participation of nucleic acids. Nucleic acids, in turn, are produced with the participation of proteins. It looks like a vicious cycle—reminiscent of the question of what came first, the chicken or the egg. In Aviezer's view, since nucleic acids are not produced without proteins and proteins are not produced without nucleic acids, the cycle "nucleic acids–proteins–nucleic acids" could not have started spontaneously. Hence, says Aviezer, such a cycle only could be started through the interference of a supernatural "guiding hand." In *In the Beginning*, Aviezer states, "[L]ife could not develop from inanimate matter because inanimate matter contains neither proteins nor nucleic acids" (p. 68). An oak's root contains neither acorns nor leaves; does this lead to the conclusion that an oak could not grow from its root?

The only conclusion that follows from the observations of the protein–nucleic acid cycle is that the cycle started in a way that is different from how it occurs now. We don't know how it happened. Aviezer, though, somehow claims to know the answer to that question and provides it in a categorical way.

The only conclusion that follows from the experimental data is that somehow, either spontaneously or through a deliberate action of a "guiding hand," the first protein appeared in the absence of a nucleic acid, or the first nucleic acid appeared in the absence of a protein. There is no information available to assert exactly how this happened. However, the opinion that the start of the cycle in question could *not* occur spontaneously has no basis in factual evidence.

There is nothing unusual in starting a cyclic process via some noncyclic event. For example, it is possible that the noncyclic event producing a protein (or a nucleic acid) required negotiating a very high potential barrier by interacting molecules. Hence, it would require very high activation energy; such a noncyclic interaction would therefore be very rare. However, rare does not mean impossible. Under certain conditions that could have occurred, even if very rarely, in the primeval atmosphere, the interacting molecules could have acquired the activation energy sufficient to overcome the potential barrier. For example, this could be a result of the presence of short-lived but powerful electromagnetic fields (perhaps as a result of intensive illumination by high-energy photons). Thus the first protein (or the first nucleic acid) could have emerged, starting a cycle wherein the necessary activation energy happened to be much lower, which made the cycle the preferred path of the reaction. Another effect with the same consequences could be a decrease of the potential barrier. The decrease of potential barriers of various interactions, for example by a shower of photons, has been observed experimentally many times (one such situation is the photoadsorption of semiconductors from colloid sols on dielectric surfaces, which normally does not occur in darkness, but is enabled by illumination, if the photon energy exceeds a certain threshold).[10] One more possibility is that the decrease of the potential barrier was due to the accidental presence of a rare strong catalyst. Then, as the first protein (or nucleic acid) was produced, the cyclic process could have become the preferred path because the necessary catalytic conditions for it become much more common.

There is no evidence that the noncyclic generation of a protein or a nucleic acid could not happen naturally. The biological sciences (including biophysics and biochemistry) have not yet identified such a process, but that does not mean science will not eventually identify it, or even reproduce it in a laboratory. Aviezer's argument in this case is based on the "God-in-the-gaps" approach,[11] which has lately been losing its attractiveness even in the eyes of the new crop of adherents of creationism.

Consider an example: let us imagine a competition in long-distance running. If the distance is, say, 10,000 meters, the runners have to circle the stadium twenty-five times. A spectator who came to the bleachers after the competition had already started observes the group of runners circling the stadium time and time again. This spectator would conclude that there was a starting point somewhere—where the runners happened to be initially. He has no direct knowledge of how the runners happened to arrive at that starting point. Maybe they descended from the bleachers, or perhaps they walked in through some gate, etc. The spectator does not know exactly how the run started, even though he realizes that it happened in a way *different* from what he now observes, where the runners instantly pass the starting point in each round, approaching it, say, from the east, and continuing their run westward. The spectator has no reason, though, to assume that the run started because of an action of a supernatural "guiding hand."

WHAT AVIEZER DOES NOT DISCUSS

It is interesting to note that in some instances Aviezer chose to keep silent in regard to some rather obvious contradictions between the Bible and science. One such example is found in his discussion of the early history of mankind, titled "After the Six Days: The Early History of Man." In this section Aviezer discusses the development of agriculture as well as the beginning of animal husbandry, which, according to archaeological findings, occurred some ten thousand years ago. In Aviezer's view, the biblical story accurately describes these milestones in mankind's history. Indeed, Aviezer tells us, Adam's family engaged in both agriculture and animal husbandry. He quotes from the Bible: "And Abel was a shepherd, and Cain was a tiller of the ground" (Gen. 4:2). Aviezer seems not to notice the chronological discrepancy between the archaeological data and the biblical account. According to the Bible, Cain and Abel were the first two sons of Adam, who himself was created by God less than six thousand years ago. Therefore, per the biblical account, the first shepherd and the first tiller of the ground lived *later* than about six thousand years ago. On the other hand, the archaeological data, referred to by Aviezer, indicate that the emergence of agriculture and animal husbandry started *several thousand years earlier*.

There are other inconsistencies in Aviezer's book. For example, he matches the creation of plants on the third day with the appearance of green plants in the Permian period 250 million years ago, whereas he matches the creation of the "big crocodiles" on the fifth day with the appearance of the Ediacra fauna in the Precambrian period—570 million years ago.

CONCLUSION

Of course, a skeptic would not fail to notice either the illogical statements or the flaws in the interpretation of probabilities utilized by Aviezer. The publication of Aviezer's book in no way changes the fact that the realms of faith and science do not intersect but rather exist independently of each other. The mental acrobatics used in publications such as those by Aviezer have no useful role. Science is a product of the human mind and human endeavors, and does not look for confirmation of its theories in the Bible or in any other religious source. Equally, faith is not based on scientific proof of what is believed to be the divine revelation; no efforts to reconcile the two areas, such as the attempt made by Aviezer, can change this situation.

NOTES

1. Nathan Aviezer, *In the Beginning: Biblical Creation and Science* (Hoboken, N. J.: KTAV, 1990); Nathan Aviezer, "The Anthropic Principle," *Jewish Action* 19 (spring 1999):9–15.

2. Raphael Falk, review of *In the Beginning*, by Nathan Aviezer (in Hebrew), *Alpai'im* 9 (spring 1994):133–142; cited in Aviezer, "The Anthropic Principle," p. 13.

3. G. N. Schlesinger, *Tradition* 23 (spring 1988):1–8; cited in Aviezer, "The Anthropic Principle," p. 13.

4. Richard Dawkins, *Climbing Mount Improbable* (New York: W. W. Norton, 1996).

5. George O. Abell, *Realm of the Universe* (Philadelphia: Saunders, 1984), p. 81.

6. See, for example, John D. Barrow and Frank J. Tipler, *The Anthropic Cosmological Principle* (Oxford: Oxford University Press, 1986).

7. A somewhat more detailed discussion of the anthropic principle can be found in Mark Perakh, "The Anthropic Principles—Reasonable and Unreasonable: And the Fallacy of the Abduction-Type Inference to a Supernatural Source of the Big Bang," Talk Reason [online], www.talkreason.org/articles/anthropic.cfm [April 1, 2002].

8. For a more detailed version, see Colin Howson and Peter Urbach, *Scientific Reasoning: The Bayesian Approach* (La Salle, Ill.: Open Court, 1993).

9. A further investigation of the logic of this proposition, based on the full form of Bayes's theorem, has been undertaken by Michael Ikeda and Bill Jefferys, "The Anthropic Principle Does Not Support Supernaturalism," Talk Reason [online], www.talkreason.org/articles/super.cfm [April 2, 2002]. Their argument

has provided an even stronger refutation of the hypothesis favoring supernatural creation of life; in fact, they seem to have shown that the more fine-tuned for life a universe is, the *less* likely it is to have had a supernatural origin.

10. Mark Perakh, Aaron Peled, and Zeev Feit, "Photodeposition of Amorphous Selenium Films by the Selor Process," part 1, *Thin Solid Films* 50 (1978):273–82; Mark Perakh and Aaron Peled, "Photodeposition of Amorphous Selenium Films by the Selor Process," parts 2 and 3, *Thin Solid Films* 50 (1978):283–92, 293–302; Mark Perakh and Aaron Peled, "Light-Temperature Interference Governing the Inverse/Combined Photoadsorption and Photodeposition of a-Selenium Films," *Surface Science* 80 (1979):430–38.

11. "God in the gaps" refers to a creationist thesis according to which gaps in scientific data have to be explained by attributing the origin of the missing part of information to the hidden hand of God. This thesis was popular in the creationist literature in the first half of the twentieth century but was gradually largely discredited. Modern creationism usually avoids using this argument, although it reappears sporadically, mostly on the fringes of the movement.

10

NOT A VERY BIG BANG ABOUT GENESIS

Gerald Schroeder Calculates the Duration of the Six Days of Creation

In this chapter I discuss three books by Gerald L. Schroeder which have gained a substantial popularity and have often been acclaimed as a very successful clarification of how to reconcile the biblical story with scientific data. The first book, *Genesis and the Big Bang: The Discovery of Harmony between Modern Science and the Bible*,[1] concisely expresses its main thrust in its subtitle. Schroeder sets out on an ambitious road aimed at demonstrating to both believers and skeptics that the contradictions between the biblical revelations and the claims of science simply stem from incorrect interpretations and that every word of the Bible is in complete agreement with the results of scientific exploration. Of specific interest to this discussion are Schroeder's views on chronology and probabilities, as well as a few other details and examples he provides.

SCHROEDER'S CHRONOLOGY

Early in *Genesis and the Big Bang*, Schroeder writes, "the discord between archaeology and theology is neither necessary nor valid. . . . My goal in this book is to explain this compatibility to expert and layperson alike" (p. 12). An important milestone on the road to Schroeder's goal is his analysis of chronological data, which until his book seemed to be irreconcilably different in the Bible and in science.

Schroeder's chronological exploration consists of two parts. The first part deals with the biblical account of the creation of the world in six days, and the second with the period of time between the end of those six days and our time. Let us consider each part of this chronology separately.

Creation of the World in Six Days According to Schroeder

The contradiction between the Book of Genesis and science has been discussed innumerable times. A notion that was often suggested (see, for example, chapters 5 and 9) was that the word *day* in the Bible was not meant to literally denote "day" as we understand it, i.e., as exactly one complete rotation of the earth on its axis, or approximately $\frac{1}{365}$ of the duration of earth's revolution around the sun. According to this interpretation, what is a billion years for us humans may be one day for God.

Despite Schroeder's assertions to the contrary, his explanation is essentially in the same vein, but with one difference: he turns to Einstein's theory of relativity in order to provide a specific clarification of the six-day creation story, which, in his view, is compatible with science (p. 27). As Schroeder explains, the theory of relativity has established, among other things, that there is no absolute time. Of course, this is true. In different frames of reference, he continues, the time interval between two events may be quite different. This is also true. What lasts six days in one frame of reference may last fifteen billion years in some other frame of reference. This is true as well.

Schroeder then gives an example of "time dilation" (p. 43). This is a well-known experimental result regarding the behavior of mu-mesons (muons). As Schroeder tells us, while 200 microseconds elapse in the frame of reference attached to the ground, only 4.5 microseconds elapse in the frame of reference of the moving muon itself. This is correct.

In his next step, Schroeder makes a leap from the case of fast-moving muons to the case of the universe's creation in six days. According to Schroeder, before creating the first man, Adam, God acted in his own frame of reference, vastly different from the frame of reference to which he would switch at the moment of Adam's creation. In the pre-Adam frame of reference—that of the Creator—what would become billions of years on the future, post-Adam clock were just days. At the moment of Adam's creation, God chose to change his frame of reference so that it would be the same for Himself and humans.

Can we assert that this idea is false? We can't. On the other hand, can we assert, based on rational considerations, that it is true? Again, we can't. Schroeder's explanation requires a leap of faith. It is fine as long as it is not intended to be an explanation based on science. There is nothing scientific in the notion that God's frame of reference may be vastly different from that of humans. As far as faith is concerned, the above assertion is not a new one, and is simply beyond any discussion in rational, scientific terms. Schroeder, though, wants readers to believe that the described explanation is somehow based on the theory of relativity. It is not.

What the special theory of relativity (STR) has established is that time indeed flows at different rates in different "inertial frames of reference." What the STR means by different inertial frames of reference is quite rigorously defined.[2] The rate of time flow is different in two frames of reference that mechanically move relative to each other with certain constant speeds. To make a period of time that is billions of years long in one frame of reference last only six days in the other frame of reference, these two frames of reference must move relative to each other at an extremely high speed.

Of course, to apply this rigorously defined situation to the creation of the universe according to the Bible requires a considerable stretch of imagination. To satisfy the requirements of STR, as per Schroeder's explanation, we have to accept first that God is a physical body; second, that He is a body that occupies a certain localized volume in space; and third, that during the six days of creation, the Creator was rushing at an enormous speed past the universe he was creating. What would then remain from the concept of the omnipresent, nonmaterial God?

In his later book, *The Science of God: The Convergence of Scientific and Biblical Wisdom*, as well as in a more recent Internet posting,[3] Schroeder has modified his interpretation of the biblical story by referring not to the special theory of relativity, but rather to Einstein's general theory of relativity (GTR). According to this theory, gravitational fields also affect the rate of time flow.[4] Schroeder builds his supposedly rational explanation of the six days of creation by calculating in detail the duration of each of the six days in terms of the human calendar, using the concept of time's dependence on gravitation.

The essence of the supposed explanation given by Schroeder is as follows: at the moment of the big bang, a "fireball" of pure energy was instantly created. This embryo of the emerging universe did not yet contain any mass, and therefore, according to Schroeder, there was no gravitation. Along with the expansion of the emerging universe, energy was transforming into mass (in accordance with Einstein's famous equation, $E = mc^2$). Accompanying the emergence of mass, gravitational forces took hold, increasing along with the appearance of the ever larger amount of mass. As the gravitational forces grew, the time flow slowed down. Schroeder suggests that what lasted just one (the first) day of creation is eight billion years on the scale of our conventional calendar. What lasted one more (the second) day of creation was four billion years on the scale of our conventional calendar, and so forth. The total duration of creation was exactly six days on the scale of time which corresponded to the levels of gravitational forces at each step of the universe's expansion, whereas on our conventional time scale it lasted about 15.75 billion years, in a good agreement with the scientific data (p. 60).

Schroeder does not explictly show a formula which would encapsulate his calculation, but it can in fact be rendered

$$T_n = \frac{1.6 \times 10^{10}}{2^n},$$

where $n = 1$ for the first day of creation, $n = 2$ for the second day, etc., and T_n is the duration of the "day" of creation number n on our time scale expressed as the number of conventional years. However, Schroeder does not explain why the rate of time flow should have gradually decreased according to this regularity rather than to any other, except for his desire to reconcile the biblical story with scientific data.

It is easy to imagine how laypersons would be impressed by this explanation, which looks so elegant on its face. I will avoid discussing some minor dubious points in Schroeder's exercise in order to concentrate instead on his most egregious misunderstanding of the matter, which renders his alleged explanation meaningless.

While supposedly utilizing the concept of relativity of time, Schroeder actually has based his discussion on the concept of an absolute time, thus contradicting the very premise of his discourse.

The concept of different rates of time flow depending on gravitation has a meaningful interpretation only in terms of at least two different bodies subjected to different gravitational forces. In each system, clocks would tick at different rates. However, to discover the difference between the clocks, there must be a way to compare them to each other.

As the amount of mass increased, as postulated by Schroeder, *all clocks* in the universe were affected equally. (There are differences between the clocks located in various parts of the universe, but this is irrelevant to Schroeder's model.) The universal clock suggested by Schroeder, namely the frequency of Penzias-Wilson radiation, is subject to that effect as well as any other imaginable clock. Since all the clocks are subject to the same effect of increasing gravitation, there is no way to discover the alleged deceleration of time flow. (The extrapolation of redshift into the past can be utilized to estimate the length of the universe's existence, but not to discover or measure the time dilation—not any more than it could be possible to lift oneself out of a swamp by grabbing one's hair with one's own hand and pulling it up without any independent point of support.) To reveal any difference in the rate of time flow, one would need an independent clock unaffected by the change of gravitation, which could be used as a reference. In other words, one would need a reference clock, which would be nothing less than a clock showing absolute time. Such a clock is not known. Since the alleged deceleration of time flow can be neither measured nor even discovered, it has no physical meaning, and therefore no effect on any physical processes, unless an absolute time, independent of frames of reference, is postulated.

Therefore, what lasted six days fifteen billion years ago lasts exactly six days now. If there were available some other observable universe that could be utilized as an independent frame of reference, then it could be possible to find out if one day in our century is different from one day fifteen billion years ago. As it is, the length of a day now is for all intents and purposes exactly the same as it was then. The length of the day is necessarily measured in arbitrary units and has no definable absolute ("actual") magnitude. Since the GTR rejects the concept of absolute time, Schroeder's idea contradicts the very essence of the theory of relativity on which he purports to have based his model.

Schroeder's Chronology between the First Six Days and Our Time

As explained earlier, Schroeder claims that at the moment the first man was created, God instantly switched from his previous frame of reference, in which the creation took six days, to a new frame of reference, which then became the same for God and humans. Since then, that frame of reference is the one that humans inhabit. Hence, everything the Bible tells us about events that occurred after Adam's appearance is chronologically precise. If the Bible says, for example, that Adam was 130 years old when his son Seth was born, then it is precisely 130 years as we understand them, i.e., the period of time it took Earth to make 130 revolutions about the sun. If the Bible says that Adam was created by God 5,750 years before the publication of *Genesis and the Big Bang*, it means precisely that—the time it took Earth to circle the sun 5,750 times, counted in the same years we conventionally mean. Schroeder tells us that this length of time is precisely compatible with the scientifically established chronology of human history: "In the year 1990, all the generations since Adam have a cumulative age of 5,750 years. This biblical date for the dawn of recorded history is closely matched by the archaeological finds of the last two centuries" (*Genesis and the Big Bang*, p. 31).

Schroeder's argument in support of this thesis is twofold. One part of his argument deals with the beginning of the Bronze Age and the other with the nature of those humanlike creatures, who, according to archaeological data, lived already thousands years earlier than about six thousand years ago.

Let us consider both parts of Schroeder's argumentation.

The Arguments Related to the Beginning of the Bronze Age

Schroeder juxtaposes the generations of the descendants of Adam's sons Seth and Cain. The Bible provides the ages of all the descendants of Seth, starting with Seth's son Enosh, all the way to Noah, who at the time of the Flood was six hundred years old. There were nine generations between Seth and Noah. Then Schroeder counts nine generations of descendants of Cain, ending with Tuval-Cain. The ages of Cain's descendants are not provided in the Bible, so Schroeder assumes that Tuval-Cain was a contemporary of Noah and from that he concludes that the time interval between Adam and Tuval-Cain was the same as between Adam and Noah, i.e., according to Schroeder's calculations, about 1,350 years. Subtracting 1,350 from 5,750, Schroeder concludes that Tuval-Cain lived about 4,400 years ago (p. 31).

This calculation, however, is wrong, and as we will see later (when discussing *The Science of God*) the author himself was forced to change his numbers. Schroeder does not say why he has changed the described data in his second book. It is sufficient, though, to read the pertinent verses (4:20–4:22) in the Book of Genesis to see the source of Schroeder's error. When counting generations between Cain and Tuval-Cain, Schroeder includes in his count Yaval (the seventh generation) and Yuval (the eighth generation), although it is clearly said in those verses that Yaval (also sometimes transliterated as Jabal) and Yuval (Jubal) were not the grandfather and the father of Tuval-Cain, but rather his half-brothers, as they all were sons of Lemach. Hence Tuval-Cain, according to the Bible, belonged to the seventh generation after Cain rather than to the ninth. This shifts Tuval-Cain's lifetime back several hundred years from Schroeder's assertion in *Genesis and the Big Bang*. (However, since we are now discussing this book, where Schroeder has based important and far-reaching conclusions on his numbers, let us accept these numbers for further discussion, especially since in his following book Schroeder did not change his conclusions based on the above dates.)

As the next step, Schroeder tells us that Tuval-Cain, according to the Bible, was the inventor of bronze. Hence, according to Schroeder, the Bible informs us that the Bronze Age started about 4,400 years ago (p. 31). This, asserts Schroeder, is precisely what science tells us. What a proof of the complete compatibility of the Bible with science!

Is it indeed? Let us check what the Bible says about that. Schroeder refers to Genesis 4:22, which, literally translated, states, "Also Zillah gave birth to Tuval-Cain, smith of all cutting tools of copper and iron."

Note that the last two words in the Hebrew text are *nekhoshet ubarzel*. Schroeder tells us that in early Hebrew the word *nekhoshet* (which he transliterates as *nhoshet*) meant both bronze and brass (p. 30).

Of course, this can only be guessed, since the literal meaning of that word is "copper," whereas the Hebrew word for "bronze" is *arad* and for "brass," *pliz*. However, the words *arad* and *pliz* are not found anywhere in the text of the Torah, which may indicate that these words appeared in the Hebrew vocabulary after the Torah was written. Therefore it is possible that the word *nekhoshet* (*nun-khet-shin-tav*) was indeed used in that context to denote not just pure copper but also copper-based alloys like bronze and brass.

However, Schroeder, in his quest for the chronological coincidence between the biblical account and scientific data, seems not to notice in the same passage about Tuval-Cain the word *barzel*, which means "iron." If Tuval-Cain made tools not only of copper or bronze, but also of iron, then we must place his lifetime at a much later date than Schroeder wants us to believe. The use of iron started roughly fifteen centuries after that of bronze. Hence, just one word omitted by Schroeder in his reference to the biblical text makes his chronological exercise collapse. Moreover, the omission of a reference to iron, which obviously could be deliberate, undermines all of Schroeder's arguments, since he could be suspected of being not quite impartial.

As we will see in *The Science of God*, where Schroeder changes his dates without explanation, the described discrepancy is only exacerbated.

The Arguments Related to Archaeological Data about Times Earlier Than 5,760 Years Ago

Having successfully, in his view, dealt with the beginning of the Bronze Age, Schroeder faces another controversy between science and the Bible. According to the Bible, Adam was created less than six thousand years ago. On the other hand, archaeological data indicate that creatures possessing many characteristics of humans lived in much earlier times. To reconcile these two viewpoints Schroeder suggests that the humanlike creatures that lived before Adam were not really human. They were, Schroeder tells us, animals that had some characteristics in common with post-Adam humans, but did not possess a human soul (in Hebrew, *neshamah*). Adam, according to that explanation, was the first real man, the first to receive from God his "neshamah," while pre-Adam humanlike animals did not have one.

Before discussing in detail Schroeder's notion about Adam being the first creature to possess the *neshamah* let us make note of the following historical evidence. If we believe Schroeder, at a certain moment in the history of men, namely less than six thousand years ago, a revolutionary event occurred which radically changed the nature of humanlike animals, converting them into real humans who were given the *neshamah*, thus separating them in a crucial manner from all other animals. However, there is historic evidence that, for example, a well-organized state existed in Egypt earlier than six thousand years ago. There is plenty of information about that state. If at a certain moment, about six thousand years ago, a revolutionary change in the nature of men took place, wouldn't we expect that such an event would be somehow reflected in the records pertaining to that time? There is no evidence in the scientifically established history of Egypt which would indicate any radical change in the nature of men at the above mentioned time.

To further discuss Schroeder's explanation, we have to define what distinguishes a man from an animal. In that endeavor I will not argue against Schroeder's contention that pre-Adam, humanlike creatures did not possess a *neshamah*, because this notion can be neither proven nor rejected based on any rationally acceptable evidence. It is a matter of faith, and in this book I argue neither in favor of nor against faith. We can, though, try to establish some verifiable criteria which would enable us to distinguish between animals and men.

It seems reasonable to assume that the following features are usual characteristics of humans while absent in animals:

- Humans possess language, necessarily spoken and, more often than not, also written. Human languages usually have a well-defined grammatical structure. No animals have a *real* language (unless the word *language* is applied to the primitive systems of sounds used by some animals, such as dolphins or chimpanzees, which lack any grammatical structure and have an extremely limited "vocabulary"). Of course, no animal uses a written language.
- Humans often form societies comprising thousands of members, structured vertically with many layers of hierarchy. The closest form of a hierarchy in the animal world is a family, say, of wolves, apes, or lions. Human society often forms a government and a system of law. There is no equivalent to government in the animal world (where the formations closest to a society are cattle herds and wolf packs).
- Humans usually develop technologies aimed at improving their standard of living. These technologies involve building living quarters, making tools for acquiring food and clothing, etc. Humans possess the capacity to improve the design of their creations and to invent new forms of technology. Some animals also build living quarters or (as beavers) even construct dams and the like, but they never change the scheme of these structures and have no evident capacity to improve and to invent.
- Humans typically develop arts, including music, dance, and visual art (such as painting). No animals are known to have any form of art.
- Finally, humans often develop some form of religion, while no animals are known to have any religious concepts or any notion of a deity (unless you believe your dog considers you to be God). I assume the development of religion can very decisively distinguish real man from any other humanlike or non-humanlike creature.

If we account for these points, then, according to Schroeder's concept, the predecessors of Adam who had not really been humans, but just humanlike animals, had no written language, no law system, no government, no art, and no religion.

Unfortunately for Gerald Schroeder, archaeological data contradict this description, indicating instead the following:[5]

- Humanlike creatures made tools of stone as early as about forty thousand years ago. Is any animal known to shape stones into tools?

- Color pictures of animals had been drawn on rocks as early as about twenty-seven thousand years ago (for example, the image of a horse in Pech Merle, France). What animal is known to engage in art?
- As early as twenty-three thousand years ago humanlike creatures embellished their looks by means of beads attached to their clothing. Is there any animal known to use artificially made clothing and, moreover, to embellish them in any way?
- As early as eighteen thousand years ago humanlike creatures used needles made of bones to sew clothing. Is there any animal known to sew clothing?
- The bow and arrow was already in use about eleven thousand years ago. Is there any animal capable of inventing and using sophisticated tools such as the bow and arrow?
- As early as about nine thousand years ago humanlike creatures already used pottery. To make the pottery, these humanlike creatures used kilns where the temperature reached about one thousand degrees. Is there any animal capable of constructing kilns and, moreover, of making and utilizing pottery?
- At about the same time, some nine thousand years ago, humanlike creatures built settlements occupying over two hectares each, where buildings were used as living quarters and as storage sheds. About eight thousand years ago, the size of such settlements was sometimes up to fifteen hectares. The inhabitants of these settlements used artificial irrigation and grew crops. They used seals, which proves the existence of some form of a writing system and of some form of documentation. Are there any animals known for building villages, using artificial irrigation, growing crops, or using a writing system and a documentation?
- Finally, as early as about eight thousand years ago—that is, some two thousand years before the date, when, according to the Bible, the first man appeared—humanlike creatures (for example, inhabitants of the Sumerian settlement at Tepe Gawra in Mesopotamia) systematically built, in the centers of their villages, religious shrines and temples. Is there an animal, however intelligent, that is known to have any religious concepts?

What we can say with reasonable confidence is that Schroeder's arguments aimed at proving the compatibility of the biblical and the scientific accounts fall apart at even a perfunctory glance.

SCHROEDER ESTIMATES PROBABILITIES

In the chapters of *Genesis and the Big Bang* dealing with probabilities, Schroeder's attitude is different from that in the chapters dealing with chronology. In the latter, Schroeder fully accepts scientific data and tries to reconcile them with the biblical story. As has been demonstrated earlier, though, Schroeder fails in his attempt.

In the chapters dealing with probabilistic estimates in regard to the origin of life, Schroeder seems to be confident in his ability to refute the scientific hypothesis in favor of the biblical story. There are no more attempts to reconcile the scientific and the biblical views, but rather an unequivocal rejection of the scientific hypothesis and an equally unequivocal acceptance of the biblical story.

The scientific hypothesis is based on the assumption that life emerged spontaneously as a result of interactions between chemicals in the primeval atmosphere of the earth several billion years ago. Another version of that hypothesis suggests that life could have been brought to the earth from some other worlds, where it originated spontaneously via this spontaneous, largely stochastic process.[6]

Schroeder bases his refutation of the scientific hypothesis on certain probabilistic considerations. For example, writing about the scientists who adhere to this hypothesis, he states, "These scientists were making assumptions without any attempt to rigorously investigate the probability of such events" (p. 111). Obviously, what Schroeder wants us to believe is that unlike those less-than-rigorous scientists, he has rigorously investigated the probability of a spontaneous emergence of life via a chance occurrence of chemical interactions. The conclusion Schroeder extracted from his rigorous investigation was that spontaneous emergence of life was impossible because the time needed for such an event to occur is so immense as to exceed the entire duration of the existence of the universe. This conclusion is far from being new and has been advanced innumerable times by the defenders of the biblical story.

This belief is in turn based on a certain interpretation of the concept of probability that Schroeder shares and tries to explain: "The probability of duplicating, by chance, two identical protein chains, each with 100 amino-acids, is 1 chance in 20^{100}, which equals the digit 1 followed by 130 zeroes" (p. 113).

While the arithmetic in the quoted segment is not exactly correct (the actual expression contains many digits other than zeros), in principle

Schroeder's statement is correct. The probability in question is indeed an extremely small number. However, as Schroeder continues, his further reasoning becomes contrary to the correct concept of probability.

This is what follows on the same page of *Genesis and the Big Bang*, "To reach the probable condition that a single protein might have developed by chance, we would need 10^{110} trials to have been completed each second since the start of the time. To carry out these concurrent trials, the feed stock of the reactions would require 10^{90} grams of carbon. But the entire mass of the Earth (all elements combined) is only 6×10^{27} grams!"

What an impressive statement so categorically demolishing the hypothesis of the spontaneous emergence of life! If only these statements and arguments were valid beyond the sheer arithmetic!

In reality, this line of reasoning misinterprets the meaning of probability. If it were correct, then we would have to accept that to win big in a California lottery, where the chance of winning the jackpot is about 1/16,000,000, one necessarily needs to play the lottery sixteen million times. We know, though, that some people have won big, having bought just one ticket the first time in their life. Such a smile from Lady Luck has happened more than once, defying the sixteen-million-to-one odds. (The concept of probability and its application is discussed in detail in chapter 13. That discussion demonstrates the lack of substantiation in Schroeder's position.)

The value of probability does not provide any information in regard to what will occur in any individual trial. If the probability of an event occurring is one out of a trillion trillions, there is nothing surprising if that event actually occurs in a particular trial, since its occurence is as likely as that of any other equally probable event. From this standpoint, the arguments based on the very small calculated probability of life emerging as a result of random chemical reactions are meaningless, since a small calculated probability in no way means the impossibility of an event.

Moreover, there are many situations where the possible events are not equally probable. As it is shown in chapter 13, a calculated probability in a certain sense reflects the level of information available about the object whose behavior is being guessed, rather than the objective likelihood of the event in question.

In regard to the random chemical reactions in the primordial atmosphere, the assumption of equiprobability has no foundation. Doubtless some chemical interactions were much more likely than some others. Certain conditions in the primordial atmosphere could have been conducive to certain interactions while inhibiting some other interactions. Under certain conditions, powerful catalytic effects must certainly have emerged, greatly enhancing the likelihood of interactions which would become steps toward the emergence of life. Schroeder's calculations of probabilities ignore such considerations and are therefore nowhere near a reasonable proof.

Life emerged at some time in the past. This was just one among a very large number N of possible (but not necessarily equiprobable) events. Its calculated probability, however small it happens to be, has by itself no cognitive value. Any other of $N - 1$ possible events could have occurred, but did not. Maybe some of those $N - 1$ events which did not happen would be even more complex and amazing than life as we know it; we will never know the answer to that. As stated earlier, the argument by Schroeder based on the calculation of probabilities of life emerging through some sequence of randomly occurring chemical reactions is meaningless. Equally unfounded are his calculations of the time allegedly necessary for such reactions to take place. If certain interactions were catalytically enhanced, with each step increasing the likelihood of the next consecutive step toward the emergence of life, the time necessary for the emergence of proteins was actually shorter by many orders of magnitude than that calculated by Schroeder.

Can these considerations serve as proof that life emerged spontaneously rather than having been created by divine design? Admittedly, they cannot. However, it shows that the hypothesis of the spontaneous emergence of life does not contradict probabilistic considerations. Moreover, Schroeder ignores the scientific theories which offer plausible scenarios regarding the spontaneous emergence of life.[7] These theories show that it is not necessary to think of the emergence of life as a single event of a very small probability but rather as a multistep process with each stage taking place with a reasonably large probability.

SOME SPECIFIC FAULTS AND ERRORS IN *GENESIS AND THE BIG BANG*

Besides the general weaknesses of Schroeder's case already discussed, *Genesis and the Big Bang* contains a plethora of small—and not so small—specific erroneous statements and arguments. Since Schroeder's books are often acclaimed for their alleged scientific sophistication, it seems worthwhile to point out at least a few of those faults in order to demonstrate the actual level of his competence on the subject.

The Bible Code Example

One of the issues Schroeder addresses is that of the so-called Bible code (pp. 182–85). However, his discourse is limited to a few primitive examples which lack any statistical significance and which testify to his insufficient familiarity with the subject he endeavors to discuss. (The Bible code controversy is discussed in detail in chapter 14 of this book, where the lack of substantiation of the claims of the Bible code proponents is shown.)

Are Weight and Mass the Same? Is Kinetic Energy Proportional to Velocity?

"Are weight and mass the same? Is kinetic energy proportional to velocity?" These questions sound especially silly considering the statements made by Schroeder, who holds a Ph.D. in physics. Unfortunately, Schroeder provides reason to raise such questions. For example, he writes, "The mass (or weight) of the object while at rest is called, in technical terms, its rest mass" (*Genesis and the Big Bang*, p. 40). I find it hard to believe that a physicist could indeed think that mass and weight are the same. I prefer to interpret this sentence as a display of sloppiness in style rather than of ignorance. Unfortunately, though, the author expresses himself in the same way more than once. The same expression "mass (or weight)" appears, for example, also on page 37, giving rise to the suspicion that Schroeder may actually believe that mass is the same as weight. Such a statement made by an undergraduate student on an exam in general physics would result in an immediate grade of F. The rest mass is a body's property, a constant independent of the frame of reference, whereas the total (or relativistic) mass is a function of velocity, and, as such, depends on the choice of frame of reference. The body's weight is a completely different quantity, reflecting the interaction of the body in question with a planet. It depends on both the mass of the body and the mass of the planet and, in the first approximation, on the squared distance from that planet's center (assuming the body in question is much smaller than the planet or has a spherical shape itself). I am sure that Schroeder studied these facts as a student.

On the same page Schroeder writes, "[A body] acquires velocity and in so doing acquires kinetic energy proportional to the velocity." Really, Dr. Schroeder! Don't you know that kinetic energy is proportional to *squared* velocity?

Another questionable usage: "The centrifugal force of the spin flattened the cloud into a disk" (p. 177). This is one more display of either a lack of sufficient understanding or of stylistic sloppiness. Centrifugal force is an example of what in physics is called "force of inertia"; it is considered to be a fictional force. The real force in this case is *centripetal* force, caused by gravitation. It is convenient in certain cases to use the concept of centrifugal force when writing certain equations (where it is sometimes referred to as D'Alembert's principle). However, saying that centrifugal force "flattened the cloud" veils the essence of the matter in a cloud of a meaningless phraseology.

Further examples of Schroeder's less than reliable statements could be given, but due to spatial considerations, we will move on.

PROBLEMS IN *THE SCIENCE OF GOD*

Schroeder's second book, *The Science of God: The Convergence of Scientific and Biblical Wisdom*, offers data which conflict with the data given in *Genesis and the Big Bang* without a word of explanation. I start with the discussion of this discrepancy.

Change of Chronology

In *Genesis and the Big Bang*, Schroeder calculated that the interval of time between Adam and Tuval-Cain was about 1,350 years, that Tuval-Cain was a contemporary of Noah, and that the Bronze Age started some 4,400 years ago (pages 31–32 and table 2). In *The Science of God*, however, the onset of the Bronze Age is said to have happened about 5,000 years ago (p. 131), i.e., about 600 years earlier). Furthermore, Schroeder claims in the second book that the time interval between Adam and Tuval-Cain was 700 years (p. 130), instead of the 1,350 years he indicated in the first book. Additionally, the date of the Flood is said in *The Science of God* to be about 4,100 years ago (p. 131), which is about 900 years later than the date of the advent of bronze given in this book. According to these altered dates, Tuval-Cain must no longer be considered Noah's contemporary.

Schroeder does not explain why he changed the dates, but it seems plausible to surmise that the alteration of dates was done because somebody had shown to Schroeder the error in his count of generations between Cain and Tuval-Cain in his first book. Still, if a writer changes his views, are the readers not entitled to at least some explanation?

Despite the change of dates, Schroeder maintains in his second book that the advent of the Bronze Age according to archaeological data precisely coincides with the biblical story (per his calculations). As in his first book, Schroeder again seems not to notice the statement in Genesis 4:22 indicating that Tuval-Cain also made tools of iron. Since in the second book the lifetime of Tuval-Cain is shifted back by some 900 years, it makes it even farther in time from the Iron Age as determined by archaeology.

Zero Time Interval in a Light Beam

The section describing the "shrinking of time" in fast-moving frames of reference according to the theory of relativity (pp. 161–64) compares two frames of reference moving with a certain constant velocity relative to each other. Two events take place in frame *A*, both occurring at the same location. Since the other frame, *B*, moves relative to *A*, the same two events occur in frame *B* at two different locations. As the theory asserts, the time interval between the two events in *A* is always shorter than in *B*. The larger the speed of motion of one frame of reference relative to the other, the larger the difference between the time intervals. If the speed approaches the speed of light, the "local" time interval (i.e., the time interval in *A*, where the events occur at the same location) approaches zero.

Schroeder considers the example of a light signal which carries information about the explosion of a supernova that occurred 170,000 Earth years ago. Since the supernova (1978A) is located 170,000 light years from the earth, the light signal took exactly that long to reach the earth. Since, though, the light signal moves with the speed of light, the flow of time in the frame of reference attached to the signal is "stopped." If there existed an observer "living" in the frame of reference of the signal (which is impossible as no physical entity other than photons can move with the speed of light) for "him" the explosion of the supernova and the arrival of the signal on Earth would have happened simultaneously.

Unfortunately, Schroeder fails to realize that there is no frame of reference which can be attached to photons. If such a frame of reference existed, photons would be at rest within it. This is impossible, as photons move with the same speed (in a vacuum) in *all* inertial frames of reference. If a frame of reference were to be attached to photons (which is impossible), any two events, such as the departure of a signal from a source (e.g., a supernova) and its arrival at the target destination (e.g., the earth) would always occur simultaneously and always at the same location. The impossibility of this event undermines Schroeder's whole amateurish exercise.

Of course, in accordance with his agenda, Schroeder tries to prove the analogy between the described "paradox" of the theory of relativity and the concept of God being "outside time." The concept of God being "outside time" belongs to the realm of faith and has nothing in common with the effect of "time stopping" in systems moving at the speed of light. Schroeder's attempt, inadvertently invoking the image of God running at the speed of light past stars and planets in order to satisfy the conditions of the theory of relativity, can only discredit his approach.

Diffraction of Waves

As part of his discourse in *The Science of God*, Schroeder explains the diffraction of waves (pp. 150–51). Up to a certain point, Schroeder's explanations are correct and explain this phenomenon, using the example of sea waves, reasonably well. Unfortunately, here, as in other cases, Schroeder fails to maintain a competent level of discourse throughout his presentation. As long as Schroeder considers diffraction on an opening whose size is close to the wavelength, his explanation is correct. However, it becomes wrong when he talks about openings whose size exceeds the wavelength. Indeed, he asserts that diffraction does not take place if the opening's size is larger than the wavelength (p. 150).

If that assertion were correct, it would be possible to build optical microscopes with unlimited magnification, because the ingress aperture of a microscope is much larger than the wavelength of visible light. If there were no diffraction on that aperture, then, utilizing a large number of consecutive lenses, one could reach a magnification power of many millions without resorting to electron microscopy. Alas, diffraction puts a limit on the resolving power of any optical device, and with it on useful magnification.

In the simplest case of diffraction on a single opening, increasing the opening's size does not eliminate diffraction but rather widens the central diffraction maximum, pushing the maxima of higher order to the opening's edges and squeezing them closer to each other. Diffraction is inevitable even if one of the opening's edges lies infinitely far, so actually rather than to speak about an opening, in this case one has to consider the wave's motion past the corner of an opaque wall. Diffraction always takes place if there is any constraint imposed on the free propagation of a wave.

This discrepancy is one of many examples of imprecision typical of Schroeder's style, casting a shadow on all of his arguments.

Miscellaneous Errors

In this section I will discuss a few more examples of the particular errors abundant in Schroeder's second book.

First, at the beginning of the section titled "The Discovery of Wave-Particle Duality" Schroeder tells the story of the discovery and interpretation of the photoelectric effect. Here, within half a page (p. 152), he manages to accumulate a long list of errors.

Schroeder starts with the statement that in 1905 Einstein "published the results of experiments that demonstrated what has become known as the photoelectric effect." In fact, Einstein did not publish any experimental data on this effect because he never performed any such experiments. The photoelectric effect was discovered and partially studied by Heinrich Hertz more than twenty-five years earlier.[8] What Einstein offered in his paper was a *theory* of that effect.[9]

Schroeder writes further, "Light, shining on certain metals, knocks free a stream of electrons." In fact, not "certain metals," but every material.

Another quote from the same page: "Einstein demonstrated that the rate at which electrons are emitted from metal is related not only to the intensity of the light beam but also to the color of the light." This is an incorrect statement. The rate of electron emission (i.e., the number of electrons emitted per second) does depend on the light intensity but does *not* depend on the "color" of the light. What indeed depends on the "color" of the light (i.e., on the wavelength) is the kinetic energy of the emitted electrons. Furthermore, Einstein did not demonstrate this, as the dependence of electron's kinetic energy, measured via the so-called stopping voltage on the wavelength of light, had already been discovered before Einstein (and then studied in detail by U.S. physicist Robert Andrews Millikan).[10]

Elsewhere, Schroeder describes experiments with a beam of atoms directed toward a plate with slits in it, writing: "Here we use a maser, a gun that can fire one atom at a time" (p. 154). Really?

A reader familiar at least with some rudimentary information in physics or electronics must wonder whether Schroeder has less understanding of atoms or of masers. It is known that the word *maser* is the acronym for "microwave amplification by stimulated emission of radiation." No maser is capable of firing atoms, either one by one or in groups. Most commonly masers are used as sources of coherent electromagnetic radiation in the microwave range. A version of a maser that works in the range of visible light is sometimes called an *optical maser*, but more often *laser*, which has by now become a ubiquitous component of many appliances, including, for example, compact disc players.

The final error I wish to discuss is found in the following statement: "The cooling effect of an expansion is logical. It is the dilution of a given amount of heat in an ever larger volume" (p. 180).

From the viewpoint of elementary thermodynamics, Schroeder's statement is nothing short of a bad joke. If Schroeder opened a college textbook on physics,[11] to the chapter on thermodynamics, he would learn that heat cannot be "diluted" in whatever volume. The expression used by Schroeder gives rise to an impression that, however hard it is to believe, he still adheres to the theory of caloric fluid, abandoned by science as erroneous since Rumford in 1796 (see chapter 5). It is a trivial knowledge in physics that heat is not a substance which is contained within a volume. (Three pages later, Schroeder repeats his meaningless statement about "heat dilution," which shows his phrasing was not a simple oversight.)

Heat is defined in thermodynamics as a quantity analogous to work and serving as a measure of energy transformation, provided such a transformation occurs via random molecular interactions. This quantity cannot experience a dilution either at expansion or in any other process. Of course, it is possible that Schroeder improperly used the term "heat," actually having in mind a component of the *internal energy* which is actually the kinetic energy of molecular motion, and which often is referred to as *thermal energy*. If this is the case, it still does not make Schroeder's notion about the reason for a temperature drop at expansion correct. According to the first law of thermodynamics, the temperature drop at expansion occurs because the work of expansion is done at the expense of the internal energy, and the temperature is the measure of that energy. Expansion causes not a "dilution" of thermal energy, but the decrease of the root-mean-square speed of molecular motion (see chapter 5 for more details).

For example, expansion into a vacuum is an isothermal process. This means that if a system expands into a vacuum, its temperature remains constant (because no work is done in such a process). Expansion itself cannot cause a drop of a system's temperature, which occurs only if, in the course of expansion, some work is done by the system.

It would be possible to continue a discussion of many erroneous and unsubstantiated statements in *The Science of God*. I feel, however, that I have already spent much more time and effort on the analysis of that opus than it deserves.

PROBLEMS IN *THE HIDDEN FACE OF GOD*

In his latest book, *The Hidden Face of God: How Science Reveals the Ultimate Truth,*[12] Schroeder mainly discusses various aspects of molecular biology, while the sections dealing with physics occupy only a small part of the book. Since I am not an expert in molecular biology, I will not argue against Schroeder's excursion into that marvelous science. However, it should be kept in mind that Schroeder is a physicist by education; hence, he is as much a dilettante in biology as I am. Any dispute between him and me in regard to biology would therefore be like a dispute between two blind men regarding the quality of a beautiful landscape painted by a famous artist. However, I am qualified to argue against Schroeder's errors when he endeavors to discuss physics. Given the sad fact that he was awarded a Ph.D. in physics by a prestigious institution, the elementary errors in Schroeder's discourse are simply stunning.

As a first example, Schroeder suggests the equation

$$h\upsilon = mc^2, \tag{10.1}$$

where h is Planck's constant, υ is the frequency of de Broglie's wave for a particle, m is the particle's mass and c is speed of light (p. 38). Unfortunately, this equation is absurd. However, it is easy to figure out how Schroeder derived it. He read somewhere about the following correct equations:

$$E = h\upsilon \tag{10.2}$$
$$E = mc^2 \tag{10.3}$$

Equation (10.2) was originally suggested by Max Planck in 1900 for the quantum of energy emitted by a black body. In 1905, Einstein applied that equation to the energy of photons regardless of whether they are emitted, are traveling, or are absorbed by a material.[13] In 1923, de Broglie suggested expanding the application of that equation to all particles, either massless as a photon or having a rest mass m.[14] Equation (10.3) is probably the most widely known equation of science, even to the general public, derived by Einstein in 1905 as part of his special theory of relativity.

In both equations (10.2) and (10.3) E denotes the energy of a particle. Obviously lacking proper understanding of these two equations, and seeing the same letter E on the left side of both, Schroeder mechanically combined equations (10.2) and (10.3) into one equation, (10.1).

Unfortunately for Schroeder, he obviously did not know that E in equation (10.2) and E in equation (10.3), while both denoting the energy of a particle, actually denote two different energies. In equation (10.2), it denotes the *variable* energy of a moving particle, related to that particle's momentum. However, in equation (10.3), it is a *constant* for a given particle, which denotes the so-called rest energy. These two types of particle energy have little to do with each other. The absurdity of Schroeder's equation, (10.1), is immediately obvious when we notice that it equalizes a variable quantity to a constant.

Planck's constant, h, equals about 6.626×10^{-34} joule-seconds while the speed of light, c, equals 2.997×108 meters per second. With this in mind, let us apply equation (10.1), for example, to the electron. The mass of an electron is close to 9.1×10^{-31} kilograms. Then the right side of equation (10.1), i.e., the electron's rest energy, is about 8.17×10^{-14} Joule, or about 5.1×10^5 electron volts (eV). Hence, if equation (10.1) were correct, all electrons in the world would always have the same energy—about 5.1×10^5 eV. To have such a level of energy, free electrons must be accelerated by a voltage a little more than half a million volts. Of course, different electrons (as well as any other particles) actually possess different energies in a wide range rather than all having the same energy of about half a million electron volts.

In fact, an equation formally identical to equation (10.1) can be legitimately written for the process of particle-antiparticle annihilation. For example, for an electron-positron annihilation an equation looking like (10.1) would be correct if m denoted the electron's (or, equally, the positron's) mass and $h\upsilon$ the energy of one of the two gamma-photons emerging as a result of those particles' mutual annihilation. The imprecision would in this case be only due to neglecting those constituents of the particle's energy which do not originate from its rest mass (such as kinetic energy of the particle's motion etc.). However, this case has nothing to do with Schroeder's discourse or with de Broglie's wave.

Equation (10.1) is not the only error in Schroeder's new book. However, it seems sufficient to limit the demonstration of the inaccuracies in Schroeder's literary production to the above examples.

Since Schroeder's insufficient competence in physics, which is his professional field, is obvious, what credibility can be given to his lengthy discourse on molecular biology, which is not his professional field? Moreover, what credibility can be given to his general thesis asserting the alleged harmony between the Bible and science?

NOTES

1. Gerald L. Schroeder, *Genesis and the Big Bang: The Discovery of the Harmony between Modern Science and the Bible* (New York: Bantam Books, 1992).

2. For elementary explanations of the special theory of relativity, see, for example, Raymond Serway et al., *Physics for Scientists and Engineers* (Philadelphia: Saunders, 1990).

3. Gerald L. Schroeder, *The Science of God: The Convergence of Scientific and Biblical Wisdom* (New York: Free Press, 1997); Gerald L. Schroeder, "The Age of the Universe" [online], www.geraldschroeder.com/age.html [May 14, 2002].

4. For an explanation of the general theory of relativity, see, for example, Clifford M. Will, "The Renaissance of General Relativity," in Serway et al., *Physics for Scientists and Engineers*, pp. 1136–45.

5. Andrew Sheratt, ed., *The Cambridge Encyclopedia of Archaeology* (New York: Crown, 1980).

6. For an explanation of scientific hypotheses of the origin of life, see, for example, a series of articles posted at The Talk.Origins Archive [online], www.talkorigins.org.

7. See, for example, Richard Dawkins, *Climbing Mount Improbable* (New York: W. W. Norton, 1997).

8. For an explanation of the photoelectric effect and its discovery by Heinrich Hertz, see, for example, Serway et al., *Physics for Scientists and Engineers*, pp. 1150–53.

9. Albert Einstein, "Über einen die Erzeugung und Verwandlung des Lichtes betreffenden heuristischen Gesichtspunkt" (On a heuristic point of view regarding the generation and transformation of light), *Annalen der Physik* 17 (1905):132–40.

10. Robert Andrews Millikan, "A Direct Photoelectric Determination of Planck's '*h*'," *Physical Review* 7 (1916):355–71.

11. For example, Halliday, Resnick, and Walker, *Fundamentals of Physics.*

12. Gerald L. Schroeder, *The Hidden Face of God: How Science Reveals the Ultimate Truth* (New York: Free Press, 2001).

13. Einstein, "Über einen die Erzeugung und Verwandlung des Lichtes betreffenden heuristischen Gesichtspunkt."

14. Louis V. de Broglie, "Ondes et quanta" (Waves and quanta), *Comptes rendus* 177 (1923):507–16.

11

A LOST CHANCE

Lee Spetner Derives Nonrandom Evolution from the Talmud

In his book titled *Not by Chance: Shattering the Modern Theory of Evolution*,[1] Dr. Lee M. Spetner offers readers interesting pieces of information from molecular biology and adjacent fields of science. Unfortunately, rather than limit his discourse to scientific arguments, Spetner's book contains passages which introduce the extraneous question of the compatibility of science with religious dogma. In the preface he writes: "I met the evolutionary theory in a serious way, and I found it hard to believe. It clashed not only with my religious views, but also with my intuition about how the information in living organisms could have developed" (p. ix).

It can be assumed from this passage that the motivation behind Spetner's effort is his desire to reconcile his religious beliefs with scientific evidence. In other words, Spetner seems to have an agenda rooted in his religious beliefs. A nonscientific agenda chosen a priori is something a scientist is supposed to avoid (although I do not wish to assert that scientists always succeed in avoiding ideological predispositions). *Not by Chance* leaves the impression that Spetner the believer was watching over Spetner the scientist, limiting his freedom to follow only the facts, wherever they might lead him.

Spetner the believer is in full force as the book starts, but becomes less visible as his story unfolds. Toward the end of the book, Spetner the scientist seems to have won the battle, as he writes about his theory, which he calls the NREH ("Non-Random Evolutionary Hypothesis"): "The NREH, on the other hand, is agnostic" (p. 212).

Immediately, however, the believer surfaces in the continuation of the quoted sentence, as follows: ". . . and poses no contradiction to creation.

The NREH, as an explanation of evolution, is in fact derivable from Talmudic sources."

While readers have no reason to doubt that Spetner himself may sincerely believe in what he says about the alleged derivation of his theory of evolution from the Talmud, the quoted statement is irrelevant as far as his theory goes. In the space of two hundred pages, Spetner offers a variety of seemingly scientific arguments against the neo-Darwinian theory (NDT) and finally suggests his own hypothesis without using any notions originating either in the Talmud or in any other religious sources or beliefs. Hence, his sudden reference to the Talmudic sources sounds like nothing short of a ransom paid to his agenda. Moreover, in such an immense compendium of commentaries and interpretations as the Talmud, one can certainly find some commentaries of this or that rabbi seemingly compatible with any views.

Thus the subtitle of the book in question, *Shattering the Modern Theory of Evolution*, seems to be misleading; in fact, Spetner's theory is itself that of evolution. Contrary to what readers could conclude from the title and subtitle, Spetner's argument is not against the theory of evolution in general. The author actually argues only against some aspects of the NDT, namely the idea that evolution necessarily included *random* variations (hence the "Non-Random Evolutionary Hypothesis").

However, this term is itself misleading. Neo-Darwinism does not maintain that evolution occurred in a purely random way; on the contrary, the Darwinian theories, in all of their variations, maintain that evolution includes both random and nonrandom elements. According to the neo-Darwinian approach, the random element of evolution entails mutations in the genome. Mutations are believed to be random—this is indeed the assumption of neo-Darwinism (although even this assumption must be qualified; see below for details). However, the evolution is rather far from being limited to mutations. Its second, equally crucial part is the concept of natural selection, which is by no means random, but rather is directed by the environment. Using Spetner's terminology, natural selection is led by *signals* from the environment.

Moreover, the adherents of neo-Darwinism (who are the target of Spetner's assault) do indeed view mutations as occurring randomly, but they qualify that concept in a substantial way. For example, one of the most distinguished defenders of Darwinism, Richard Dawkins—who, unlike Spetner, is a professional biologist—offers five aspects in which mutation is not completely random, adding, "There are, in truth, many respects in which mutation is not random." A few pages later, Dawkins continues, "The Darwinian says that variation is random in the sense that it is not directed toward improvement, and that the tendency toward improvement in evolution comes from selection."[2]

In light of these perspectives, the very name Spetner chose for his hypothesis seems to be misdirected. The author appears to be fighting against a straw man. What is, then, the real difference between neo-Darwinism, which Spetner rejects, and his NREH?

Spetner's hypothesis shifts the nonrandom step of evolution from natural selection to mutations. According to Spetner, random mutations followed by natural selection could not ensure the rate of evolution rapid enough for the appearance of the enormous multitude of the existing species. Hence, insists Spetner, to explain the actual rate of the evolution of the species, we must assume that mutations are not random but rather caused by signals from the environment.

Spetner's hypothesis is by no means a novel one. Similar ideas have been suggested before and rejected by the overwhelming majority of biologists for a number of reasons, mainly because they are not supported by evidence. Spetner's predecessors are sometimes referred to as "mutationists." For example, Dawkins provides a rather detailed criticism of the views of such "mutationists."[3]

Maybe, unlike his predecessors within the ranks of "mutationists," Spetner came up with some hitherto unknown arguments in favor of nonrandom mutation that is triggered by a "signal" from the environment and directed toward improvement? Unfortunately, this is not the case. There is no positive evidence in Spetner's book that would support his hypothesis. All his argumentation is of a negative character, wherein, rather than offering evidence in favor of his hypothesis, Spetner tries to show that the neo-Darwinian theory entails serious faults. Of course, even if Spetner's critique of neo-Darwinism were well substantiated, this in itself would not signify the validity of his alternative hypothesis.

Although Spetner is entitled to suggest whatever hypothesis he chooses, in order to be accepted in science his mislabeled Nonrandom Evolutionary Hypothesis would require, first, an explanation of the nature and mechanism of the supposed environmental signals that cause necessary, useful mutations. Second, his theory would require empirical evidence to support his hypothesis of directed rather than random mutations. Unfortunately, Spetner's discourse meets neither of the two requirements. His hypothesis is an unsubstantiated assumption which lacks both empirical foundation and explanatory details that are necessary to lead from his ideas to a real scientific theory. No wonder Spetner published his work as a popular book for a broad audience rather than in a peer-reviewed scientific journal.

Spetner's critical discussion of various aspects of neo-Darwinism comprises three parts: probabilities, information, and randomness. All three elements of Spetner's discourse will be discussed in subsequent sections of this chapter.

It is hard to disagree with Spetner when he says that his hypothesis is agnostic. Every scientific hypothesis is. However, this statement, made at the end of the book, contradicts his agenda as could be understood from his statements at the beginning.

As has been pointed out in previous chapters, questions of faith cannot and must not be handled by means of scientific research, which has no tools to either confirm or reject religious claims, except for cases of deliberate fraud that can be debunked by an impartial test. Spetner's hypothesis provides no arguments either in favor of or against any religious beliefs, however strongly some defenders of religious views—including Spetner himself—may wish to derive such conclusions from that hypothesis.

It is interesting to mention that *Not by Chance*, which its author intended to use as a weapon defending religious beliefs, was subjected to rather strong criticism in *Jewish Action* magazine, which serves the Orthodox Jewish community. In his review of Spetner's book, Dr. Carl Feit concentrates on the very heart of Spetner's discourse, namely the rejection of randomness in the process of evolution.[4] Feit is mainly interested in showing that randomness is not contrary to the biblical story. While the religious reasons to disagree with Spetner are beyond the scope of my discussion, there are certain points in Feit's review which could very well be accepted in a scientific analysis of Spetner's writing. As Feit indicates, Spetner's premise can be characterized as the "God-in-the-gaps" approach to the conflict between religion and science. The adherents of this approach look for "gaps" in scientific theories, i.e., for points either insufficiently explained or not explained at all by the existing theories, and insist that the only explanation for such points is the presence of a supernatural being. Feit, who is a believer, rejects such an approach because the ongoing development of science can eventually lead to a scientific explanation of points that are still obscure, thus forcing religion to retreat with humiliation. Feit also indicates that Spetner has left out many important areas of modern biological research such as "complexity theory, neutral mutations, self-organization in complex systems, artificial life, common attractors, the modular domain structure of modern proteins."[5] Because Feit is an expert in biology, this part of his critical remarks casts a shadow on the very aspects of Spetner's book which purportedly constitute its strongest points. Since I am not a biologist, I leave it to Feit and other experts in biology to discuss the biological aspects of *Not by Chance*, discussing instead some points which are closer to my own expertise.

SPETNER CALCULATES PROBABILITIES

In two chapters of *Not by Chance*, Spetner devotes much space to the discussion of probabilities. In chapter 4, "Is the Deck Stacked?" Spetner discusses probabilities of random mutations and of other events that, according to the Darwinian approach, were the steps of evolution. Spetner repeats here the same well-known calculations of probability which have been copied, with slight variations, from book to book, written by opponents of the theory of evolution but actually irrelevant for the problem of the origin of life. (These calculations are discussed at length in chapter 13, and, to some extent, in chapters 9 and 10.) I will not repeat here all these arguments, but will only discuss briefly a few points in regard to Spetner's calculations.

From the cognitive viewpoint, the concept of probability first and foremost reflects the level of ignorance about a situation. If we possessed the full knowledge of the latter, we would deal with certainties rather than probabilities. The extremely low values of probabilities of various events, such as the spontaneous emergence of life and the like (which are often quoted by the opponents of evolution), primarily reflect the paucity of information about the events which would lead, for example, to useful surviving mutations or to the emergence of life. This lack of knowledge forces us to assume an enormous number of possible competing events thus resulting in the very low probabilities cited by Spetner, Aviezer (see chapter 9), and others.

How is the probability of an event estimated to be, say, $1/N$, where N is a very large number? This estimate is tantamount to the assumption that there could be N possible events, of which, say, the emergence of life is just one. Assuming further that all those N possible events are equally probable, we divide 1 by N and arrive at the very low probability $1/N$, which turns out to be, say, $1/10^{27}$ (as in Spetner's example). The conclusion derived by the antievolutionists is that the cited probability is so small that the event in question is all but impossible.

This way of thought, common to many discussions by the opponents of evolution, is flawed as it misinterprets the meaning of probability. Indeed, if the probability of an event (say, of a useful mutation) is calculated as $1/N$, it means that we assumed the competition of N equally probable events. One of those N equally probable events must necessarily happen (even though we don't know in advance which one). If any of these events is viewed as impossible, then by the same token each of them must be viewed as impossible, as all those events have exactly the same chance of happening or not happening. This conclusion, based on the assumption of impossibility of any one of N events, is absurd. Hence the premise must be wrong.

If a certain event, A (for example, a certain mutation) whose probability was calculated as $1/N$ did not actually happen, it only means that some other event, B, whose probability was equally small, happened instead. Why could event B happen but the equally probable event A could not? From the standpoint of probability, there is no difference between all these N events, even if one of them is very special from some nonmathematical viewpoint (for example, the spontaneous emergence of life).*

Discussing the analogy with a lottery, Spetner considers a favorite example of the opponents of evolution, that of multiple wins, and states correctly that the probability of cheating seems to be much larger than the probability that a person wins a lottery twice in a row. Similarly, he states that in a poker game the simultaneous occurrence of two straight flushes indicates fraud by a much larger probability than its happening by chance. Writing about a cowboy who was shot to death for holding straight flushes in two consecutive hands, he comments, "As we have seen in the story of the poker-playing cowboy, too much luck might be not good for you. So too in nature, if we see the occurrence of the event with exceedingly low probability, we must suspect the event was not random" (p. 94). This last claim is unsubstantiated. Indeed, fraud can be more probable than two simultaneous straight flushes in a poker game, or a double win in a lottery. However, this example is irrelevant in regard to nature, where there is no analogy of fraud. In nature an event is possible despite its exceedingly small *calculated* probability, because, first, the very small value of the calculated probability is mainly due to the paucity of information about the situation, and, second, because the event whose

*Readers may notice that this issue has already been addressed (see chapters 2, 9, and 10), but I believe it is worthwhile to return to it in each case, because this aspect of probability is repeatedly discussed (incorrectly) by many authors with creationist sympathies; I'd prefer to be accused of beating a dead horse rather than omitting a relevant counterargument. This point will surface again in later chapters as well.

probability was calculated as very small is actually as probable as any other of N events assumed to be possible.

Chapter 4 of *Not by Chance* contains many other calculations of probabilities, but all of them are fraught with similar fallacies that are discussed and dismissed throughout this book, especially in chapter 13. This includes Spetner's argument against the theory of random evolution, which, based on his calculations of probabilities in "Is the Deck Stacked?" is not convincing.

In chapter 6, titled "The Watchmaker's Blindness," Spetner returns again to the estimation of probabilities, this time using it to refute the ideas suggested by Dawkins. Here, again, Spetner offers calculations of extremely small probabilities of such events as encountering a "perfect bridge hand." According to that calculation, the probability of getting a perfect hand in one deal is $1/10^{28}$. This number, although correct from a formal viewpoint, is irrelevant. Spetner accuses Dawkins of not understanding the meaning of probability, but actually, such an accusation could be directed toward Spetner himself, who obviously knows how to calculate probabilities but ascribes to them properties these quantities cannot possess. The value of probability does not predict the outcome of any particular event. Despite the extremely low calculated probability of an event, it can well happen on the very first trial, whereas an event whose calculated probability is much higher may not happen even in hundreds of thousands of tests. When such problems as the possibility of the spontaneous emergence of life are discussed, their calculated probability is largely irrelevant and cannot be used as proof of any opinions on that subject. (This question will be discussed at length in chapter 13.)

Moreover, if the problem under discussion is the probability of a particular mutation, the actual probability is immensely larger than that calculated by Spetner (or, say, Nathan Aviezer), as they artificially limit the number of possible pathways to unrealistically few choices (often just one choice), while in fact the number of possible pathways leading to *useful mutations* is very large. In the case of the origin of life, biologists do not adhere at all to the hypothesis that the first living matter emerged as a result of a chance encounter of all the necessary components; therefore all these calculations of minuscule probabilities of such chance events are in fact addressing a straw man–type problem.

SPETNER DISCUSSES RANDOMNESS

While the NDT is based on the assumption that evolutionary changes occur via random mutations followed by nonrandom natural selection, Spetner's NREH is based on the idea of nonrandom mutations triggered by the demands the environment imposes on species. Obviously, since the difference between the NDT and the NREH essentially boils down to one between random and nonrandom chains of events, the concept of randomness becomes germane for the discussion of these two hypotheses. Consequently, we could expect that Spetner would provide some definition of randomness, as it is the fundamental concept of his hypothesis. Strangely, Spetner seems not to be worried about the precise meaning of a term he uses so frequently in his discourse. As a result, in different contexts, he uses the terms *random* and *randomness* in different ways, seemingly not noticing the vagueness this sloppiness of usage imparts to his discourse.

The following quote provides a good example:

> The motion of these genetic elements about to produce the above mutations has been found to be a complex process and we probably haven't yet discovered all the complexity. But because no one knows why they occur, many geneticists have assumed they occur only by chance. I find it hard to believe that a process as precise and as well controlled as the transposition of genetic elements happens only by chance. Some scientists tend to call a mechanism *random* before we learn what it really does. If the source of the variation for evolution were point mutations, we could say the variation is random. But if the source of variation is the complex process of transposition, then there is no justification for saying that evolution is based on random events. (p. 44)

In this passage, we see that Spetner used the term *random* twice, once in relation to a "mechanism," and once to "events." He apparently assumes that the term in question is universally understandable without explanation, in both contexts in which he uses it. Let us guess what Spetner means by *random*. From the quoted text it seems to follow that, according to Spetner, random is an event that, first, occurs just by chance, and, second, is simple, comprising only one step. As for *random mechanism*, this term seems to mean a combination of consecutive random events.

If my interpretation of the meaning Spetner implies in the term *random* differs from what Spetner himself would say, it only means that he should have defined his usage of this term in a less ambiguous way.

As another example, Spetner writes, "Darwinian theory asks that the mutations be both spontaneous and random" (p. 46). It is not clear whether Spetner intended this sentence as a quasi-direct quotation from some writings by NDT adherents, or it is given as his own formula, but in any case he uses the expression "spontaneous and random" without any indication that he might view it as imprecise or misleading. Does that expression mean that events (and mechanisms?) can be random but not spontaneous, spontaneous but not random, or both spontaneous and random?

The necessity for a stringent definition of the these terms and for following that definition consistently is crucial because the concept of randomness (and related to it, concepts of complexity and spontaneity) are fundamental for his hypothesis.

Spetner gives no indication that he is familiar with the mathematical definitions of randomness and complexity (a question that was discussed in chapters 1 and 2 of this book).

At the core of Spetner's hypothesis is the suggestion that the variations leading to evolution are triggered by forces of environment and are directional rather than random. To substantiate that suggestion, he must first clearly understand in what way the variations in question are not random. To this end he must clearly define what is random and what distinguishes nonrandom from random. Without first building the foundation in terms of random versus nonrandom, Spetner's idea remains too vague for a scientific hypothesis.

Spetner may believe that the variations he admits to being the steps in evolution are nonrandom, but he actually has no knowledge that allows him to clearly establish that the variations in question are indeed nonrandom. So far his classification of events as random or nonrandom remains a matter of personal preference rather than an established fact.

The closely connected concepts of randomness and complexity are strictly defined in the algorithmic theory of probability (ATP). Let us recall these definitions. The complexity of a system (or a process), which is often referred to as *Kolmogorov complexity*, is defined in the ATP as the minimal size of an algorithm (or a program) which can encode the system (or the process) in question. From this definition follows the definition of randomness as follows: a system (or a process) is random if its complexity approximately equals the size of the system (or process) itself (usually measured in bits).

The term "approximately" appears in the above definition because, as the ATP shows, randomness is a matter of degree. The closer the value of a system's (or a process's) complexity to its own size, the closer that system (process) is to perfect randomness. This shows that the concept of randomness is more complicated than it may seem at first glance. The demarcation between random and nonrandom events or mechanisms is diffuse—events or mechanisms can be more random or less random. Unfortunately, this feature is absent in Spetner's discussion.

To base any hypothesis on the concept of random versus nonrandom requires a quantitative approach, one that accounts for the mathematical definitions of randomness and complexity cited here. These definitions are universal and applicable to all systems and processes. However, there are no signs of such an approach in Spetner's book. Therefore when the author tells us that certain events are random while some others are nonrandom, we cannot verify his statements in order to judge his line of thought rationally.

According to his biographical data, Spetner has extensive experience in research related to communication systems, and also, according to his book, must be well versed in information theory. Therefore he is expected to also be familiar with the mathematical concept of randomness as well, even more so because he has based his hypothesis on the distinction between random and nonrandom events. Strangely, though, there is not a sentence in Spetner's book which would reveal his knowledge of the pertinent mathematical concepts. This gives his suggestion of supposedly nonrandom sources of evolution the flavor of speculation on a popular rather than on a scientific level.

SPETNER DISCUSSES INFORMATION

In chapter 5 of *Not by Chance*, "Can Random Variation Build Information?" one of Spetner's main arguments is that random variations never, or almost never, lead to the buildup of information. Since this concept is one of the foundations of his NREH, one would expect that the fundamental concept of that hypothesis, information, would be presented in an unambiguous, rigorous manner.

Information theory is the scientific basis of modern communication technology. Spetner is introduced to the reader as an expert in signal processing, a field largely based on information theory. Hence the readers are entitled to a good, professional discussion by Spetner of matters related to information in its application to the NREH. Unfortunately, this is not the case.

Like with probabilities and randomness, Spetner uses the term *infor-*

mation vaguely, as if this concept has no quantitative measure. Therefore, when he maintains that certain variations do or do not add information to a system, these statements remain unsubstantiated. Having accused some other authors (e.g., Dawkins) of lacking calculations which would support their theories or hypotheses, Spetner himself does not offer quantitative estimates of the change in the amount of information caused by the variations he discusses. Without such calculations his assertions that this or that variation do or do not add information to a system remain unverifiable. His statements in regard to particular variations that allegedly add no information are based not on factual evidence but only on his intuition, which may or may not lead to the right conclusion.

One such ambiguous situation is Spetner's example of a streptomycin molecule fitting into a specific location in a ribosome (see figures 5.3 and 5.4 in *Not by Chance*). There is a steric match between the streptomycin molecule and the bacterium's ribosome. The streptomycin molecule fits a specific site in the ribosome as a key fits into the proper lock, inhibiting the bacterium's normal functioning and eliminating its detrimental effect on the human body. A random mutation can cause a change in the shape of the pertinent site in the ribosome, making it unsuitable for the attachment of a streptomycin molecule. Such a mutated bacterium acquires immunity against streptomycin, thus improving its chance for survival. Spetner maintains that the described mutation leads to a loss rather than an increase of information.

However, this assertion is not substantiated. The described change in the ribosome's shape may either increase or decrease the amount of information associated with that ribosome. To decide whether it is a loss or a gain in information, a detailed calculation is necessary, based on the detailed knowledge of a particular shape's change. As long as such knowledge is not available, my guess as to whether the information increased or decreased is as good as yours—or as Spetner's.

The author's attempt to substantiate his assertion that the amount of information is decreased by the described mutation because this mutation makes the ribosome less specific is itself unsubstantiated. The ribosome may become less specific in relation to streptomycin, but may instead become more specific in relation to some other substance. Since information about such a possibility is absent, there is no reason to assert that *specificity*, in Spetner's sense, has indeed dropped. Therefore Spetner's assertion that the mutation in question resulted in a decrease of information is pure speculation with no evidentiary value.

WHAT REMAINS?

In his effort to substantiate his hypothesis, Spetner uses calculations of probabilities, the question of randomness versus nonrandomness, and the buildup of information as his three foundation stones. Since we have dismissed the calculations of probabilities as irrelevant, the notion that random events are not capable of causing evolution as uncertain, and the alleged absence of information buildup as unproven, the natural question to be asked is, "What then remains of the book in question to be taken seriously?"

My answer to that question is, almost nothing. There is no chance Spetner's book might be taken seriously by biologists. Indeed, well-substantiated critiques of *Not by Chance* from a biological standpoint have been offered, for example, by Gert Korthof and Ian Musgrave.[6] Of course, adherents of intelligent design may praise Spetner for supposedly showing that evolution cannot generate information, but this thesis has been promoted by other ID creationists at a much higher level of sophistication, and all of their arguments were rejected by the overwhelming majority of both biologists and information theorists.

Regardless of whether Spetner's hypothesis is correct or not, it entails no religious consequences, even if that might be contrary to Spetner's own intention. Spetner's attempts to interpret his hypothesis as proving certain religious concepts were, first, not useful, and, second, unsuccessful, since his hypothesis can be viewed equally as either confirming or contradicting those concepts, which in fact have no relation to the veracity of this hypothesis. Another weakness of Spetner's book is his attempt to substantiate his hypothesis by invoking probability, randomness, and information, but having done so in a way that does not meet the requirements of scientific discourse.

Spetner had a chance to write a serious discussion of an important problem, suggesting his view of that problem in a nonsensational way, and to submit it to a peer-reviewed professional journal. Instead, he chose another path—to appeal to the public's interest in sensational claims, regardless of their degree of rigorously scientific substantiation. One might say that the title of his book, *Not by Chance*, would have been more appropriate if it had instead read *A Lost Chance*.

NOTES

1. Lee M. Spetner, *Not by Chance: Shattering the Modern Theory of Evolution* (New York: Judaica Press, 1998).

2. Richard Dawkins, *The Blind Watchmaker* (New York: W. W. Norton, 1986), pp. 305–308; the quotations are from pp. 306 and 308.

3. Ibid., pp. 305–309.

4. Carl Feit, "Not by Chance! The Fall of Neo-Darwinian Theory by Dr. Lee M. Spetner," *Jewish Action* (spring 1999):87–89.

5. Ibid., p. 86.

6. Gert Korthof, "Could It Work? Does Evolution Work by Accumulation of Random Mutations as Neo-Darwinism Claims or by 'Adaptive' Mutations as Spetner Claims?" Was Darwin Wrong? [online], home.planet.nl/~gkorthof/kortho36.htm [June 23, 2003]; Ian Musgrave, "Spetner and Biological Information," Talk Reason [online], www.talkreason.org/articles/spetner_v2.cfm [July 22, 2003].

PART 3

TWO BITS OF A GENERAL DISCUSSION AND ONE TELLTALE EXAMPLE

12

SCIENCE IN THE EYES OF A SCIENTIST

In the preceding chapters, either those discussing books and papers devoted to the most recent incarnation of creationism disguised as the intelligent design theory, or those purporting to assert the harmony between science and the Bible, the question repeatedly arose of what constitutes legitimate science and how to distinguish it from bad science or pseudoscience. Obviously, this distinction is possible only if a certain definition of science has first been agreed upon. Unfortunately, however, it seems that a rigorous definition of science is hard to offer—perhaps even impossible, for reasons which may become clear from the following discourse. In a nonrigorous and most general way, science can perhaps be defined as a human endeavor consciously aimed at acquiring knowledge about the world in a systematic and logically consistent manner, based on factual evidence obtained by observation and experimentation.

As should become clear from the following discourse, it is hard to adhere to the above definition of science consistently, because the latter seems to inevitably encompass elements beyond the limitations of that definition.

The subject of this chapter is often discussed within the framework of the philosophy of science. I would like to stress from the very beginning, though, that this chapter is not a piece on philosophy of science. My intention is to look at the problem from a standpoint of a scientist, mainly stemming from my own personal experience. I envision this as a glance from the "inside" of science rather than from a philosophical "outside," since I am not a philosopher and do not pretend to be one.

There seems to exist a whole industry of philosophy of science with its own jargon, definitions and multiple cross-references comprehensible only to its practitioners. Though there are serious discourses in that vast literature, sometimes its priesthood excels in providing a seemingly sophisticated analysis of science by people who themselves have never seen the inside of a research lab.

Although philosophers of science meet at conferences where they exchange their views on science and publish books and articles in many journals which are fully or partially devoted to philosophy of science, all this lively activity seems to occur outside of science itself and seems to have little impact on the everyday work of practicing scientists. Most scientists seem to have no interest in philosophy of science, which therefore seems to exist for its own sake. Since universities and research establishments are willing to pay salaries to philosophers of science, and their literary output finds readers willing to pay for the privilege, obviously there is a consensus that philosophy of science has its use, and I have no intention to argue against that. However, readers must be warned that they will not find in this chapter either many references to the publications of philosophers of science or attempts to enter the temple where the discourse of philosophy of science occurs.

Whereas philosophers may avoid discussing the specific nuts and bolts of scientific work, a glance from within science requires analyzing specific details of the process of scientific inquiry. If I become too technical, I may lose those readers who are not professionally versed in science. If I resort to oversimplifications, I may inadvertently seem to be an amateur philosopher, which is something I wish to avoid. I don't know if I can succeed in such a balancing act between the two approaches and probably will not be able to completely eschew the trap of coming across as an amateur philosopher. If that is the case, I ask the reader to believe it is inadvertent.

I also realize that it may be impossible to analyze the essence of science as though all of its branches were characterized by the same set of features. Science is so multifaceted that an analysis which is quite adequate for some fields of science may be irrelevant in others. My approach will be relevant first and foremost to such fields as physics, chemistry, engineering disciplines, and biology, while many aspects of my discourse may be less applicable to other areas of knowledge.

One of the first questions in the discussion of science concerns its driving force. While different scientists may be driven by different motivations, I submit that the main reason for the existence of scientific research is the curiosity which seems to be inherent in human nature. This statement does not imply that every man and woman necessarily possesses such curiosity; on the contrary, the vast majority of people lack it. As the great

Russian poet Aleksandr Pushkin said (he meant only the Russian people, but his words can probably be applied to humanity as a whole), "We are lazy and not curious."[1] However, there has always existed a fraction of people differing from others in that they are driven by insatiable curiosity. Apparently, some ethnic groups, at different historical periods and for unknown reasons, have had a larger fraction of such people than others.

Over the course of millennia, science underwent multiple modifications, remaining for many centuries indistinguishable from philosophy. Modern science, which is quite distinct from philosophy, developed largely during the last few centuries.

In a certain sense, the relationship between science, philosophy, and religion can be represented by three concentric circles (necessarily simplifying the actual situation). The inner circle encompasses scientific knowledge; its radius is constantly increasing, as more and more knowledge is accumulated. Beyond that inner circle of scientific knowledge is a concentric circle with a larger radius; the area between the two circles represents the domain of philosophy. Philosophy reigns where science has not yet acquired sufficient knowledge, leaving room for speculation limited only by logic. Finally, an even larger concentric circle surrounds the first two circles; this is the domain of religion, where neither science nor philosophy has so far managed to offer good explanations of the world. The expansion of science pushes philosophy into a gradually narrowing domain where it still has the freedom to pose logical explanations not based on rigorously collected factual data. This in turn pushes religion even farther to the margins of the world's comprehension. The enormous power of survival obviously inherent in religion is due not to its gradually shrinking explanatory abilities but rather to its ability to satisfy human emotional needs.

FEATURES OF GOOD SCIENCE, BAD SCIENCE, AND PSEUDOSCIENCE

Philosophers of science often discuss the distinction between good science, bad science, and pseudoscience. I have no intention to delve into the analysis of the distinctive features of these three categories, leaving it instead to philosophers. I believe bad science usually becomes pseudoscience if it is stubbornly pursued despite evidence to the contrary.

A case of bad science can be exemplified by the story of so-called cold fusion. The sensational discovery of this phenomenon was claimed in 1989 by Martin Fleischmann and B. Stanley Pons, working together, and by Steven Jones independently. The story of cold fusion is well known so I will not repeat it here.[2] What is relevant to my discussion is that the results of these researchers could not be reproduced by other scientists. Nevertheless, despite the disappointment in the evident irreproducibility of the initial results by Fleischmann and Pons, some research in that direction is still going on, and regular conferences devoted to that subject are being held. Despite all that effort, no actual cold nuclear fusion has so far been found; therefore most scientists are inclined to brand all this work as pseudoscience. However, although the claims of Pons and Fleischmann were almost certainly the results of experimental sloppiness, and cold fusion seems highly unlikely, pursuit of this evasive phenomenon may be useful. The famous physicist Ernest Rutherford was once asked why he allowed one of the researchers in his lab to conduct an obviously futile study. Rutherford is reported to have answered something like, "Let him do it. If he will not find what he is looking for, maybe he'll find something else which happens to be useful." The effort of those researchers who continue working on the phenomenon referred to as cold fusion, even if there seems to be little chance they will indeed discover a real cold nuclear fusion process, still may be part of good science; in the course of their work, novel methods of experimentation are being developed, and new, useful data, often having little relation to cold fusion per se, are being obtained.

Whereas bad science, under certain circumstances, may sometimes reconstruct itself into good science, pseudoscience is often an endeavor which essentially is not real science but rather something else disguised in quasi-scientific clothes for the sake of a certain agenda—usually having nothing to do with science and in fact often hostile to it. It may be harmful because of its ability to disguise itself as genuine science and thereby to force on scientists a defensive effort that wastes valuable human resources.

Without trying to define what distinguishes good science from bad science or pseudoscience, let us review a few examples of what clearly is pseudoscience. Probably the most egregious example of pseudoscience that had tragic consequences on an enormous scale is that of Marxism. Created in nineteenth-century Europe mainly by Karl Marx and Friedrich Engels, it was so successfully disguised as science that it won a large number of adherents despite its obvious contradiction to facts. Marxism comprised several parts, the first of which can roughly be defined as a philosophy of the so-called dialectical materialism, the second as the historical analysis of society's development (so-called historical materialism), and the third as

Marxist political economy. Dialectical materialism was an eclectic philosophical theory combining elements of Ludwig Feuerbach's materialism and Friedrich Hegel's dialectics. Neither Marx nor Engels contributed much to the development of these philosophical ideas, but they did reformulate those ideas into a neat set of simple statements which could be much easier comprehended by nonphilosophers (such as a vulgarized version of Hegel's principles of the transition of quantity into quality, or the negation of negation, etc.). (It is interesting to note that a frequent feature of pseudoscientific theories is that they often suggest a neat and simple scheme allegedly representing complex reality.) The core of Marxism is, though, its theory of the class struggle, whose essence is succinctly expressed by the maxim that the history of mankind is the history of the struggle of classes. Marxism reduces all the complexity of human history to this one factor. Marx and Engels did not seem to be worried by the fact that their representation of history was contrary to the histories of such countries as, for example, China or India, and reflected even the history of Europe only in a rather strained way. Such absurdities as interpreting religious wars as the results of the struggle of economic classes did not seem to make the creators of "scientific Marxism" pause and allow for the role of any factors besides class struggle in human history. They arbitrarily chose a very narrow subset of data from the much wider set of available data in order to fit the data into their preconceived theory. This is an example of pseudoscience, whose main fault is its theory is built accounting for only a tiny, deliberately selected fraction of the available factual evidence. However, since Marxism had all the appearance of a scientific theory, its predictions won wide popularity, with one of the consequences being the bloody revolution in Russia in 1917. Although the actual revolution and its consequences obviously did not fit the blueprint predicted by Marxism, the theory acquired the status of a godless religion, believed with a fanatical stubbornness despite its obvious futility and the bloodbath it caused. Indeed, as in many religions, one of the persistent claims by Marxism was that it was omnipotent because it was true.

Another example of a pseudoscience is Trofim Lysenko's pseudobiology, which was imposed by decree of the Soviet Communist rulers as the only true biology compatible with Marxism-Leninism. Lysenko's theory was hogwash, with no basis in factual data. Its imposition also had tragic consequences, as many genuine scientists were arrested and perished in the Gulag because they tried to refute Lysenko's pseudoscience by pointing to the numerous data which it contradicted.

One more example—this time rather comic—was the supposedly revolutionary theory of viruses suggested in 1949 in the USSR by a semiliterate veterinary physician named Georgiy M. Boshian.[3] According to that theory, which was officially approved by the Communist Party, viruses constantly convert either into bacteria or into crystals and back into viruses. Boshian's theory denied Louis Pasteur's acclaimed scientific achievements. Actually, Boshian's theory was based on fictional data. When, after many unsuccessful attempts, a commission of scientists finally gained access to Boshian's secret laboratory, all they found, instead of samples of the alleged virus-bacteria colonies, was plain dirt.

Other examples of pseudoscience include "creation science," with its distortion and misuse of facts, as well as its more recent and more refined reincarnation under the label of the intelligent design theory. As discussed in several previous chapters, this theory, promoted by a large group of writers, including many with scientific degrees from prestigious universities and with long lists of publications, and propagated at various levels of sophistication, has all the appearance of scientific research, as it offers definitions, hypotheses, laws, models, and theories like a genuine science. What is absent in the intelligent design theory, though, is evidence. No relevant data which would support its hypotheses, laws, models, or theories are found in the articles and books written by proponents of intelligent design—only unsubstantiated assumptions. Therefore it can justifiably be viewed as pseudoscience.

THE BUILDING BLOCKS OF SCIENCE

There are seven principal building blocks of science:

- Methods of experimentation
- Data
- Bridging hypotheses
- Laws
- Models
- Cognitive hypotheses
- Theories

This list may create the impression that science can be presented by a neat scheme, as a combination of clearly distinguishable building blocks. However, no such neat scheme exists. The building blocks of science can overlap, intersect, and emerge in an order different from the list presented above. However, the division of the body of science into supposedly separate building blocks, besides providing convenience in analyzing this complex and multifaceted form of human endeavor, also reflects its real composition, even though the boundaries between its constituents may sometimes be diffuse.

Methods of Experimentation

The statement that methods of experimentation form the engine of the progress of science seems to be trivial. Although this statement is especially transparent in physics, chemistry, biology, and engineering science, it is valid as well for any other science, even if it may sometimes be not as obvious.

When Italian astronomer Galileo Galilei built his telescope, it provided an enormous impetus to the development of astronomy. When the Hubble telescope was put in orbit, astronomy underwent another powerful push forward. Similarly, when Dutch scientist Antonie van Leeuwenhoek built his microscope, it revolutionized biological science. When the first electron microscope was built, it again revolutionized several fields of science, leading to amazing modern achievements such as actually seeing and even manipulating individual atoms.

These well-known examples are only the tip of the iceberg, because the progress of science is pushed forward by the multiple everyday inventions of various ingenious methods of experiment or observation.

In thousands of research laboratories, scientists whose names are usually unknown outside the narrow circle of their colleagues, apply their ingenuity and inventiveness every day to creating new, ever more subtle methods of questioning nature. Without these often very ingenious methods and sophisticated experimental setups, there would be no progress of scientific penetration into objective reality, which often jealously guards its secrets.

For example, in the late seventies, at one of the foremost research institutions in the United States, a group of scientists was developing certain types of magnetic memory for computers. In one part of that study a magnetic wire was pulled at a certain speed through a tube filled with an electrolyte, and electrochemical processes on the surface of the wire were investigated. The researchers encountered a problem: as the electrolyte was pumped through the tube, the friction between the electrolyte and the walls of the tube caused the pressure to drop along the wire, distorting the data.

Considerable effort was invested in finding a way to eliminate or at least alleviate the described effect, but the effect persisted. Then a guest scientist, who had never before dealt with the experiment in question, was invited to attend a research meeting (which I attended). At the very first meeting where the detrimental effect of the pressure drop was discussed, the guest scientist looked at the diagram of the tube and the wire moving through it and said, "Why, instead of pumping the electrolyte through the tube, eliminate the pump, make a rifle-type spiral groove on the inner surface of the tube, and put the tube into rotation. The groove will push the liquid, whose pressure will be automatically kept constant all along the wire."

This is an example of how a fresh look at a problem may suddenly solve a seemingly difficult problem in a simple way, but this example is typical of what often happens in thousands of research institutions.

The unstoppable development of experimental methods, from designing giant accelerators of elementary particles to small improvements in measuring and observational methods, is what underlies the progress of science.

Therefore, the statement that a revolution in science starts when the experimental technique proceeds to the next digit after the decimal point (increasing sensitivity tenfold) reflects an important aspect of the progress of science.

Data

Definition

In the words of the famous Russian philosopher Ivan Pavlov, facts are the air a scientist breathes. The term *data* is synonymous with *facts*. It is also synonymous, in a certain sense, with *evidence*. Without data there is no science, only pseudoscience. There are two main types of data, *observational* and *experimental*. The demarcation between observation and experiment is not quite sharply defined, since observation may be a natural part of a designed experiment.

Observation, and even some primitive form of experimentation, may also occur in nonscientific activities. In a popular comic Russian poem for children, a Russian Orthodox priest counts crows sitting on trees, just for the heck of it. This activity is an observation and is accompanied by a measurement, but this fact does not make the priest's activity scientific. Scientific observation or experimentation, while driven mainly by curiosity, always has a conscious or subconscious purpose—to establish facts which may shed light on the intrinsic structure and functioning of the real world.

A crucial element of a scientific experiment is measurement. There is a theory of measurement which teaches us about the precautions necessary to ensure the reliability of the measured quantities and about the proper estimate of unavoidable errors of measurement. This theory is beyond the scope of this chapter, but we have to realize that, on the one hand, data obtained via a properly conducted experimental procedure are reasonably reliable, but on the other hand they are always only approximately true and cannot be relied upon to an extent exceeding that determined by the properly estimated margin of error.

Astronomy presents an example of a science wherein observation is by far more prevalent than experimentation. A classic example of valuable observational data is the tables of the planets' positions compiled by Danish astronomer Tycho Brahe over the course of many years of painstaking observations accompanied by meticulous measurements. These tables, after they wound up in the possession of German astronomer Johannes Kepler, served as the data the latter used to derive his famous three laws of planetary motion. I will discuss later the transition from data via hypotheses to laws, and from laws to theories. My point now is to illustrate how reliable data, gained via a proper observational procedure accompanied by measurement, became a legitimate part of a scientific arsenal.

A classic example of the experimental acquisition of data is the discovery, in the early years of the twentieth century, of superconductivity by a group of researchers in Leiden, the Netherlands, headed by Dutch physicist Kamerlingh Onnes. These researchers systematically measured the electric resistivity of various materials at lower and lower temperatures. In a specially designed experiment, samples of material were gradually lowered into a Dewar flask on whose bottom was a puddle of liquid helium while their electric resistance was measured. Analysis of the data obtained led to the conclusion that at certain very low temperatures, the electric resistivity of certain materials dropped to zero. This unexpected result seemed contrary to the understanding of electric resistance prevalent at that time; however, data take precedence over theories. The scientists of Leiden, however puzzled and astonished by their data, came to believe in the reliability of their data and thus announced the discovery of superconductivity. Their data were then reproduced by other scientists, and a new law asserting the fact of superconductivity was formulated. The formulation of a law did not mean a theory of superconductivity was forthcoming—it took half a century for a theory of superconductivity to be developed. Neither the law nor the theory of this phenomenon would have been possible without first acquiring reliable data on the behavior of superconducting materials.

As we will see, the path from data acquisition to a law can be arduous and prolonged. It involves steps requiring imagination and inventiveness, because no law automatically follows from data. This is even truer for the path from data, via law, to theories. However, only those laws and theories which stem from reliable data are constituents of genuine science.

To illustrate this last statement, I would like to cite an article written in 1994 by three Israelis, Doron Witztum, Eliyahu Rips, and Yoav Rosenberg (collectively referred to as WRR), for the mathematical journal *Statistical Science*.[4] This article caused a prolonged and heated discussion. As a result of a thorough analysis of WRR's methodology, the overwhelming majority of experts in mathematical statistics concluded that WRR's data were obtained through a procedure which in many respects was contrary to the rules of proper statistical analysis. In other words, the community of experts concluded that WRR's claim stemmed from *bad* data. Therefore, WRR's work was rejected as *bad* science. It was not, though, originally rejected as *pseudo*science, because WRR based their conclusion on certain data to which they applied a statistical treatment (although this was partially flawed as well). However, despite the scientific community's almost unanimous rejection of WRR's work as based on unreliable data, two of them (Witztum and Rips), as well as a small circle of their adherents, stubbornly continued to insist that they made a genuine discovery, an insistence that converts their theory from bad science to pseudoscience. Because their alleged Bible code is quite educational with regard to how pseudoscience appears and persists, it is discussed at length in chapter 14.

On the other hand, claims such as those by Immanuel Velikovsky have been rejected from the very beginning as pseudoscience because they did not stem from any data but only from arbitrary assumptions. In his book *Worlds in Collision*,[5] Velikovsky offered a bunch of wild theories supposedly explaining many mysteries that contemporary science could not explain. For example, one of his suggestions was that when, according to the biblical story, Yehoshua (Joshua) stopped the sun in the sky, the earth indeed stopped its rotation. Another hypothesis by Velikovsky postulated a near-collision of Venus and Mars with Earth in order to explain numerous biblical miracles. Of course, there exists no data to serve as evidence for such theories; therefore they were justifiably relegated to pseudoscience. While Velikovsky acquired substantial notoriety and was compared in nonscientific publications to Newton, Einstein, and other great scientists, no scientific magazine accepted his papers because they, while plainly contradicting Newtonian mechanics, did not offer a shred of evidence to support his claims.

Reproducibility and Causality

To be legitimately useful in science, data must meet several requirements, one of which is reproducibility. Neither the reputation of the scientists claiming certain experimental results nor the impressive appearance of data seemingly conforming to the strict requirements of a properly conducted experiment are sufficient for the data to be accepted as a contribution to science. Data become a part of science only after they have been reproduced by other scientists. Indeed, the rejection of the original cold fusion hypothesis by the majority of physicists was due precisely to the fact that other groups of researchers could not reproduce the data claimed by Pons and Fleischmann. Likewise, the data claimed by Witztum and Rips could not be reproduced by other scientists, which was the main reason their theory was rejected by the scientific community.

The requirement of reproducibility is based on the assumption of causality as a universal law of nature. This assumption presupposes that reproducing certain experimental conditions must necessarily lead to reproducing the outcome of the experiment. This supposition is of course a philosophical principle to be accepted a priori, with an ancient origin; it was already discussed in detail by Aristotle, who introduced the concept of a hierarchy of four causes: the so-called material causes, the efficient causes, the formal causes, and the final causes. In more recent times, the principle of causality known as the *principle of determinism* was formulated by French scientist Pierre Laplace and has been universally accepted in science for at least two centuries.

The advent of quantum mechanics seemed to shatter that principle. The very fact that many outstanding scientists were prepared to abolish a principle that had been a foundation of experimental science for so long testifies against the claims by adherents of intelligent design theory who assert that scientists are dogmatically adhering to "icons" of metaphysical concepts rather than keeping open minds.

However, the interpretation of quantum-mechanical effects as a breach of causality is not the only one being discussed. Another interpretation is that the seeming breach of causality may simply testify to the insufficient understanding of submicroscopic processes wherein causality appears to be absent.

First of all, even in the case of quantum-mechanical events, causality is obviously present as long as the macroscopic manifestations of those events are observed. For example, causality seems to be a problem in experiments with microscopic particles passing through slits in a partition. If only one slit is open, on the screen behind the partition a diffuse image of the slit is observed. If, however, two slits are open, the image on the screen is a set of fringes. On the macroscopic level, the outcome of the experiment is predictable and consistent with the principle of causality. Indeed, the image on the screen is always the same if only one slit is open, and can be reproduced at any time, anywhere in the world. If two slits are open, the picture on the screen is different from the case of only one opened slit, but it is predictable and can also be reproduced at any time, anywhere in the world.

However, the validity of causality becomes an issue when the details of the event are considered on a microscopic level. If an electron is a particle, it cannot pass simultaneously through two slits. Therefore we could expect that in the case of two opened slits, the image on the screen would be two separate diffuse images of slits rather than a set of fringes. Indeed, how can an electron passing a particular slit "know" if the other slit is open? However, the behavior of individual electrons is different, depending on whether the other slit is open.

Thus, the results of the described experiments are sometimes interpreted as indicating that the same microscopic conditions lead to different outcomes, depending on the variations in macroscopic conditions; thus causality is absent.

Quantum mechanics tells us that the explanation for this phenomenon is that electrons are not particles like those we can see, touch, etc. They are very different entities that, under certain conditions, behave like waves rather than like particles. It does not seem possible to visualize an electron, because it is unlike anything we can interact with by means of our senses in our macroscopic world. Of course, it is well known that macroscopic waves (for example, sea waves) can very well pass several openings simultaneously. Therefore, the two-slit experiment is not necessarily an indication of the absence of causality on a microscopic level, but possibly only an indication of our inability to visualize a subatomic particle.

Physicist Richard Feynman has said that nobody understands quantum mechanics.[6] To interpret this statement, we have first to agree what the term *understand* means. We can't visualize an electron, and in this sense we may say that we don't understand its behavior. However, we can describe an electron's behavior reasonably well using the mathematical apparatus of quantum mechanics, and we can successfully predict many features of that behavior. In this sense, we can say that we do indeed understand quantum mechanics.

Since we can't visualize the intrinsic details of an electron's behavior, we are actually uncertain of whether, on the microscopic level, its behavior is indeed nondeterministic, or we simply don't know what causes this or that seemingly random path of the electron. And since we are uncertain, it seems unnecessary to assert categorically that the principle of causality breaks down at the microscopic level.

Let us look at another situation wherein causality is often claimed to be absent. If we have a lump of radioactive material, we can easily measure the rate of its atoms' decay. This rate is a constant for a specific material. For example, we can find experimentally that, in the course of a prolonged experiment, the fraction of the atoms that decayed every second was, on the average, 0.01. Thus, if initially the lump consisted of N_0 atoms, one second later the number of atoms of that material was $N_1 < N_0$, one more second later, the number of atoms was $N_2 < N_1$ etc., the ratios N_1/N_0, N_2/N_1, etc., averaged over the duration of the experiment, would be 0.99. This result is quite reproducible; thus on the macroscopic level causality again seems to be intact. However, if we discuss the phenomenon on a microscopic level, the question arises of why at any particular moment a particular atom decays while other atoms do not. All atoms within the lump of material are, from a macroscopic viewpoint, in the same situation. There is a certain probability of an atom's decay, which is the same for every atom in that lump of material; however, at each moment some atoms decay, whereas others do not. Therefore, the argument goes, causality does not seem to be at work in that process.

The problem of possible "hidden parameters" affecting such microscopic processes is the subject of a rather heated discussion among scientists, but so far no consensus has been reached. For example, Feynman maintains that an explanation referring to "hidden parameters" is impossible, because nature itself "does not know in advance" which atom will decay at which moment, or which path a particular electron will choose at a given moment.[7] While Feynman's view seems to be the most widely shared by physicists, there also are dissenters who are inclined to assume the existence of "hidden parameters." I will not express here my personal view of the problem because it seems to be beyond the main theme of my discussion. For the purpose of this discourse it seems sufficient to assert that causality seems to be always in effect on a macroscopic level, thus allowing the requirement of reproducibility of experimental data to remain valid in any legitimate research.

Ceteris Paribus

The next question to be discussed in relation to data as a foundation of science is the so-called principle of *ceteris paribus,* a Latin expression meaning "everything else equal." The idea of that principle is as follows. Imagine that in a certain experiment the behavior of a quantity x is studied. This quantity is affected by a number of factors, referred to as A, B, C, D, etc. Some of these factors may be known, while some others may not be. From a philosophical standpoint, everything in nature is interconnected, and therefore the number of factors affecting x is immensely large, so we never can account for all of them. However, an overwhelming majority of this multitude of factors have only a very minor effect on x, and therefore, in practical terms, most of them are of no consequence for the study at hand. Still, there is a set of factors whose effect on x cannot be ignored if we want to gain meaningful data from our study.

The principle of ceteris paribus is a prescription regarding how to conduct the study. According to this principle, the process of research must be divided into a number of independent series of experiments. In each series only one of the factors—A, or B, or C, etc. has to be controllably changed, while the rest of the factors must be kept unchanged. For example, if A is being changed, B, C, D, etc. have to be kept constant, and the behavior of x has to be recorded for each value of A. Then another series of experiments has to be conducted, wherein only factor B is being changed and x recorded while A, C, D, etc., are kept constant.

Of course, the application of the described procedure is based on the assumption that factors A, B, C, etc., are independent of each other, so that it is possible to change at will any one of these factors without causing the rest of them to change simultaneously.

Unfortunately, there are two problems inseparable from this approach: One is related to errors of measurement, and the other to the very core of the ceteris paribus principle, i.e., to the assumption that each factor is independent of the others.

The problem of errors stems from the fact that no measurement can be absolutely precise and accurate (the difference between precision and accuracy is defined in the theory of measurement and is beyond the scope of this discourse). That is to say, even if factors A, B, C, etc. are independent of each other, it is impossible to guarantee that while one of them is being changed, the rest of them remain indeed absolutely constant, because the values of *B*, *C*, *D*, etc., while in principle remaining constant, cannot be measured with absolute precision and accuracy. Moreover, the values of factor *A* itself, which are supposed to be changed in a controlled way, are never guaranteed to have precisely the desired values, and, finally, the target of the study, *x* itself, cannot be measured with absolute precision and accuracy.

The problem with the principal foundation of ceteris paribus methods stems from the fact that in many systems, factors *A*, *B*, *C*, etc. are in principle *not* independent of each other. Hence, as factor *A* is being changed, it may be impossible *in principle* to suppress the simultaneous changes of factors *B*, *C*, *D*, etc., which are supposed to remain constant. Such systems are sometimes referred to as "large," "complex," "diffuse," or "poorly organized," the study of which can only be conducted beyond the confines of ceteris paribus.

In general terms, all systems are complex, such that factors *A*, *B*, *C*, etc. are always interdependent. However, the extent of their interdependence may be insignificant to a degree which makes the assumption of ceteris paribus reasonable. For example, assume we study the behavior of ice/water in the vicinity of 273 K, which is, of course, 0°C or 32°F. The factor we change is temperature. Besides temperature, the behavior of ice/water is affected by a multitude of other factors. Most of them are insignificant and can be ignored because they have only a very minor effect. However, at least two factors have a pronounced effect—pressure and the purity of the water. In this case it is relatively easy to ensure ceteris paribus: we can keep a sample of ice in a vessel where pressure is automatically controlled and kept at a predetermined level, regardless of changes in temperature. Likewise, the purity of ice can be maintained at a constant level at all values of temperature investigated. In such a study we find that ice containing, say, not more than 0.01 percent impurities, at a pressure of about 10^5 pascal, melts at about 273 K.

Errors

We have to distinguish between various types of errors. Sometimes errors stem from a fundamentally wrong approach to scientific inquiry. It may be due to an insufficient understanding of the subject by a researcher, which results in an improperly designed experiment; to a preconceived view or belief; or to a strong desire on the part of a researcher to confirm his preferred theory. The researcher may subconsciously ignore results contrary to his expectations and inadvertently choose from the set of measured data only those which jibe with his expectations. Errors of this type, while by no means uncommon, should be viewed as pathological; they produce *wrong* data, one of the constituents of bad science. An example of such bad data is the case of the alleged Bible codes discussed in chapters 6 and 14.

However, even in the most thoroughly designed experiments, errors inevitably occur, despite the most strenuous effort on the part of a researcher to avoid them. These *honest* errors are due to the inevitable imperfection of experimental setups and to the equally inevitable imperfection of the researcher's performance.

Until recently, a common notion in the theory of experiment was the distinction between systematic and random errors. While this distinction is logically meaningful and often can be made quite rigorously, lately it has become evident that on a deeper level the demarcation between these two types of error is rather diffuse.[8] Moreover, the error can often be viewed as either systematic or random, depending on the formulation of the problem. For example, an error can be viewed as systematic if we always conduct measurements in laboratory *A*. However, the same error can be viewed as random if we have a choice among many laboratories, *A*, *B*, *C*, etc., and choose laboratory *A* to conduct the measurement on a particular day, by chance. The deeper analysis of that question, which is a proper subject of the theory of experiment, is beyond the scope of this chapter.

The question of experimental errors is, though, related to the problem of causality. If data are a legitimate component of science only if they are reproducible, the immediate question is: What are the legitimate boundaries of reproducibility? It is an indisputable fact that the expression "precise data," if interpreted literally, is an oxymoron. If an experiment is repeated many times, in each run the measured data will be to a certain extent different. There is no way to ensure absolute reproducibility of data. The conventional interpretation of that fact is not that it negates the principle of causality but rather that in each experiment there are factors that uncontrollably change, despite the most strenuous effort on the part of the experimenter to keep all conditions of the experiment under control.

Therefore the principle of causality is not a direct product of experimental evidence but rather a metaphysical principle borne out by the total body of science. Science is based on the assumption of causality, without which most of it simply would make no sense.

Therefore the inevitable errors of measurement are something science has to live with, making, though, a very strong effort to distill objective truth from the chaos of measured numbers. Data that form a legitimate foundation of laws and theories are therefore the result of a complex process through which the wheat is separated from the chaff and regularities have to be extracted from the error-laden experimental numbers in order to build a bridge from data to laws.

In connection with the question of experimental errors, the notion of *noise* seems to be relevant. The term *noise* entered science from information/communication theory. This branch of science studies the transmission of information (in the form of "signals") inevitably accompanied by noise in the elements of the transmission channel. The noise is superimposed on the useful signal and has to be filtered out if the signal is to be reliably interpreted. The term *noise* has gradually gained a wider use, referring to any unintended extraneous input to the experimental setup. Normally, to extract useful data from the experimental output the desired data have to be distilled from the overall output by excising noise. However, noise has happened to be the source of scientific discoveries more than once. For example, when, at the end of the nineteenth century, physicists studied the processes in a cathode ray tube, these processes were always accompanied by the emission of a certain type of radiation from the walls of the tube. Some researchers did not notice these rays; others did but viewed them just as noise and ignored them. Then German physicist Wilhelm Röntgen, who specialized in the study of dielectrics, and was generally highly meticulous in his research, looked at the "noise" as a phenomenon interesting in itself. He thoroughly studied the phenomenon and thus discovered a new type of radiation which he called X rays.

Another well-known example of how a phenomenon which was viewed as noise led to an important discovery is that of background cosmic radiation discovered in 1965 by Arno Penzias and Robert Wilson.[9]

Generally speaking, if an experiment yields results which could be predicted, this adds little to the progress of science. By contrast, the matter becomes really interesting if the results of an experiment look absurd. If a researcher encounters an absurd result that persists despite efforts to clean up the experiment and to eliminate possible sources of error, there is a good chance the researcher has come across something novel and therefore interesting. Absurd outcomes may often be attributed to noise, but sometimes that noise carries information about a hitherto unknown effect or can be utilized in an unconventional manner for a deeper study of the phenomenon.

Let me give an example from my personal experience. Many years ago I studied the process of adsorption of various molecules on the surface of metallic and semiconductor electrodes. One of the methods of that study was the measurement of the so-called differential capacity of the electric double layer on the electrode surface. A couple of my students built a setup enabling us to measure the differential capacity at various values of the electric potential imposed on the electrode. After having worked with that setup for several weeks, they became frustrated by the instability of the measured capacity. Each time they changed the value of the electrode potential, the differential capacity started shifting, and they had to wait for many hours and sometimes days until a new, uncertain equilibrium seemed to set in. They viewed this unpredicted "instability" of the measured differential capacity as noise preventing them from performing the desired measurement. They complained to me about their frustrating experience. I remember that moment when a kind of sudden light exploded in my mind and I shouted, "We got an unexpected result. The capacity creeps—and this means that if you measure the curve of capacity versus time, this curve will contain information about the kinetics of the adsorption process!" I set out to develop equations reflecting the capacity's relaxation. A new method of scientific study we named potentiostatic chronofaradometry was thus born. Instead of trying to eliminate the supposed noise and measure the supposed equilibrium values of the differential capacity, now my students concentrated on measuring those very relaxation curves which they angrily considered to be annoying noise. Various data on the kinetics of the adsorption/desorption process were extracted from that noise, assisted by the equations derived for that process.[10]

Complex Systems

For over two hundred years, one of the underlying hypotheses of science, often unspoken, was that every system that is a subject of a scientific study is "well organized," in the sense that it was always possible to separate a few phenomena or processes of similar nature, dependent on a limited set of important factors. The system's behavior could be then studied by isolating factors one by one, finding the functional dependencies between pairs of factors, and attributing to those functional dependencies the status of laws. In the twentieth century the described approach started encountering serious difficulties.[11] In many systems the separation of various factors turned out to be impossible. Science encountered the ever-increasing number of what are referred to as "large," "poorly organized," or "diffuse" systems, etc., these labels reflecting various aspects of such systems' behavior.

It should be noted that the concept of a large or poorly organized system is not equivalent to the concept of *stochastic* systems. This type of system was well known and successfully dealt with in nineteenth-century physics research. An example of a stochastic system is a gas occupying a certain volume. It consists of an enormous number of molecules (for example, at atmospheric pressure and room temperature, one cubic centimeter of gas contains about 10^{19} molecules). The behavior of each molecule is determined by the laws of Newtonian mechanics. However, because of the immense number of molecules, it is impossible to analyze the behavior of a gas by solving the equations of mechanical motion for each molecule. Therefore, nineteenth-century scientists came up with very powerful statistical methods which enabled them to analyze the behavior of a gas without a detailed analysis of the motion of each individual molecule. (There are theories according to which a system already becomes stochastic, i.e., not treatable by studying the behavior of its individual elements, if the number of these elements exceeds about thirty.)

However, stochastic systems such as gases behave stochastically only on a microscopic level. On a macroscopic level they can be treated using the ceteris paribus approach. Macroscopic properties of a gas can be studied by isolating and changing these properties in a controlled manner one by one while keeping the rest of the properties constant. This is possible because the immense number of elements of such systems does not translate into a large number of macroscopic properties. A gas containing a quadrillion quadrillion microscopic molecules, all of them identical, macroscopically is characterized only by a few parameters, such as pressure, temperature, and volume.

On the other hand, large or poorly organized systems are characterized by a large number of macroscopic parameters, many of which cannot be individually varied without simultaneously changing some other parameters affecting the system.

Historically, one of the situations whose study mightily served the realization of the ineliminable interdependence of various parameters in a poorly organized system is emission spectral analysis.[12] In this process, a sample of material to be studied is placed between two electrodes into which a high voltage is fed. The breakdown of the gap between the electrodes causes a spark discharge wherein the temperature reaches tens of thousands of Kelvin. The density of energy at certain locations on the electrodes reaches a huge level. An explosive evaporation occurs, creating a cloud of evaporated matter between the electrodes in which various processes including diffusion, excitation, and radiation take place simultaneously. Besides, oxidation-reduction processes occur on the electrodes' surface, and diffusion of elements occurs in their bulk, dependent on the temperature gradient, phase composition of electrode material, defects, etc. Attempts to analyze this extremely complex combination of phenomena by means of separating various factors one by one turned out to be in vain. The system is too complex, forcing scientists to give up attempts to proceed using the ceteris paribus approach.

To study large or complex systems, the science of the twentieth century developed two main approaches: one is the multidimensional statistical approach, and the other is the cybernetic (or computer modeling) approach.

The statistical approach was to a large extent initiated by the British mathematician Ronald A. Fisher. It is often referred to as multidimensional mathematical statistics (MDMS). Essentially, MDMS is a logically substantiated formalization of such an approach to the study of large systems wherein the researcher deliberately avoids a detailed penetration into intricate mechanisms of a complex phenomenon, resorting instead to its statistical analysis using multiple variables.

The application of MDMS requires solving a number of problems involving the appropriate strategy of an experiment—the proper choice of the essential parameters—since accounting for too many parameters may make the task beyond the available intellectual and computational resources, while accounting for too few parameters may reduce the solution to a triviality. The effectiveness of MDMS has been drastically improving with the availability of ever more powerful computer hardware and software. Statistical models of complex systems have been successfully employed for a multitude of problems which could not be tackled by the traditional methods of direct measurement.

The cybernetic approach, whose origin can be attributed to U.S. mathematician Norbert Wiener, is in some respects very different from the statistical one, although it also has become really useful only with the advent of powerful computers. While MDMS and computer modeling are principally different, both are successfully applied, sometimes to the study of the same complex system.

One of the frontiers of modern science is the field of artificial intelligence, which exemplifies the cybernetic approach to a complex system—in this case, human intellect—by using a computer-generated model to describe it. (Biological systems are mostly large or poorly organized in the sense relevant to this discussion.)

Another example of a large system is the economy of any country. In every country there is a multitude of factories, farms, companies, and individuals each pursuing their own particular economic goals. All these elements of the economic system interact with each other in an immensely complicated web of transactions, changing every moment such that this enormously complex system is in a constant state of flux. Trends in such systems are usually not obvious. It is impossible to account for each feature of that immensely complex game, not only because of its sheer size and the huge number of its constituent elements, but also because of its unstable character. Before the available data about the state of economy have been digested and interpreted, they have already changed. No computer, however powerful, can follow all the nuances of the economic game. Therefore the economy cannot in principle be "scientifically managed" as it supposedly was in the allegedly socialist "planned economy" of the former USSR. This was one of the reasons for its abysmal failure—it was no more scientific than it was really socialist. The actual economic system in the former USSR was state capitalism, with all the drawbacks of an extreme form of a monopolistic capitalist system but without the advantages of free enterprise. The attempts to sustain a planned economy allegedly based on scientific analysis of the available resources, demands, and supplies resulted in a seemingly controlled but actually chaotic economy wherein stealing and cheating became the only possible means of survival for the powerless slaves of the state.

A classic example of a large system, the economy of the modern world functions in a cyclic manner, following its own poorly understood complex statistical laws wherein its oscillatory motions have an immense inertia, such that attempts to steer it in this or that direction (for example, by regulating the interest rate or changing the tax laws) have only a marginal effect on the waves of prosperity and recession replacing each other as if by their own will. Analytical approaches such as those outlined in this section are the most potent tools we have for gaining a deeper understanding of such complexity.

Bridging Hypotheses

Perhaps the title of this section should more appropriately be "Bridging Hypothesis," in the singular, because this hypothesis is essentially always the same. This type of hypothesis constitutes a "bridge" from the raw data to the postulating of an objective law.

In the case of simple or well-organized systems, which can be studied under the ceteris paribus condition, the bridging hypothesis appears in its most explicit form, while in the case of large or poorly organized systems it can often be hidden within the complex web of the data themselves.

Imagine an experiment performed with the assumption of ceteris paribus wherein a set of values of a target quantity x has been measured for a set of values of a parameter a, so that for each value a_1, a_2, a_3, etc., corresponding values x_1, x_2, x_3, etc., have been measured and also the margin of error $\pm\Delta$ has been estimated so that every value of x_i is believed to be within the margin of $x_i \pm \Delta$.

Every set of x and a, however extensive, is always still only a selected subset of all possible values of x and a. Reviewing the sets of measured numbers, we try to discern a regularity connecting a and x. To do so, we have necessarily to *assume* that the set of actually measured numbers indeed reflects an objective regularity. There is never unequivocal proof that such a regularity indeed objectively exists; we have to postulate it *before* a law can be formulated.

More often than not, such a hypothesis is implicitly present without being explicitly spelled out. In scientific papers we usually see statements introducing a law without mentioning the bridging hypothesis. The researchers reporting their results routinely assert that, for example, "As our data show, in the interval between a_1 and a_k, x increases proportionally to a." This statement is that of a law. In fact, though, stating such a law might not be legitimate without first postulating the very objective existence of a law connecting x to a. The bridging hypothesis according to which the observed sets of values of x correspond to the selected sets of values of a, not as an accidental result of an experiment conducted under a limited set of conditions but because of an objectively functioning law, must necessarily precede, if often subconsciously, the statement of a law.

The bridging hypothesis—which is necessarily present, most often implicitly, in the procedure of claiming a law—is never more than a hypothesis and cannot be proven. Fortunately it happens to be true, at least as an approximation, in the vast number of situations, thus constituting a reliable building block of science. In fact, if sets of values of a and x seem to match each other according to some regularity, more often than not such a regularity does indeed exist, although sometimes it may also happen to be illusory.

Of course, the bridging hypothesis is not applied if no law is postulated. A researcher may obtain sets of values of the measured quantity x corresponding to a set of values of parameter a that may look like a display of regularity, but for various reasons she may reject the bridging hypothesis and avoid an attempt to spell out a pertinent law. The reasons for that may be, for example, some firmly established theoretical concepts making the law in question very unlikely, or doubts regarding the elimination of some extraneous factors which could distort the data, thus creating the false appearance of a regularity—or any number of other reasons.

More often, though, a scientist may be just uncertain whether to accept or reject the bridging hypothesis rather than flatly rejecting it. Often such uncertainty is based simply on insufficient data. The researcher is uncertain whether or not the available sets of data do indeed properly represent the entire multiplicity of possible data. If that is the case, the proper remedy is to resort to a statistical study of the phenomenon in question. The proper tool for such a study is the part of mathematical statistics called hypotheses testing.

Hypotheses testing involves introducing two competing hypotheses, one called the *null hypothesis*, and the other the *alternative hypothesis*. The null hypothesis is the assumption that the available set of data does not represent a law. The alternative hypothesis is in this case a synonym for what I called the bridging hypothesis, i.e., the assumption that the available set of data reflects a law. As a result of a proper statistical test, the researcher compares the likelihood of the null hypothesis versus that of the alternative hypothesis. The hypothesis whose likelihood is larger is accepted, while the other hypothesis is rejected (see chapter 14 for hypotheses testing as applied in the Bible code controversy). This choice is never assured to be ultimately correct, since discovery of additional data may change the likelihood of the competing hypotheses. In this sense, laws of science are usually held to be tentative.

Fortunately, the rigorous self-verification inherent in proper scientific procedure makes a vast number of laws of science work very well despite their tentative character. In the next section we will discuss laws in greater detail.

Laws

Whereas the bridging hypothesis discussed in the previous section is simple and uniform in that it simply assumes that the set of experimental data reflects a law, the specific formulation of a law itself is far from being simple and uniform. While in every scientific procedure the same bridging hypothesis is present, this in itself contains no indication of how the particular law has to be spelled out.

Formulating a law is not a mechanical process of stating the evident behavior of the studied quantities; it involves an interpretive effort on the part of the scientist, who makes choices among various alternative interpretations of the assumed regularity that have to be extracted from arrays of numbers defined within certain margins of error.

In the simple case of a functional dependence between two variables, the scientist is confronted with a set of experimental points which seem to display certain regularity, often rather nebulously because of inevitable experimental errors. Hence, besides the bridging hypothesis, which simply postulates the existence of a law, the scientist must then postulate the specific law itself.

Consider an example. In Coulomb's well-known law of interaction between point electric charges, the force of interaction is assumed to be inversely proportional to the squared distance between the charges.[13]

This assumption is, however, not a direct conclusion from the experimental data. One of the reasons for this is the inevitability of errors occurring in every procedure of measurement. If the dependence of Coulomb's force on the distance between the point charges is measured, it is assumed that 1) while the distance is being changed, the interacting charges are indeed being kept constant; 2) no other charges which could distort the results are anywhere close to the measuring contraption; 3) the distance itself is measured with sufficient accuracy; etc. All of this is, of course, just an approximation. Let us imagine that in the course of an experiment in which Coulomb's force was measured, the distance between the supposedly *point* charges was given the following values (in whatever units of length): 1, 2, 3, 4, 5, and 6. The total of six points at which these values were determined with an error not exceeding, say, ±5 percent. This means that when the distance was assumed to be 2, it could actually be anything between 1.9 and 2.1, and similarly for every other value of distance. Imagine further that the measured Coulomb's force (in whatever units of force) happened to have the following set of values, averaged over many repeated measurements: 1097, 273.87, 121.64, 68.97, 43.12, and 31.01. Repeating the measurement many times, the experimentalist evaluated the margin of error for the force to be ±10 percent. Reviewing the set of numbers for the values of the measured force, the researcher notices that these numbers are rather close to a set of numbers obtained by dividing the maximum force of 1097 (measured for the distance of 1) by the squared distances. Indeed, dividing 1097 by the squared values of distances 2, 3, 4, 5, and 6, one obtains the following set: 274.25, 121.89, 68.56, 43.88, and

30.47. Every number in the experimentally measured set differs from a corresponding number in the second, calculated set, but by not more than 10 percent, which is within the margin of error for that measurement. Although none of the measured numbers exactly equal the numbers calculated upon the assumption that the force drops inversely proportional to the distance, the scientist *postulates* that the difference between the measured and calculated sets is due to experimental error and that the real law, hidden behind the measured set of numbers, is indeed the inverse proportionality between the force and the *squared* distance.

What is the foundation of that postulate? If instead of the power of 2, the distance in the same formula appeared with the power of, say, 2.008 or 1.9986, such a formula would describe the experimental data as well as Coulomb's law, where the power is exactly 2. Of course, the precision and accuracy of the measurement can be improved so that the margin of error is substantially reduced. However, it can never be reduced to zero. With a much improved technique we may be able to assert that Coulomb's force changes inversely proportional to the distance to the power being, say, between 1.9999999987 and 2.0000000035, but we can never assert, based just on the experimental data, that the power is exactly 2. We have to *postulate* that the power is exactly 2 rather than any other number between 1.9999999987 and 2.0000000035.

Such a postulate is often made on nonscientific grounds. It is often based on some metaphysical consideration or some philosophical principle. We actually have no scientific grounds to prefer the power of 2 to, say, the power of 2.0000986345, since both numbers fit the experimental data equally well. We choose 2 not because the data directly point to that choice but because this choice seems to be either simpler (i.e., using Ockham's razor[14]), or more elegant, or more convenient, or maybe just favored by a particular scientist for purely personal reasons.

Furthermore, even the principal form of the law—the power function—is not logically predetermined by the data. There are many formulas differing from the simple power function which would yield numbers within the margin of error of the experimental data. For example, it is always possible to construct a polynomial expression whose coefficients are chosen in such a way as to make the numbers calculated by that expression match the measured data within the margin of the experimental error. Hence, not only the power of 2 in the formula of Coulomb's law, but even the form of the law itself is the result of a scientist's assumption.

From the above discussion it follows that the laws of science are necessarily postulates. We don't *know* that the distance in Coulomb's law must be squared, we *postulate* it.

The choice of a postulate is limited by two considerations. One is the requirement not to contradict experimental data, so we may not arbitrarily assume that the distance in the formula in question must have the power of, say, 2.75, because it is contrary to evidence. The other requirement is not to contradict the overall body of scientific knowledge. Otherwise any number which is within the margin of the experimental error has, in principle, the same right to appear in the formula of law, and the choice of 2 instead of, say, 1.99999867 is a postulate based on metaphysical grounds.

The application of some mathematical apparatus such as, for example, least square fit, does not principally change the situation, because the output of the mathematical machine is only as good as the input.

This discourse may create an impression of arbitrariness of the laws of science. Some philosophers of science think so. Are the laws of science indeed arbitrary?

They are not—the fault of the discourse presented here is that it concentrates on data obtained in a particular experiment as if these data exist in a vacuum. In fact, every experiment adds only a very small chunk of new information to the vast arsenal of scientific knowledge, and the interpretation of any set of data must necessarily fit in with the entire wealth of science.

If we review the example of Coulomb's law, we will see that the choices both of the power function for distance between the interacting point charges and of the exact value of the power as 2, although resulting from a postulate, were not really arbitrary if the results of Coulomb's experiment are viewed from a broader perspective, accounting for information stemming from outside of the experiment itself. Support for the choice of that particular form of Coulomb's law historically came later than Coulomb's law was suggested. Hence Coulomb indeed had to postulate the form of the law without having support from any independent source. However, the chronological order in which different steps of scientific inquiry may occur is of secondary importance. Coulomb's postulate, seemingly based only on the simplicity and elegance of his law, got strong independent support after the concept of the electric field entered science. At that stage of the development of the theory of electricity the force of electric interaction was redefined as the effect of an electric field on an electric charge. A simple geometric consideration shows that the electric field of a lone electric point charge drops inversely proportional to the squared distance from that charge, simply because the area over which the field acts increases proportionally to the squared distance. This conclusion is not the result of a postulate stemming from experimental data but a mathematical certainty. (Of course, the mathematical certainty in itself is based on the postulate of space being Euclidean.[15])

The convergence of the indisputable geometric facts with Coulomb's experimental data provides an extremely reliable conclusion ascertaining the validity of Coulomb's law in its originally postulated form.

Therefore the notion, often discussed in the philosophy of science, that researchers have unlimited flexibility in drawing an infinite number of curves through any number of experimental points (or, more precisely, through the margin of error surrounding each such point),[16] while correct in an abstractly philosophical way, is of little significance for the actual scientific procedure of stating a law. While theoretically one can indeed draw any number of fancy curves through any set of points on a graph, in fact a scientist is limited in his choice of legitimate curves connecting his experimental points, just as he is limited in postulating a law, based on the two factors listed earlier. When facing the choice among various possible ways to connect experimental points by a continuous curve, the researcher not only matches the particular set of data to other sets of related data, but is also guided by reliable theoretical concepts.

To summarize, we can state that laws of science are both postulates and approximations. Fortunately, as the great successes of science and the technology based on it prove, those laws which have become firmly accepted as part of science are, without exception, *good* postulates and *good* approximations.

Models

The term *model* may have various meanings; for example, a *model husband* is not the same as a *model airplane*. Even in science, this term may be used for different concepts. For example, a *mathematical model* of a phenomenon is a term denoting a set of equations (often differential equations) reflecting the interconnections among various parameters affecting a given system's behavior. A *computer model* is a program designed to *simulate* a certain situation or process.

In this section, I will discuss what is sometimes referred to as a *physical model*. I will refer to it simply as "model." To explain the meaning of this term as it will be used in this section, we first have to state that every model, in a certain limited sense, is always assumed to represent a real object. A model is an imaginary object which has a limited number of properties in common with a real object, while the real object represented by the model in question also has many other properties which the model does not possess.

It can be said that at the level of scientific theories, science actually deals only with models rather than with the real objects these models represent. The reason for the replacement of real objects with models is that real objects are usually so immensely complex that they cannot be comprehended in all of their complexity by means of a human theoretical analysis. Out of necessity, science resorts to simplifications in its effort to study and understand the behavior of real systems. To do so, science replaces, in the process of a theoretical interpretation of laws, real objects with simpler representations of the latter which we call models.

There are many classical examples of the use of models. Recall again the story of Brahe and Kepler. The former had meticulously accumulated numerous data regarding the positions of the planets of the solar system. The latter reviewed these data, postulated that these data reflected a law (i.e., introduced a bridging hypothesis) and further postulated the three specific laws of planetary motion.

The next step in science was to offer a theory explaining Kepler's laws, i.e., to develop a new science—celestial mechanics. This task was brilliantly performed by Newton, who derived theoretical equations explaining Kepler's laws based on his own theory of gravitation and his general laws of mechanics. In doing so, Newton wrote equations in which planets were assumed to be *point masses*. A point mass is a model of a planet, sharing with the real object it represents—a planet—only one property, the planet's mass. A real object such as a planet has an enormous number of properties besides its mass. Replacing a real object with its substantially simplified model, Newton made it possible to derive a set of elegant equations which provided an excellent approximation of the actual planetary motion.

As another example, the branch of applied science called *seismometry* studies the propagation of seismic waves in the crust of the earth. The real object of study in seismometry is the same as in celestial mechanics—the planet. Obviously, using a point mass as a model would be useless for the purposes of seismometry. The model used in that science is the so-called *elastic half-space*. In that model, only three properties are common with the real object—the modulus of elasticity, Poisson's coefficient, and the sheer modulus. The mass of the planet, which is the property of the model of the planet in celestial mechanics, is absent in the model of the same planet in seismometry, whereas the moduli and coefficients of the seismometric model are absent in the model of the same planet in celestial mechanics. The two models of the same real object are vastly different in the two mentioned branches of science, to such an extent that it is hard to find anything else in common between them.

Both models are highly successful. Choosing these models has enabled scientists to build good theories successfully—in one case, of planetary motion, and in the other, of the propagation of mechanical waves in the crust of a planet.

How is a model chosen? The choice is to be made of those properties of the real object which are of primary importance for the studied behavior of the system in question, while all other properties of the real object are deemed to be of negligible significance and are therefore not attributed to the model. There are no known specific criteria telling the scientist how to distinguish between those properties of the real object which are crucial for the problem at hand and those which are insignificant for that particular theory. The scientist must make the choice based on his intuition, his experience, and his imagination. If the model has been chosen successfully, it paves the way to developing a good theory of the phenomenon. If some crucial properties of the real object are ignored in the model, it leads to a defective theory. If some insignificant properties of the real object are attributed to the model, its theoretical treatment may turn out to be too unwieldy, and the ensuing theory too awkward. The proper choice of a model, which sometimes is made subconsciously, is an extremely important step in a scientific endeavor. This is the step where the deviation from good to bad science occurs most often.

Let us now review an example of a bad model. In chapter 2, a critical review of *Darwin's Black Box*, by Michael Behe, was offered. Various facets of Behe's concept of "irreducible complexity" of micromolecular systems were criticized. Now I will look at Behe's concept from the viewpoint of his choice of a model. An alleged example of irreducible complexity used by Behe in several of his publications is a mousetrap. This device, which, unlike a biological cell, is quite simple, consists of five parts, each of which according to Behe are necessary for the proper functioning of the mousetrap. If any of these parts were absent, the mousetrap would not work. Therefore, maintains the author, it must have been designed in a one-step process, as a whole rather than having been developed in stages from simpler versions. Behe asserts that a similar situation exists in biological cells, which contain a very large number of proteins all working together so that the removal of even a single protein would render the cell incapable of performing a certain function necessary for the organism's well-being. Since all proteins are necessary for the proper functioning of a cell, Behe argues, the cell could not evolve from some simpler structures because such structures would be useless and hence could not be steps of evolution.

It is clear that Behe uses a mousetrap as a model of a cell. This is an example of how bad a model may happen to be.

Looking at Behe's mousetrap model, let's assume first that the author is correct in stating that removing any part of it would render it dysfunctional. Even with this assumption, the fundamental fault of that model is that it ignores some crucial properties of cells and therefore is inadequate for the discussion in question. The first fault of Behe's model is that his model (the mousetrap) unlike a cell, cannot replicate itself. The mousetrap (like Paley's watch)[17] has no ancestors and will have no descendants. Therefore the mousetrap certainly could not evolve from a more primitive ancestor. Biological systems do replicate, though, and therefore are capable of evolving via descent with modification. The second fault of Behe's model is that it ignores the redundancy of biological structures. For example, the DNA molecule contains many copies of identical formations, some of which serve as genes and can continue to perform a certain useful function while their copies are free to undergo mutations and thus lead to new functions. The mousetrap lacks such redundancy and therefore is a bad model which does not illustrate the supposed irreducible complexity of microbiological structures.

Moreover, Behe's five-part mousetrap is not irreducibly complex at all. Out of its five parts, four can be removed one by one and the remaining four-part, then three-part, then two-part, and finally one-part contraption would still be usable for catching mice. Alternatively, the five-part mousetrap can be built up step by step from a one-part contraption, making it gradually two-part, three-part, four-part, and finally five-part contraptions, each of them capable of performing the job of catching mice, this capability slightly improving with each added part. (Nice versions of simpler mousetraps with one, two, three, and four parts were demonstrated by John McDonald and discussed in chapter 1.)

If the mousetrap were capable of reproduction, it certainly could have evolved from a primitive but still usable one-part device via descent with modification governed by natural selection. Hence, the model chosen by Behe not only fails to support his thesis but actually can serve to argue against it.

Behe's supporters often refer to his mousetrap example as a good illustration of irreducible complexity, obviously not noticing the inadequacy of this model. In particular, William Dembski tried to justify Behe's use of a mousetrap as a model of irreducible complexity.[18] Dembski's assertions are critically discussed in chapter 1.

Cognitive Hypotheses

The choice of a cognitive hypothesis is closely related to the choice of a model. Sometimes these two steps merge into one. In other cases, though, the choice of a model is a step separate from the choice of a cognitive hypothesis.

A cognitive hypothesis is the necessary first step toward developing a theory. A law postulated on the basis of experimental data is always of the phenomenological character. It describes the regularity which has been assumed to govern the observed data, without explaining the mechanism of the process which results in those data. A theory of a phenomenon, on the other hand, has to be of explanatory nature, i.e., it must suggest a logically consistent and plausible explanation of the mechanism in question. In order to develop a theory, a cognitive hypothesis must be first chosen. The cognitive hypothesis is a basic idea serving as a foundation for the theory's development.

Choosing a cognitive hypothesis is often a subconscious procedure, based essentially on the scientist's imagination. Experience certainly may play a significant role in a scientist's ability to come up with a hypothesis which would provide a basis for the development of theory. However, many cases are known in which a young scientist with little or no experience suggested ideas which escaped the minds of much more experienced colleagues. One of the most famous examples is Einstein's theory of photoelectric effect, for which he was awarded the Nobel Prize. Einstein was only twenty-six at that time and had no record of preceding scientific achievements. The photoelectric effect, which was discovered by German physicist Heinrich Hertz more than twenty-five years earlier, had a number of puzzling features, and none of the leading experienced scientists could offer any reasonable explanation of those features. In a brief article Einstein offered a brilliant interpretation of the data obtained in the experiments, which immediately solved the puzzle in a consistent and transparent way.[19] The cognitive hypothesis offered by Einstein maintained that electromagnetic energy is emitted, propagates, and is absorbed by solids in discrete portions (later named photons). Based on that hypothesis, Einstein developed a consistent theory which solved the puzzle of the photoelectric effect in a plausible way.

Many scientists who have suggested various hypotheses often assert that they cannot explain how these hypotheses formed in their mind. Let me give an example from my own experience. Many years ago, I studied internal stress in metallic films. At an early stage of that study I had already found that in the process of a film's growth two types of stress, tensile and compressive, emerge simultaneously, each apparently due to a different mechanism. Reviewing the multitude of experimental data, I tried to imagine what the mechanisms for the emergence of each of the two types of stress could be. I will refer here only to the hypothesis I offered for tensile stress (although I also offered another hypothesis for compressive stress). I cannot give any explanation of why and how I came up with the hypothesis according to which tensile stress in growing films appeared because of the egress of a certain type of crystalline defects, the so-called dislocations, to the surface of the crystallites. At that time, the very existence of dislocations was disputed by some scientists (although a theoretical model of dislocations had already been developed).[20] One day, without consciously realizing why, a vivid picture of dislocations moving away from each other until they reached the surface of crystallites emerged in my mind, and I had a feeling that I could somehow see the microscopic world of crystals.

The model I had in mind was a crystal containing a certain concentration of dislocations that all repelled each other, causing their motion toward the surface of crystals. The vision of the egress of these dislocations causing tensile stress in films originated purely in my imagination. On the basis of that vision I set out to derive formulas which would enable me to calculate stress originating from the egress; i.e., I proceeded from the cognitive hypothesis to a theory.[21] Years later, when improved methods of electron microscopy were developed, not only were dislocations directly observed, vindicating the dislocation theory itself, but also, a little later, the very egress of dislocations that I had hypothesized. Thus the hypothesis which originally was born simply out of my imagination was justified.

Although my theory of tensile stress in films was, of course, only a very small contribution to science, I believe that the process is similar also in the case of much more important scientific theories.

Sometimes more than one hypothesis can be offered for the same data; i.e., different models of the same object or phenomenon can compete with each other. The competition between hypotheses often extends into a competition between the theories developed on the basis of those hypotheses, the next step in the scientific process.

Theories

Although the development of science is a very complex process, it is more or less a common situation wherein the steps already discussed are present in the development of a particular chunk of science. The final step is usually developing a theory, although these steps may partially merge, may be taken subconsciously, and/or may occur out of the order described here.

Each of these steps requires specific qualities from a scientist. For example, an accomplished experimentalist may happen to be a poor theoretician, and vice versa. Along the same lines, offering a cognitive hypothesis requires imagination, and different scientists may possess it to a different degree. Development of a theory, on the other hand, requires the pertinent skill and is facilitated by experience.

There is a distinction between particular theories designed to explain the results of a specific experiment and theories of a general character, embracing a whole branch of science. Theories of the first type are often suggested by thc same researchers who have conducted the experiment. Theories of the second type usually are developed by theoreticians who rarely or never conduct experiments. This division of tasks between participants of the scientific process is necessary because, first, modern science has reached a very high level of complexity and sophistication, thus necessitating a specialization of scientists in narrow fields, and second, each of the steps of scientific inquiry requires a different type of talent and skill.

A scientific theory is an explanation of a phenomenon that is systematic, plausible, and compatible with the entirety of the available experimental or observational evidence.

As Einstein is quoted to have said, data remain while theories change with time. Each theory is only as good as the body of data it is compatible with. If new data are discovered, it may require reconsideration of a theory or even its complete replacement by another theory. Therefore it is often said that every theory can only make one of two statements: either *no* or *maybe*, but never an unequivocal *yes*. In other words, it is often said that all scientific theories are tentative because there is always the possibility that new facts will be discovered which would disprove the theory.

While the above assertion certainly contains much truth, I believe that it requires amendment. Reviewing the theories accepted in science, it is easy to see that there are two types. Most of the theories are indeed only tentative, because, as is often stated, the experimental or observational data "underdetermine" the theory. However, there are certain theories whose reliability is so firmly established that the chance of them ever being disproved is negligible.

An example of such a theory is the periodic system of elements as posited by Russian chemist Dmitry Mendeleyev. There is hardly a chance it will ever be disproved. This statement holds not only for the phenomenological form of the periodic system as was suggested by Mendeleyev, but also for the explanatory theory of the periodic system of elements, based on quantum-mechanical concepts.

Since, as I stated in the beginning of this chapter, I am not a philosopher and am quite reluctant to engage in a philosophical discourse, I will not discuss here the ideas suggested by philosophers such as Thomas Kuhn and Karl Popper that are so popular in publications on philosophy of science; I will only say that, in my view, the theories of these two quite famous figures have little relevance to the practice of science in any substantial way.

SCIENCE AND TECHNOLOGY

The amazing progress of various modern technologies is commonly attributed to input from science. While this attribution is true, the relationship between science and technology is not unidirectional. Historically, while many innovations in technology have resulted from direct applications of scientific discoveries, some branches of science themselves have received a strong impetus from technological products of an inventor's intuition. Moreover, even in cases where an invention is made without a direct connection to a specific scientific discovery or theory, the inventor has the advantage of standing on a firm foundation of the entire body of contemporary science. Therefore it can be said that the progress of technology, directly or indirectly, has been ensured by the progress of science.

A good example of the emergence of a new technology as a direct result of the progress of science is the explosion in the use of semiconductors starting in the late 1940s. Until the mid-1930s the very term *semiconductor* was almost completely absent in the scientific literature. Solid materials were usually divided into only two classes—conductors and dielectrics. The materials which eventually were classified as the third separate type of solids, semiconductors, had been viewed simply either as bad conductors or as bad dielectrics. In the 1930s, the quantum-mechanical theory of conductivity was developed (the so-called energy bands theory),[22] which identified three distinctive classes of solids from the standpoint of the behavior of their electrons. This theory predicted the peculiar features of the behavior of semiconductors as a separate class distinct from conductors and dielectrics.

Semiconductor technology was practically nonexistent at that time, although some rather primitive devices were already in use, such as electric current rectifiers in which the semiconductor copper sulfide was utilized. The theoretical explanation of the mechanism of electric conductivity laid the foundation for the development of various semiconductor devices, but it did not happen overnight. Although the scientific basis for semiconductor technology had been established, it took about fifteen years before that technology started taking off. Since then, however, the development of semiconductor technology directly stemming from the progress of physics of semiconductors has been proceeding at an amazing, ever-increasing rate, resulting in a revolution in communication, automation, and many other fields.

An example of the opposite sequence of the science-technology interaction is the story of Swedish engineer Carl Gustaf de Laval, the inventor of the steam turbine. In the process of designing his turbine, Laval encountered many difficult problems, but he possessed an inordinate degree of engineering intuition and inventiveness. Among several inventions Laval made when designing his turbine, two inventions stand alone as amazing insights into phenomena that had not yet been studied and understood by science at that time. They are known as Laval's flexible shaft and Laval's nozzle.

Laval's turbine was designed to rotate at a very high speed. The turbine's wheel sits on a rapidly rotating shaft. At certain values of the shaft's rotational speed that are referred to as critical speeds, a phenomenon called resonance occurs when the shaft experiences sharply increasing bending forces. This may cause the breakdown of the shaft. According to common sense, it would seem reasonable to increase the shaft's strength by increasing its diameter. However, this also causes the bending force to increase. Instead of increasing the shaft's diameter, Laval made his shafts very thin, hence very flexible. Amazingly, instead of making the rapidly rotating shafts weaker, the decrease of their diameter led to a peculiar phenomenon, unforeseen by the science of mechanics of rotating bodies. As the speed of rotation reached the critical value, the bending force reversed its direction, straightening the shaft instead of bending it. Nobody knows why and how the idea of a flexible shaft occurred to Laval. The scientific explanation of the behavior of the flexible shafts at critical speeds was developed years after Laval's turbine was introduced, becoming a separate chapter in the branch of science called technical dynamics. Laval's insight provided a strong impetus for the development of that science, which, in turn, later provided powerful input into the development of various technologies.[23]

Laval's genius and intuition were equally amazing in his invention of Laval's nozzle. Without going into the details of that extraordinary invention, let me state that the subsequent development of a theory explaining the behavior of Laval's nozzle gave a mighty push to the emergence of a new scientific field—gas dynamics, which in turn has become a foundation of important modern technologies.

While paying proper tribute to Laval's genius and recognizing that his amazing insights were not based on any specific scientific discoveries of his time, we must also realize that Laval, however inventive and inordinately talented he was, could not have made his inventions if he had lived, say, a hundred years earlier. Whereas he did not derive the ideas of the flexible shaft and a novel type of a nozzle from any specific scientific theory, his entire mental makeup and way of thinking were formed as products of his time, with its already high level of scientific knowledge in physics, chemistry, and many other disciplines.

One more feature of the interaction between science and technology is that it often becomes hard to distinguish where science ends and technological development starts. This convergence of scientific research and the development of new technologies is often referred to by the not precisely defined term *applied science*.

Research divisions of large high-tech and pharmaceutical companies as well as university departments financed by industry are engaged in intensive studies wherein the distinction between pure and applied science is sometimes hard to see. For example, two of the foremost research institutions in the world—Bell Labs and IBM's research division—recently announced the independent successful development of new types of non–silicon-based transistors of molecular-size dimensions. This impressive breakthrough in nanotechnology was made possible only because in both institutions research was not limited to the narrow task of technological advance but involved intensive research of the type traditionally considered pure science.

Here is another example, once again from my own experience (of course, much more modest than the examples discussed in this section so far). In 1973 I had a student working toward a Master of Science degree who was studying the effect of illumination on the adsorption behavior of semiconductor surfaces. When I was preparing instructions for this student, I had a hunch that the effect of the so-called photoadsorption, known to exist on solid semiconductor surfaces, must exist as well for small semiconductor particles if the latter were suspended in water. Together with a chemist who worked in my lab, we prepared a colloid solution containing small spherical particles of the semiconductor selenium. I placed a transparent Plexiglas plate on the solution's surface, ran to the building's roof and exposed the solution to the rays of the sun. To my excitement, a smooth film of amorphous selenium grew rapidly on the plate right in front of my eyes, confirming my hunch that high-energy photons must lower the potential barrier for electrons on suspended semiconductor particles. Soon a few students joined the project. Besides the theory I developed for what was labeled "inverse photoadsorption," the project also resulted in a technology for photodeposition of semiconductor films.[24]

SCIENCE AND THE SUPERNATURAL

In this section I will briefly discuss the question of whether or not the attribution of a phenomenon to a supernatural source may be legitimate in science. This question is separate from the problem of the relationship between science and religion. The latter problem is multifaceted and includes a number of questions such as the compatibility of scientific data with religious dogma (in particular with the biblical story of creation), the roles of science and religion for the society as a whole and for individuals, the impact of science and religion on the moral fiber of the society, and many others. In this section, however, my topic is much narrower, and relates only to the question of whether or not science may legitimately consider possible supernatural explanations of the laws of nature and of specific events.

Obviously, the answer to that question depends on the definition of science. If science is defined as being limited only to "natural" phenomena, then obviously anything which is supernatural is beyond scientific discourse. Such definitions of science have been suggested by scientists and even produced as outcomes of legal procedures.

While I understand the reasons for the limitations of legitimate subjects for scientific exploration only to "natural" causes and, moreover, emotionally am sympathetic to them, I think there is no real need to apply such limitations, and a definition of science should not put any limits on legitimate subjects for the scientific exploration of the world. Indeed, although science has so far had no need to attribute any observed phenomena to a supernatural cause, and in doing so has achieved staggering successes, there still remain unanswered many fundamental questions about nature. Possibly such answers will be found someday, if science proceeds on the path of only "natural" explanations. Until such answers are found, nothing should be prohibited as a legitimate subject of science, and excluding the supernatural out of hand serves no useful purpose.

It seems relevant to discuss at this point the distinction between rational and irrational approaches to problems. It seems a platitude that to be legitimately scientific, a discourse must necessarily be rational, while religious faith does not *have* to be based on a rational foundation. If this is so, what makes a discourse rational?

For the purpose of this discussion, I suggest that in order to be viewed as rational, a discourse must meet the following requirements: (1) it must distinguish between facts and the interpretation of facts; and (2) it must distinguish among (*a*) undeniable facts proven by uncontroversial direct evidence, (*b*) plausible but unproven notions, (*c*) notions agreed upon for the sake of discussion, (*d*) notions that are possible but not supported by evidence, and (*e*) notions contradicting evidence.

If we adopt this viewpoint, we must also accept that a rational *conclusion* is such that is based solely on *facts* proven by uncontroversial direct evidence. Statements and notions offered without meeting the criteria listed above I will view as irrational, regardless of how many adherents they may have.

(Note that this definition does not contain criteria designed to assert that a notion, conclusion, or view in question is *true*. In other words, my definition of *rational* does not necessarily coincide with a definition of *true*.)

If science is defined as a human endeavor aimed at acquiring knowledge about the world in a systematic and logically consistent way, based on factual evidence obtained by observation and experimentation, then there is no need to exclude anything from the scope of scientific exploration simply because it does not belong in science *by definition*.

However, removing the prohibitions that limit the legitimate boundaries of science to whatever can be defined as "natural" does not mean that in science anything goes. To be a proper part of science, the process of developing an explanation of phenomena must necessarily be rationally based on factual evidence, i.e., on reliable data.

So far, no attribution of any events to a supernatural source has ever been based on factual evidence. To clarify my thesis by example, let us recall the so-called Bible code, which was briefly mentioned in preceding sections and which is discussed in detail in chapters 6 and 14. Recall that in a paper by three Israeli authors published in the journal *Statistical Science*, a claim was made, based on a statistical study of the text of the Book of Genesis, that within this text there exists a complex "code" predicting the names as well as dates of birth and death of many famous rabbis who lived centuries after it was written.[25] If the claim in that paper were true, it would mean that the author of the Book of Genesis knew about the events which would occur centuries after compiling his text. This would constitute a miracle and hence must have a supernatural explanation (although at least one writer suggested that the alleged code was inserted into the text of the Bible by visitors from other planets). Despite the obvious religious implications of the claim made by the three authors, the editorial board of the journal, which comprised twenty-two members, approved the publication of the article. This was a convincing display of scientists keeping an open mind, contrary to the persistent assertions by the adherents of intelligent design such as Johnson (see chapter 3), who accuses scientists of having closed minds.

There was nothing seemingly nonscientific in the paper; hence, the claim about the "code" became a legitimate subject of scientific discussion, despite its supernatural implications.

If the claim by the three Israelis were confirmed and hence accepted by the scientific community as a result of properly collected data and of their proper analysis, their results would have become part of science, despite their religious implications. Moreover, the proof of the existence of a real code in the Bible would constitute a strong scientific argument for the divine origin of the Bible and hence support the foundation of the Jewish and Christian religions. This is a way the supernatural may legitimately enter the realm of science, but it requires reliable factual evidence.

The Bible code, however, was shown rather swiftly to be far from proven. It was convincingly shown that the claim in question was in fact based on bad data (see chapter 14). Moreover, a number of serious faults were identified in the statistical procedure used by the three authors. Hence, the story of the alleged Bible code, rejected by the scientific community, did not change the fact that so far no miracles have ever been proven by scientific methods.

Therefore, although in my view the reference to a supernatural source of the events should not be excluded from science out of hand—i.e., simply by definition—accepting the claim of a miracle into science can only take place if it is based on reliable evidence, meeting the requirements of scientific rigor.

If the past is any indication of the future, the chance that an actually observed phenomenon will be successfully attributed to a supernatural source in a scientifically rigorous manner is very slim indeed.

CONCLUSION

In 1948, a meeting of biologists took place at the Academy of Agricultural Science in Moscow, where the scientists who objected to Trofim Lysenko's pseudoscience were verbally abused, derided, and insulted as a prelude to their subsequent dismissal from all universities and research institutions, as well as their arrest and imprisonment. Officially the meeting was supposed to be devoted to a scientific discussion of the problems of modern biology and agriculture. However, at the last session of this meeting, Lysenko's speech included the lines, "The question is asked in one of the notes handed to me, 'What is the attitude of the Central Committee of the Communist Party to my report?' I answer: the Central Committee of the Party examined my report and approved it." Everybody in the audience stood up, and a lengthy applause followed.

So, we see that sometimes truth in science is decreed by the ruler of a country. Is this a normal process of establishing the validity of a scientific theory? Of course not—this was an example of a pathological usurpation of the power of science by dictatorial rulers of a totalitarian state.

How, then, is the validity of a scientific theory established? There is no Central Committee empowered to decide which theory is correct. Of course, there are numerous organizations, such as the Royal Society of London; academies of science in various countries; and scientific societies, both national and international, which award prizes and honorary fellowships for achievements in science, thus bestowing an imprimatur of approval on certain theories and discoveries, but these official or semiofficial signs of recognition serve more as tributes to human vanity than as real proof of the validity of scientific theories. The most prestigious of all these prizes is perhaps the Nobel Prize, awarded by the Swedish Academy of Science. Whereas the majority of the Nobel Prize winners are indeed outstanding contributors to science, even this enormously prestigious prize could in certain cases invoke the following line from a play by Aleksandr

Griboedov, popular in Russia, titled *The Grief because of Wisdom*: "Ranks are given by men and men are prone to err." Prizes and fellowships, however prestigious, are not a legitimate means for establishing the validity and importance of this or that theory.

Is, then, the validity of a theory established by a vote of scientists, so that the choice between theories is made based on the majority of votes? Despite the indisputable fact that often a single scientist may turn out to be right, even if all the rest of scientists think otherwise, the acceptance of theories is indeed actually determined by an unwritten consensus among scientists. Such a consensus is usually not formed by a vote at a single meeting of the experts in a given field, but is rather achieved in a haphazard way through a piecemeal exchange of information among specialists in that field. Sometimes it happens swiftly, when scientists working in a given area of research overwhelmingly come to believe in a theory because of its strong explanatory power and solid foundation in data—an example of such a theory is the theory of quarks suggested by Murray Gell-Mann.[26] In some other cases a theory may wait for a prolonged time before becoming generally accepted—for example, the theory of dislocations, mentioned earlier. Suggested as a hypothesis in the early 1920s by three scientists working independent of each other (in England, Russia, and Japan), and developed rather quickly into a mathematically sophisticated branch of physics of crystals, it had indisputable explanatory power and elucidated in a very logical way many effects which seemed puzzling at that time. Nevertheless, for many years it could not win universal acceptance because the very existence of dislocations assumed by the theory was not experimentally proven. Only with drastic improvements in electron microscopy were the dislocations experimentally observed and thus the theory vindicated and universally accepted.

In that sense, it can be said that scientific theories gain the imprimatur of approval through a "vote" by scientists. Similarly, they fall out of favor by the same procedure.

This means that science is largely immune to attacks by amateurs and pseudoscientists, regardless of whether they try to undermine science under the banner of an undisguised creationism or by masquerading as legitimate alternatives to mainstream science, as intelligent design proponents are doing nowadays.[27] I believe that their attack is doomed to fail, although they can be expected to continue generating annoying noise, distracting scientists from fruitful research in the process.

An illustration of this viewpoint can be seen in the following story. The leading creationist organizations in the United States, such as the Discovery Institute's Center for Science and Culture, the Institute for Creation Research, and Answers in Genesis all have triumphantly published lists of "scientists who are critical of the Darwinian evolution theory." Such a list published by the Discovery Institute, for example, comprises exactly one hundred names (among which are actually some nonscientists). Of course, scientific disputes are not resolved by popular vote, so these lists of Darwin's deniers are irrelevant to the creation-evolution controversy. However, the National Center for Science Education (NCSE) has come up with a tongue-in-cheek response to the creationist antievolution lists, having collected over four hundred signatures of scientists in support of evolution against creationism, with one peculiar twist: all the signatories, besides being all *qualified scientists*, all have the same first name—Stephen (or its variations such as Steven, Stefan, and the like). The choice of the first name was to honor the late prominent defender of evolution Stephen Jay Gould. Since it is estimated that approximately 1 percent of all scientists bear this name, this list means that there are over forty thousand qualified scientists supporting evolution against creationism.[28] Forty thousand versus one hundred—isn't this an illustration of the fallacy of the frequent claims by neocreationists such as Johnson or Dembski, who maintain that Darwinism is a "dying theory," to be replaced imminently by intelligent design? Darwinism is part of science, and science is robust enough to withstand the attack by pseudoscientists of the kind represented by Dembski, Johnson, and their cohorts.

The discussion in this chapter reflects the view of science held by one particular scientist. Other scientists may have a very different vision of their chosen profession. Additionally, philosophers of science may be expected to dismiss this chapter as twaddle by an amateur. So be it.

One thing, however, seems quite certain. Whatever the definition of science may be, and despite the misuse of its achievements by all kinds of miscreants, and despite its multiple shortcomings and failures, it is probably the most magnificent of the human endeavors.

NOTES

1. Aleksandr Sergeyevich Pushkin, "The Journey to Arzrum during the Campaign of 1829" (in Russian), in *Polnoye Sobraniye Sochineniy* (Complete works), vol. 6 (Moscow: Nauka, 1964), p. 668.

2. But see, for example, Alan Cromer, *Uncommon Sense: The Heretical Nature of Science* (New York: Oxford University Press, 1993), pp. 160–66.

3. This example is presented in greater detail in Mark Popovskiy, *Upravlyaemaya Nauka* (Manipulated science) (London: Overseas Publications Interchange, 1978), pp. 113–16.

4. Doron Witztum, Eliyahu Rips, and Yoav Rosenberg, "Equidistant Letter Sequences in the Book of Genesis," *Statistical Science* 9, no. 3 (1994):429–38.

5. Immanuel Velikovsky, *Worlds in Collision* (New York: Macmillan, 1950).

6. Richard Feynman, *The Nature of Physical Law* (New York: Random House, 1994), p. 123.

7. Ibid., p. 141.

8. Vasiliy V. Nalimov, *Teoriya Eksperimenta* (Theory of experiment) (Moscow: Nauka, 1971), p. 77.

9. For details, see Steven Weinberg, *The First Three Minutes* (New York: Basic Books, 1977).

10. M. Ya. Popereka (a.k.a. Mark Perakh), "Methods of Quantitative Investigation of Adsorption on Solid Electrodes," in *Twenty-fifth Meeting of the International Society of Electrochemistry: Extended Abstracts* (Brighton, U.K.: University of Southampton Press, 1974), pp. 305–307; M. Ya. Popereka (a.k.a. Mark Perakh) and V. N. Lebedeva, "Potentiostatic Chronofaradometry: Method for a Quantitative Study of Adsorption on Solid Electrodes" (in Russian), part 1, in *Adsorbtsionnye plyonki* (Adsorption films), ed. A. I. Baranov, M. Ya. Popereka, and V. M. Rudyak, Fizika plyonok 2 (Kalinin, USSR: Kalinin University Press, 1972), pp. 3–25; M. Ya. Popereka (a.k.a. Mark Perakh) and V. M. Romanov, "Potentiostatic Chronofaradometry: Method for a Quantitative Study of Adsorption on Solid Electrodes" (in Russian), part 2, in *Adsorbtsionnye Plyonki*, pp. 26–30.

11. Nalimov, *Teoriya Eksperimenta*, pp. 7–9.

12. A detailed discussion of emission spectral analysis is given in Nalimov, *Teoriya Eksperimenta*, p. 8.

13. Coulomb's law is explained in David Halliday, Robert Resnick, and Jearl Walker, *Fundamentals of Physics*, extended ed. (New York: John Wiley and Sons, 1993), pp. 639–42.

14. Ockham's (or Occam's) razor is a medieval rule of economy named for English theologian and nominalist philosopher William of Ockham (ca. 1285–ca. 1347/49): "Plurality should not be posited without necessity."

15. A brief but very clear explanation of the distinctions between Euclidean and non-Euclidean geometry is given in Jan Culberg, *Mathematics from the Birth of Numbers* (New York: W. W. Norton, 1997), pp. 381–83.

16. See, for example, Del Ratzsch, *Nature, Design, and Science: The Status of Design in Natural Science* (New York: State University of New York Press, 2001), pp. 79–82.

17. William Paley's "watchmaker argument," discussed in detail in his *Natural Theology: Or, Evidences of the Existence and Attributes of Deity Collected from the Appearances of Nature* (1802; reprint, Boston: Gould and Lincoln, 1852), is one of the earliest and perhaps best-known versions of the argument from design.

If a watch were found on a heath, wrote Paley, it would point to the existence of a watchmaker who designed the watch. The universe is like a giant, extremely complex watch; this, in Paley's view, points to the existence of a designer of the universe, i.e., God. The arguments of the new crop of creationists are often just variations of the same watchmaker argument, offered in quasi-scientific jargon.

18. William A. Dembski, *No Free Lunch: Why Specified Complexity Cannot Be Purchased without Intelligence* (Lanham, Md.: Rowman and Littlefield, 2002), pp. 261–67.

19. Albert Einstein, "Über einen die Erzeugung und Verwandlung des Lichtes betreffenden heuristischen Gesichtspunkt" (On a heuristic point of view regarding the generation and transformation of light), *Annalen der Physik* 17 (1905):132–40.

20. Boris Gruber, *Theory of Crystal Defects* (New York: Academic Press, 1964).

21. M. Ya. Popereka (a.k.a. Mark Perakh), "On the Origin of Internal Stresses in Electrolytically Deposited Materials" (in Russian), *Fizika metallov i metallovedeniye* 20 (1965):753–62; *Vnutrenniye napryazhenia elektrolitichski osazhdayemykh metallov* (Internal stress in electrolytically deposited films) (Novosibirsk, USSR: Zapadno-Sibirskoye Knizhnoye Izdatelstvo, 1966); English translation published for the National Bureau of Standards and the National Science Foundation by the Indian National Scientific Documentation Centre (New Delhi, 1970); "A Theory of Internal Stresses in Electrolytically Desposited Materials" (in Russian), in *Trudy Tret'ego Mezdunarodnogo Kongressa po Korrozii Metallov* (Proceedings of the Third International Congress on Metallic Corrosion), vol. 3 (Moscow: Mir, 1965), pp. 350–56.

22. Charles Kittel, *Introduction to Solid State Physics* (New York: John Wiley and Sons, 1976), pp. 183–204.

23. Some fifty years after Laval's death, certain features of the phenomenon Laval had intuitively foreseen were still subject to theoretical analysis; see, for example, M. Ya. Popereka (a.k.a. Mark Perakh), "Investigation of Critical Speeds of Rapidly Rotating Shafts," *Doklady Akademii Nauk Tajikistana* (Reports of the Academy of Sciences of Tajikistan), no. 10 (1954):81–85.

24. Mark Perakh, Aaron Peled, and Zeev Feit, "Photodeposition of Amorphous Selenium Films by the Selor Process," part 1, *Thin Solid Films* 50 (1978):273–82; Mark Perakh and Aaron Peled, "Photodeposition of Amorphous Selenium Films by the Selor Process," parts 2 and 3, *Thin Solid Films* 50 (1978): 283–92, 293–302; Mark Perakh and Aaron Peled, "Light-Temperature Interference Governing the Inverse/Combined Photoadsorption and Photodeposition of a-Selenium Films," *Surface Science* 80 (1979):430–38.

25. Witztum, Rips, and Rosenberg, "Equidistant Letter Sequences in the Book of Genesis."

26. For an explanation of the theory of quarks, see, for example, Raymond Serway et al., *Physics for Scientists and Engineers* (Philadelphia: Saunders, 1990), pp. 1428–31.

27. Mark Perakh and Matt Young, "Is Intelligent Design Science?" in *Why Intelligent Design Fails: A Scientific Critique of the Neo-Creationism*, ed. Matt Young and Taner Edis (Piscataway, N.J.: Rutgers University Press, forthcoming).

28. For the detailed description of "Project Steve," see "Resources," National Center for Science Education [online], www.ncseweb.org/article.asp?category=18 [August 11, 2003]. On this site there are also links to creationist sites where the lists of anti-Darwinians are posted.

13

IMPROBABLE PROBABILITIES

Assorted Comments on Some Uses and Misuses of Probability Theory

The calculation of probabilities is routinely used in books and articles wherein the authors attempt to prove the impossibility of the spontaneous emergence of life, or of the universe as a whole. Likewise, the probabilistic approach is quite common in various treatises relating to narrower issues, for example the so-called Bible code.

The estimates of probabilities in the books and articles discussed in the preceding chapters are often wrought with errors and improper interpretations of the concepts of probability theory. Therefore it is desirable to discuss probability in general, in as simple a way as possible in order to make it comprehensible for nonexperts.

This chapter contains a critical review of misuses of probability as well as a discussion, designed for nonexperts, of some seminal concepts of that theory.

Probability theory is at the core of at least two important scientific disciplines: mathematical statistics and statistical physics. Statistical physics seems to be immune to misuses by dilettantes; mathematical statistics, however, has been routinely abused by many amateurs who try to utilize it to substantiate various pseudoscientific claims, some of which were discussed in the preceding chapters.

While an incorrect application of mathematical statistics may involve any part of this science, a large portion of such errors occurs already at the stage when its seminal quantity, probability, is miscalculated or misinterpreted. One example of an incorrect application of the probability concept is the attempts by the proponents of the so-called Bible code to calculate the probability of occurrence of certain letter sequences in various texts (see chapter 14). Another example is the calculation of the probability of the spontaneous emergence of life on earth. There are, of course, many other examples of improper uses of the probability calculation.

There are many good textbooks on probability theory;[1] usually they make use of a rather sophisticated mathematical apparatus. This chapter is not meant to be one more discussion of probability on a rigorously mathematical level; rather, probability will be discussed almost without resorting to mathematical formulas or to the axiomatic foundation of probability theory.[2] I will try to clarify this concept by considering examples of various situations in which different facets of probability manifest themselves and can be viewed in as simple a way as possible. Of course, since probability theory is essentially a mathematical discipline, it is possible to discuss it only to a very limited extent without resorting to some mathematical apparatus. Hence, this chapter will stop at the point where the further discussion without mathematical tools would become too crude.

ESTIMATION OF PROBABILITY IS OFTEN TRICKY

Calculation of probabilities is sometimes a tricky task even for qualified mathematicians, not to mention laypersons. Here are two examples of rather simple probabilistic problems whose solution often escaped even some experienced scientists.

The first problem is as follows. Imagine that you watch buses arriving at a certain bus stop. After watching them for a long time, you determine that the interval between the arrivals of any two sequential buses is, *on the average*, one minute. The question is, How long should you expect to wait for the next bus if you start waiting at an arbitrary moment in time? Asked to answer that question, many people would confidently assert that the average time of waiting is thirty seconds. This answer would be correct if all the buses arrived at exactly the same interval of one minute. However, the situation is different in that one minute is just the *average* interval between any two consecutive bus arrivals. This number—one minute—is a *mean* of a distribution over a range from zero to a maximum which is larger than one minute. Therefore, the average waiting time, regardless of when you start waiting, is one minute, rather than thirty seconds.

The second problem was used in a popular television game show hosted by Monty Hall, wherein the players were offered a choice among three closed doors. Behind one of the doors, there was a valuable prize, while behind the two other doors there was nothing. Obviously, whichever door a player chose, the probability of winning the prize would be 1/3. However, after the player chose a certain door, the host, who knew where the price was, would open one of the two doors not chosen by the player, and show that there was nothing behind it. At that point, the player would be given a choice, either to stick with the door he had already chosen or to choose instead the remaining closed door. The problem a player faced was to estimate whether or not changing his original choice would provide a better chance of winning. Most people, including some trained mathematicians, answered that the probability of winning is exactly the same regardless of whether the participant sticks to the originally chosen door or switches to the other, yet unopened door. Indeed, at first glance it seems that the chance of the prize being behind either of the two yet unopened doors is the same. However, this is wrong. Changing the choice from the originally chosen door to the other yet unopened door actually doubles the probability of winning.

To see why this is so, note that at the beginning of the game, there was only one winning door and two losing doors. Hence, when a player chose arbitrarily one of the doors, the probability of his choosing the winning door was 1/3, while the probability of his choosing the losing door was 2/3, i.e., twice as large. Now, if the player luckily chose the winning door, he would win if he did not change his choice. This situation happens, on the average, in one-third of games played, if the game is played many times. If, though, the player happened to choose the losing door, he would have to change his choice in order to win. This situation happens, on the average, in two-thirds of games played, if they are played many times. Hence, to double her chance to win, the player should change her original choice.

If many people, including trained mathematicians, are often confused by these rather simple probabilistic situations, the misuse of probabilities in many more complex cases could (and does) happen quite often, which demonstrates that extreme caution is necessary if probabilities are used to arrive at important conclusions.

SOME ELEMENTARY CONCEPTS

Consider some games of chance. Each time we toss a coin it can result in either *tails* or *heads* facing up. If we toss a die, it can result in any of six numbers facing up, namely 1, 2, 3, 4, 5, or 6. If we want to choose one card out of a deck of fifty-two cards scattered on a table face down, and turn its face up, it can result in any one of those fifty-two cards facing up.

Let us now introduce certain terms commonly found in discussions of probability. Each time we toss a coin or a die, or turn over a card, this will be referred to as a *trial*. In the case of a coin, the trial can have one of two possible *outcomes*, tails (T) or heads (H). With the die, each trial can result in any of six possible outcomes, 1, 2, 3, 4, 5, or 6. In the case of fifty-two cards, each trial can result in any one of fifty-two possible outcomes, e.g., the five of spades, or the seven of diamonds, etc.

Now assume we conduct the game in sets of several trials. For example, one of the players tosses a die five times in a row, resulting in a set of five outcomes, for example 5, 3, 2, 4, and 4. Then his competitor also tosses the die five times, resulting in some other combination of five outcomes. The player whose five trials result in a larger sum of numbers wins. The set of 5 (or 10, or 100, or 10,000, etc.) trials constitutes a *test*. The combination of 5 (or 10, or 100, or 10,000, etc.) outcomes obtained in a test constitutes an *event*. Obviously, if each test comprises only a single trial, the terms *trial* and *test* as well as the terms *outcome* and *event* become interchangeable.

For the further discussion we have to introduce the concept of an "honest coin" (also referred to as a "fair coin"). It means we postulate that the coin is perfectly round, its density is uniform all over its volume, and in no trial do the players consciously attempt to favor either of the two possible outcomes. If our postulate conforms to reality, what is our estimate of the *probability* that the outcome of an arbitrary trial will be, for example, T (or H)?

First, it seems convenient to assign to something that is certain a probability of 1 (or, alternatively, 100 percent). It is further convenient to assign to something that is impossible a probability of 0. Then the probability of an *event* that is not certain will always be between 0 and 1 (or between 0 and 100 percent).

Now we can reasonably estimate the actual value of some probabilities in the following way. For example, if we toss a fair coin, the outcomes H and T actually differ only in the names we give them. Thus, in a long sequence of coin tosses, H and T can be expected to happen almost equally often. In other words, outcomes H and T can be reasonably assumed to have the same probability. This is only possible if each has probability of 1/2 (or 50 percent).

Since we use the concept of probability, which by definition is not certainty, it means that we *do not* predict the outcome of a particular trial. We expect, though, that in a large number of trials the number of occurrences of H will be roughly equal that of T. For example, if we conduct one million trials, we expect that in approximately half of them (i.e., in close to five hundred thousand trials) the outcome will be T and in about the same number of trials it will be H.

Was our postulate of an honest coin correct? Obviously it could not be absolutely correct: no coin is perfect. Each coin has a certain imprecision of shape and mass distribution, which may make the T outcome slightly more likely than the H outcome, or vice versa; a player may inadvertently favor a certain direction, which may be due to some anatomical peculiarities of his arm; there may be occasional wind affecting the coin's fall; and so forth. However, for our theoretical discussion we usually ignore the listed possible factors and assume an honest coin. We will return later to the discussion of possible deviations from a perfectly honest coin.

We see that our postulate of an honest coin led to another postulate, that of the equal probability of possible outcomes of trials. In the case of the coin there were two different possible outcomes, T and H, equally probable, but in other situations there can be any number of possible outcomes. In some situations those possible outcomes can be assumed to be all equally probable, while in others the postulate of their equal probability may not hold. In each specific situation it is necessary to establish whether or not the postulate of equal probability of all possible outcomes is reasonably acceptable, or else it must be dismissed. Ignoring this requirement has been the source of many erroneous considerations of probabilities. I will discuss specific examples of such errors later on.

Now consider one more important feature of probability. Suppose we have conducted a coin trial and the outcome was T. Suppose we proceed to conduct one more trial, tossing the coin once more. Can we predict the outcome of the second trial given the known outcome of the first trial? Obviously, if we accept the postulate of an honest coin and the postulate of equal probability of outcomes, the outcome of the first trial has no effect on the second trial. Hence, the postulates of an honest coin and of equal probability of outcomes lead us to a third postulate, that of independence of tests. This postulate is based on the assumption that in each test the conditions are exactly the same, which means that after each test the initial conditions of the first test are exactly restored. The applicability of this postulate must be ascertained before any conclusions can be made in regard to the estimation of probabilities. If the independence of the tests cannot be ascertained, the probability must be calculated differently from the situation when the tests are established to be independent.[3]

A discussion analogous to that of the honest coin can be also applied to cases where the number of possible outcomes of a trial is larger than two—be it three, six, or ten million. For example, if instead of a coin, we deal with a die, the postulate of an honest coin has to be replaced with the similar postulate of an honest die, while the postulates of equal probability of all possible outcomes (of which there now are six instead of two) and of independent tests have to be verified as well before calculating the probability.

The postulate of an honest coin or its analogs are conventionally implied when probabilities are calculated. Except for some infrequent situations, this postulate is usually reasonably valid. However, some writers who calculate probabilities do not verify the validity of the postulates of equal probability and of independence of tests. This is not an uncommon source of erroneous estimation of probabilities. Pertinent examples will be discussed later on.

PROBABILITIES OF EVENTS

Suppose we conduct our coin game in consecutive sets of ten trials each. Each set of ten trials constitutes a test. In each ten-trial test the result is a set of ten outcomes, constituting an event. For example, suppose that in the first test the event comprised the following outcomes: H, H, T, H, T, T, H, T, H, and H. Hence, the event in question included six heads and four tails. Suppose that the next event comprised the following outcomes: T, H, T, T, H, H, H, T, T, and T. This time the event included six tails and four heads. In neither of the two events the number of Ts was equal to the number of Hs, and, moreover, the ratio of H to T was different in the two tests. Does this mean that, first, our estimate of the probability of H as 1/2 was wrong, and, second, our postulate of equal probabilities of H and T was wrong? Of course not.

We realize that probability does not predict the outcome of each trial and hence does not predict particular events. What is, then, the meaning of probability?

If we accept the three postulates introduced earlier (the honest coin, equal probability of outcomes, and independence of tests) then we can define probability in the following manner. Let us suppose that the probability of a certain event A is expressed as $1/n$, where n is a positive number. For example, if the event in question is the combination of two outcomes of tossing a coin, the probability of each such event is 1/4, where $n = 4$. It means that in a large number x of tests event A will occur, *on the average*, once in every n tests. For this prediction to hold, x must be much larger than n. The larger the ratio x/n is, the closer the number of occurrences of event A will be to the probability value, i.e., to one occurrence in every n tests.

For example, as we concluded earlier, in a test comprising two consecutive tosses of a coin, the probability of each of the four possible events is the same 1/4, so $n = 4$. It means that if we repeat the described test x times, where x is much larger than 4 (say, one million times) each of the four possible events—namely HH, HT, TT, and TH—will happen, on the average, once in every four tests.

We have now actually introduced (not quite rigorously) one more postulate, sometimes referred to as the law of large numbers. The gist of this law is that the value of probability can be some accidental number unless it is determined over a large number of tests. The value of probability does not predict the outcome of any particular test, but in a certain sense we can say that it "predicts" the results of a very large number of tests in terms of the values averaged over all the tests.

If any one of the four postulates is not held (i.e., the number of tests is not much larger than n; the coin or die etc. is not honest, the outcomes are not equally probable, or if the tests are not independent) the value of probability calculated as $1/n$ has *no meaningful interpretation.*

Ignoring this statement is often the source of unfounded conclusions from probability calculations.

Later I will discuss situations where some of the four postulates do not hold (in particular, the postulate of independence), but the probabilities of events each comprising several trials can be reasonably estimated nevertheless.

The definition of probability we have established at this point is sometimes referred to as the "classical" definition. However, there are in probability theory also some other definitions of probability, which overcome certain logical shortcomings of the classical definition and generalize it.[4] In this chapter I will not use those more rigorous definitions explicitly (even though they may sometimes be implied) since the classical definition is sufficient for our purpose.

Let us now discuss the calculation of the probability of an event. Remember that an event is defined as the combination of outcomes in a set of trials. For example, what is the probability that in a set of two trials with a coin the event will be "TH," i.e., that the outcome of the first trial will be T and of the second trial will be H? We know that the probability of T in the first trial was 1/2. This conclusion stemmed from the fact that there were two equally probable outcomes. The probability of 1/2 was estimated dividing 1 by the number of all possible equally probable outcomes (in this case 2). If the trials are conducted twice in a row, how many possible equally probable events can be imagined? Here is the obvious list of all such events: TT, TH, HH, and HT. The total of four possible results, all equally probable, covers all possible events. Obviously, the probability of each of those four events is the same—1/4. We see that the probability of the event comprising the outcomes of two consecutive trials equals the product of probabilities of each of the sequential outcomes. This is one more postulate, which is based on the independence of tests: the rule of probabilities multiplication. The probability of an event is the product of the probabilities of the outcomes of all sequential trials constituting that event.[5] As we will see later, this rule has certain limitations.

SOME SITUATIONS WITH SEEMINGLY UNEQUAL PROBABILITIES

Let us discuss certain aspects of probability calculations which have been a pivotal point in the dispute between "creationists" (those who assert that life could not have emerged spontaneously but only via a divine act by the Creator) and their opponents (those who adhere to a theory asserting that life emerged as a result of random interactions between chemical compounds in the primeval atmosphere of our planet or of some other planet). In particular, creationists maintain that the probability of the spontaneous emergence of life is so negligibly low that it would not happen.[6]

Lest I be misunderstood, I would like to point out that in this chapter I am not discussing whether the creationists or their opponents are correct in their assertions in regard to the origin of life; this question is very complex and multifaceted, and the probabilistic argument often employed by the creationists is only one aspect of their view. What I will show is that the probabilistic argument itself as commonly used by many creationists is unfounded and cannot be viewed as proof of their views, regardless of whether those views are correct or incorrect.

The probabilistic argument often used by the creationist can be rendered as follows. Imagine tossing a die with six sides. Repeat it one hundred times. There are many possible combinations of the six numbers (we would say there are many events possible, each comprising one hundred outcomes of individual trials). The probability of each event is exceedingly small (about $1/10^{77}$) and is the same for each combination of numbers, including, say, a combination of one hundred "4s," that is, 4, 4, 4, 4, 4, 4, 4, 4, 4, 4, . . . etc. (an outcome of 4 one hundred times in a row). However, say the creationists, the probability that the set of one hundred numbers will be some random combination is much larger than the probability of one hundred 4s, which is a unique, or "special," event. Likewise, the spontaneous emergence of life is a special event whose probability is exceedingly small; hence, it could not happen spontaneously. Without discussing the ultimate conclusion about the origin of life, let us discuss only the example with the die.

Indeed, the probability that the event will be some random collection of numbers is much larger than the probability of rolling all 4s, but that does not mean anything. The larger probability of random sets of numbers is simply due to the fact that it is a *combined probability of many events*, while for "all 4s" it is the probability of only one particular event. From the standpoint of the probability value, there is nothing special about "all 4s"; it is an event exactly as probable as any other individual combination of numbers, be it all 6s, half 3s and half 5s, or any arbitrary, disordered set of one hundred outcomes 1 through 6, like 2, 5, 3, 6, 1, 3, 3, 2, etc. The probability that one hundred trials result in any *particular* set of numbers is always less than the combined probability of all the rest of the possible sets of numbers, exactly to the same extent as it is for all 4s. For example, the probability that one hundred consecutive trials will result in the disordered set of numbers 2, 4, 1, 5, 2, 6, 2, 3, 3, 4, 4, 6, 1, etc.—which is not a "special" event—is less than the combined probability of all other about 10^{77} possible combinations of outcomes, including all 4s, to the same extent as this is true for the "special" event of all 4s itself.

Creationists usually proceed to assert that the "special" event whose probability is extremely small simply would not happen. However, this argument can be equally applied to any competing event whose probability is equally extremely small. In the case of a set of one hundred trials, every one of about 10^{77} possible events has the same exceedingly small probability. Nevertheless, one of them must necessarily take place. If we accept the probabilistic argument of the creationists, we will have to conclude that *none* of the about 10^{77} possible events could have happened—which is an obvious absurdity.

Of course, nothing in probability theory forbids any event to be special in some sense, and spontaneous emergence of life qualifies very well for the title of a special event. However, being special from our human viewpoint in no way makes this or any other event stand alone from the standpoint of probability estimation. Therefore probabilistic arguments are simply irrelevant when the spontaneous emergence of life is discussed.

Let us look once more, by way of a very simple example, at the argument based on the very small probability of a special event versus nonspecial ones. Consider a case when the events under discussion are sets of three consecutive tosses of a coin. The possible events are as follows: HHH, HHT, HTT, HTH, TTT, TTH, THH, and THT. Let say that for some reasons we view events HHH and TTT as "special," while the rest of the possible events are not. If we adopt the probabilistic arguments of creationists, we can assert that the probability of a special event, say, HHH (which is in this case 1/8), is less then the probability of the event "not HHH" (which is 7/8). This assertion is true. However, it does not at all mean that event HHH is indeed special from the standpoint of probability. Indeed, we can assert by the same token that the probability of any other of the eight possible events; for example, HTH (which is also 1/8) is less than the probability of event "not HTH" (which is 7/8). There are no probabilistic reasons to see event HHH as happening by a miracle. Its probability is not less than that of any of the other seven possible events. This conclusion is equally applicable to situations in which not eight but billions of billions of alternative events are possible.

Thus the "all 4s" type of argument has no bearing on the question of the spontaneous emergence of life. (It must be added that the probabilistic argument of creationists also fails for another reason: scientific theories of the emergence of life do not assume that life emerged in one step as a result of a purely chance encounter of suitable molecules; the process of the emergence of the first "replicator" molecules, which served as the beginning of living matter, is viewed as comprising many steps, the probability of each being immensely larger than the instant emergence of, say, a protein or an RNA molecule in one fell swoop.)[7]

I will return to the discussion of supposedly special versus nonspecial events in a subsequent section of this chapter.

TESTS WITHOUT REPLACEMENT

Now I will discuss situations in which the probabilities calculated before the first trial cannot be directly multiplied to calculate the probability of an event.

Imagine a box containing six balls identical in all respects except for their colors. Let one ball be white, two balls red, and three balls green. We randomly pull out one ball. (The term *random* in this context is equivalent to the previously introduced concepts of an honest coin and an honest die.) What is the probability that the randomly chosen ball is of a certain color? Since all balls are otherwise identical and are chosen randomly, each of the six balls has the same probability of 1/6 to be chosen in the first trial. However, since the number of balls of different colors varies, the probability that a certain color is chosen is different for white, red, and green. Since there is only one white ball available, the probability that the chosen ball will be white is 1/6. Since there are two red balls available, the probability of choosing a red ball is 2/6 = 1/3. Finally, since there are three green balls available, the probability that the chosen ball happens to be green is 3/6 = 1/2.

Assume first that the ball chosen in the first trial happens to be red. Now, unlike in the previous example, let us proceed to the second trial without replacing the red ball. Hence, after the first trial there remain only five balls in the box, one white, one red and three green. Since all these five balls are identical except for their color, each of them has the same probability, 1/5, of being randomly chosen in the second trial. What is the probability that the ball chosen in the second trial is of a certain color? Since there is still only one white ball (W) available, the probability of choosing that ball randomly is 1/5. There is now only one red ball available, so the probability of choosing a red ball (R) randomly is also 1/5. Finally, for a green ball (G) the probability is 3/5. So if in the first trial a red ball was randomly chosen, the probabilities of balls of different colors to be randomly chosen in the second trial are 1/5 (W), 1/5 (R), and 3/5 (G).

Assume now that in the first trial not a red, but a green ball was randomly chosen. Again, adhering to the no-replacement procedure, we proceed to the second trial without replacing the green ball in the box. Now there remain again only five balls available, one white, two red, and two green. What are the probabilities that in the second trial balls of specific colors will be randomly chosen? Each of the five balls available has the same probability of being randomly chosen, 1/5. Since, though, there are only one white, two red, and two green balls available, the probability that the ball randomly chosen in the second trial happens to be white is 1/5, while for red or green balls it is 2/5.

Hence, if the ball chosen in the first trial happened to be red, then the probabilities in the second trial would be 1/5 (W), 1/5 (R), and 3/5 (G). If, though, the ball chosen in the first trial happened to be green, then the probabilities in the second trial would change to 1/5 (W), 2/5 (R), and 2/5 (G).

The conclusion: in the case of trials without replacement, the probabilities of outcomes in the second trial depend on the actual outcome of the first trial; hence in this case the tests are *not* independent.

When the tests are not independent, the probabilities calculated separately for each of the sequential trials cannot be directly multiplied. Indeed, the probabilities, calculated before the first trial, were as follows: 1/6 (W), 2/6 = 1/3 (R) and 3/6 = 1/2 (G). If we multiplied the probabilities as in the case of independent tests, the probability, for example, of the event RR would be 1/3 × 1/3, which equals 1/9. Actually, though, the probability of that event is 1/3 × 1/5, which is 1/15. Of course, probability theory provides a way to deal with the no-replacement situation, using the concept of so-called conditional probabilities. However, some writers utilizing probability calculations seem to be unaware of the distinction between independent and nonindependent tests; ignoring this distinction has been a source of crude errors in their work.

One example of such erroneous calculations of probabilities is how some proponents of the so-called Bible code estimate the probability of the appearance in a text of certain letter sequences (see also chapter 14).

The letter sequences in question are the so-called equidistant letter sequences (ELSs; recall them from chapter 6). For example, in the preceding sentence the word *question* includes the letter *s* as the fourth letter from the left. Skip the preceding letter *e* and there is the letter *u*. Skip again the preceding letter *q* and the space between the words (which is to be ignored) and there is the letter *n*. The three letters, *s*, *u*, and *n*, separated by "skips" of two letters, constitute the word *sun* if read from right to left. This is an ELS with a "skip" of –2. There are many such ELSs, both from right to left and from left to right, in any text.

As mentioned in several previous sections, there are people who are busy looking for arrays of ELSs in the Hebrew Bible, believing these arrays have been inserted into the text of the Bible by the divine Creator and constitute a meaningful code. As one of the arguments in favor of their beliefs, the proponents of the code attempt to show that the probability of such arrays of ELSs happening in a text by sheer chance is exceedingly small, and therefore the presence of those arrays of ELSs must be attributed to divine design.

There are a few publications in which attempts have been made to apply an allegedly sound statistical test to the question of the Bible code. The methodology used by Witztum, Rips, and Rosenberg (WRR),[8] for example, has been thoroughly analyzed in a number of critical publications and shown to be deficient (see chapter 14). This methodology goes further than the application of probability theory, making use of some tools of mathematical statistics, and therefore is not discussed here since this chapter is only about probability calculations. However, besides the article by WRR and some other similar publications, there are many publications where no real statistical analysis is attempted, but only "simple" calculations of probabilities are employed.[9] There are common errors in these publications, one being the multiplication of probabilities in cases when the tests are not independent. (There are also many Internet publications in which a supposedly deeper statistical approach is utilized to prove the existence of the Bible code. These calculations purport to determine the probability of appearance in the text not just of individual ELSs, but of whole clusters of such. Such analysis usually starts with the same erroneous calculation of probabilities of individual words as examined in the following paragraphs.) Usually the authors start by choosing a word whose possible appearance as an ELS in the given text is being explored. When such a word has been selected, its first letter is therefore determined. The next step is estimating the probability that the letter in question will appear at arbitrary locations in a text. The procedure is repeated for every letter of the chosen word. After having allegedly determined the probabilities of occurrence of each letter of a word constituting an ELS, the proponents of the code then multiply the calculated probabilities, claiming to have determined the probability of the occurrence of the given ELS.

Unfortunately, however, such multiplication is illegitimate. Indeed, a given text comprises a certain set of letters. When the first letter of an ELS has been chosen (and the probability of its occurrence *anywhere* in the text has been calculated) this makes all the sites in the text occupied by that letter inaccessible to any other letter. Let us assume that the first letter of the word in question is X, and it happens x times in the entire text, whose total length is n letters. The proponents of the code calculate the probability of X occurring at any arbitrary site as x/n. Already at this step, the calculation is erroneous. It would be correct only for a random collection of n letters, among which letter X happens x times. However, for a meaningful text this calculation is wrong.[10] Since we wish at this time to address only the question of the test's independence, though, let us accept the described calculation for the sake of discussion. As soon as letter X has been selected, and the probability of its occurrence at *any* location in the text allegedly determined, the number of sites accessible for the second letter in the chosen word decreases from n to $n - x$. Hence, even if we accept the described calculation, then the probability that the second letter (let us denote it *Y*) appears at an arbitrary, still accessible site is now $y/(n - x)$, where y is the number of occurrences of letter *Y* in the entire text. It is well known that the frequencies of various letters in meaningful texts are different. For example, in English the most commonly occuring letter is *e*, whose frequency (about 12 percent) is about 180 times larger than that of the least common letter, *z* (about 0.07 percent).[11]

Hence, depending on which letter is the first one in the chosen word, i.e., on what the value of x is, the probability of the occurrence of the second letter, estimated as $y/(n - x)$, will differ.

Therefore we have in the described case a typical situation without replacement, where the outcome of the second trial (the probability of *Y*) depends on the outcome of the preceding trial (which in turn depends on the choice of *X*). Therefore the multiplication of calculated probabilities performed by the code proponents as the second (as well as the third, the fourth, etc.) step of their estimation of ELS probability is illegitimate and produces meaningless numbers of alleged probabilities.

The probabilities of various individual letters appearing at an arbitrary site in a text are not very small (mostly between about 1/8 and 1/100). If a word consists of, say, six letters, the multiplication of six such fractions will result, though, in a very small number, which is then considered to be the probability of an ELS but is actually far from the correct value of the probability in question.

Using $y/(n - x)$ instead of y/n, thus correcting one of the errors of such calculations, would not suffice to make the estimation of the probability of an ELS reliable. The correct probability of an ELS could be calculated based on certain assumptions in regard to the text's structure that distinguish meaningful texts from random conglomerates of letters. However, there is no mathematical model of meaningful texts available, and therefore the estimations of the ELS probability, even if calculated accounting for interdependence of tests, would have little practical meaning until such a mathematical model is developed.

Finally, the amply demonstrated presence of immense numbers of various ELSs in both biblical and any other texts, in Hebrew as well as in other languages (see chapter 14), is the simplest and most convincing proof that the allegedly very small probabilities associated with ELSs, as calculated by the proponents of the Bible code, are of no evidential merit.

COGNITIVE ASPECTS OF PROBABILITY

So far I have discussed the quantitative aspects of probability. I will now discuss probability from a different angle, namely its cognitive aspects. This discussion will be twofold. One side of the cognitive meaning of probability is that it essentially reflects the amount of information available about possible events. The other side of the cognitive aspect of probability is the question of what the significance of this or that probability value essentially is.

Imagine that you want to meet your friend, who works for a company with offices in a multistory building downtown. Close to 5 P.M. you are on the opposite side of the street, waiting for your friend to come out of the building. Let us imagine that you would like to estimate the probability that the first person coming out will be male. You have never been inside that building, so you have no knowledge of the composition of the people working in that building. Your estimate most likely will be that the probability that the first person coming out will be male is 1/2, and the same probability for a female. Let us further imagine that your friend who works in that building knows that among the people working there, about 2/3 are female and about 1/3 are male. His estimate will then be that the probability that the first person coming out will be male is 1/3 rather than 1/2. Obviously, the objective likelihood of a male coming out first does not depend on who makes the estimate: it is 1/3. The different estimates of probability are due to a factor that has no relation to the subject of the probability estimation: the different level of information about the subject possessed by you and your friend. Because of a very limited knowledge about the subject, you have to assume that the two possible events—a male or a female coming out first—are equally probable. Your friend, however, knows more; in particular, he knew that the probability of a female coming out first was larger than that of a male coming out first.

This example illustrates an important property of calculated probability: it reflects the level of knowledge about a subject. If we possess full knowledge of the subject, we know exactly, in advance, the outcome of a test, so instead of probability we deal with certainty.

One case in which we have full knowledge of a situation is when an event has actually occurred. In such a situation the question of the probability of the event is meaningless. After the first person has actually come out of the building, the question of the probability of that event becomes moot. Of course we still can calculate the probability of that event, but in doing so we necessarily deal with an imaginary situation, assuming the event has not yet occurred.

Serving as a reflection of the level of knowledge about a subject is the most essential feature of probability, from the viewpoint of its cognitive essence.

What about the examples with a coin or a die, where we thought we possessed full knowledge of all possible outcomes, and where all those possible outcomes definitely seemed to be equally probable? We did not possess such knowledge! Our assumption of the equal probability of either heads or tails, or of the equal probability of each of the six possible outcomes of a trial with a die was due to our limited knowledge about the actual properties of the coin or of the die. No coin or die is perfect; therefore, in tests with a coin, either heads or tails may have a slightly better chance of occurring. Likewise, in tests with a die, some of the six sides of the die may have a slightly better chance of facing upward. For example, in tests conducted by British mathematician Karl Pearson with a coin in 1921, after it was tossed 24,000 times, heads occurred in 12,012 trials, tails in 11,988 trials.[12] Generally speaking, a slight difference between the numbers of heads and tails is expected in a large sequence of truly random tests. On the other hand, we cannot exclude the possibility that the described result was due, at least partially, to a certain imperfection in the coin used or in the procedure employed.

Since we have no knowledge of the particular subtle imperfections of a given coin or die, we have to postulate the equal probability of all possible outcomes.

In the tests with a die or a coin, we at least know all possible outcomes. There are many situations in which we have no such knowledge. If that is the case, we have to assume the existence of some supposedly possible events that are actually impossible but that we simply cannot rule out.

For example, assume we wish to estimate the probability that upon entering a property at 1236 Honey Street, the first person we meet will be an adult man. We have no knowledge of that property. In particular, we don't know if it is occupied by a group of monks, meaning that there will be no women or children at that address. If we also don't know the percentage of women, men, male children, and female children in that town, we have to guess that the probability of encountering each of the four supposedly possible types of a person is the same—1/4. If we knew that this residence was indeed occupied by monks, we would estimate the probability of encountering an adult female or children of either sex as being very small. Then the probability of meeting an adult man would be calculated as close to one, i.e., to certainty.

Quite often, very small calculated probabilities of certain events are due to the lack of information and hence to an exaggerated number of supposedly possible events, many of which are actually impossible. One example of such a greatly underestimated probability of an event is that of the spontaneous emergence of life. The calculations in question are based on a number of arbitrary assumptions and deal with a situation whose details are largely unknown. Therefore, in such calculations the number of possible events is greatly exaggerated, and all of them are assumed to be equally probable, which leads to extremely small values of calculated probability. Actually, many of the allegedly possible paths of chemical interactions may be impossible, and those possible are by no means equally probable. Therefore (and for some other reasons as well) the extremely small probability of the spontaneous emergence of life must be viewed with the utmost skepticism.

Of course, it is equally easy to give an example of a case in which insufficient knowledge of the situation results not in an increased but rather in a decreased number of supposedly possible outcomes of a test. Imagine that you made an appointment over the phone to meet John Doe at the entrance to his residence, but you have never before seen his residence. When you arrive at his address you discover that he lives in a large apartment house which seems to have two entrances at the opposite corners of the building. You therefore have to watch both entrances. Your estimate of the probability that John would exit from the eastern door is 1/2, the same as from the western door. The estimated probability, 1/2, results from your assumption of the equal probability of John's choosing either of the exits and from your knowledge that there are two exits. However, what if the building has also one more exit in the rear, which you are not aware of? If you knew that fact, your estimated probability would drop to 1/3 for each of the doors. Insufficient knowledge (you only knew about two possible outcomes) led you to an increased estimated probability compared with one accounting for all three possible outcomes.

Now let us discuss the other side of the cognitive aspect of probability. What is the real meaning of a calculated value of probability if it happens to be very small?

Consider first the situation when all possible outcomes of trials are supposedly equally probable. Assume the probability of an event, A, was calculated as $1/n$, where n is a very large number, making the probability of the event very low. Often, such a result is interpreted as an indication that the event in question should be considered, for all intents and purposes, as practically impossible. However, such an interpretation, which may be psychologically attractive, has no basis in probability theory. The actual meaning of that value of $1/n$ is just that—the event in question is one of n equally probable events. If event A has not occurred, it simply means that some other event, B, has occurred instead. But event B had the same very low probability of occurring as event A. So why could the low-probability event B actually occur but event A, which had the same probability as B, could not occur?

An extremely low value for a calculated probability has no cognitive meaning in itself. Whichever one of n possible events actually occurs, it necessarily has the same very low probability as the others but occurs nevertheless. (Therefore, as mentioned earlier, the assertion of impossibility of such events as the spontaneous emergence of life, based on its calculated very low probability, has no merit.)

If possible events are actually not equally probable, which is a more realistic approach, a very low calculated probability of an event has even less of a cognitive meaning, since its calculation ignores the possible existence of preferential chains of outcomes which could ensure a much higher probability for the event in question.

This discourse may give an impression in the minds of some readers that my thesis was to show that the concept of probability is really not very useful since its cognitive content is very limited. However, this was by no means my intention. When properly applied and if not expected to produce unrealistic predictions, the concept of probability may be a very potent tool for shedding light on many problems in science and engineering. When applied improperly and if expected to be a magic bullet to produce predictions, it often becomes misleading and a basis for a number of unfounded and sometimes ludicrous conclusions. The real power of properly calculated and interpreted probability is not in the calculations of probability of this or that event, when it is indeed of a limited value, but when the probability is utilized as an integrated tool within the much more sophisticated framework of either mathematical statistics or statistical physics.

PSYCHOLOGICAL ASPECTS OF PROBABILITY

Scientific theories often seem to contradict common sense. When this is the case, it is usually common sense that is deceptive, while the assertions of science are correct. The whole science of quantum mechanics, one of the most magnificent achievements of the human mind, seems to be contrary to common sense based on everyday human experience.[13]

One good example of the above contradiction is related to the motion of spacecraft in orbit about a planet. If two spacecraft are moving in the same orbit, one behind the other, what should the pilot of the second craft do if she wishes to overtake the first? Common sense tells us that the pilot has to increase the speed of her craft along the orbital path. Indeed, that is what we do when we wish to overtake a car that is ahead of us on a road. However, in the case of an orbital flight, common sense is wrong. To overtake the first spacecraft, the pilot of the second craft must *decrease* rather than increase her speed. This theoretical conclusion of the science of mechanics has been confirmed in multiple space flights despite its seeming contradiction to the normal experience of car drivers, pedestrians, runners, and horsemen. Likewise, many conclusions of probability theory may seem to contradict common sense, but nevertheless probability theory is correct, while common sense in those cases is wrong.

Consider an experiment with a die, where events are sets of ten trials each. We assume an honest die as well as independence of outcomes. If we toss the die once, each of the six possible outcomes has the same chance of happening, 1/6. Assume that in the first trial the outcome was, say, 3. Then we toss the die the second time. It is the same die, tossed in the same way, with the same six equally probable outcomes. To get an outcome of 3 is as probable as any of the five other outcomes. The tests are independent, so the outcome of each subsequent trial does not depend on the outcomes of any of the preceding trials.

Now toss the die in sets of ten trials each. Assume that the first event is as follows: *A* (3, 5, 6, 2, 6, 5, 6, 4, 1, and 1). We are not surprised in the least since we know that there are 6^{10} (that is, 60,466,176) possible, equally probable events. Event *A* is just one of them and does not stand alone in any respect among those over sixty million events, so it could have happened in any set of ten trials as well as any other of those sixty million variations of numbers. Let us assume that in the second set of ten trials the event is *B* (6, 5, 5, 2, 6, 2, 3, 4, 1, and 6). Again, we have no reason to be surprised by such a result since it is just another of those millions of possible events, and there is no reason for it not to happen. So far the probability theory seems to agree with common sense.

Assume now that in the third set of ten trials the event is *C* (4, 4, 4, 4, 4, 4, 4, 4, 4, and 4—the "all 4s" event described earlier in the chapter). I am confident that in such a case everybody would be amazed, and the immediate explanation of that seemingly "improbable" event would be the suspicion that either the die has been tampered with or that it was tossed using some sleight of hand.

While cheating cannot be excluded, this event does not necessarily require such an assumption. Indeed, what was the probability of event *A*? It was one in over sixty million. Despite the exceedingly small probability of *A*, its occurrence did not surprise anybody. What was the probability of event *B*? Again, only one in over sixty million but we were not amazed at all. What was the probability of event *C*? The same one in over sixty million, but this time we are amazed.

Why does "all 4s" seem amazing? Only for psychological reasons. It seems easier to assume cheating on the part of the dice-tossing player than the never-before-seen occurrence of all 4s in ten trials. What is not realized is that the overwhelming majority of events other than this one were never seen, either. There are so many possible combinations of ten numbers, composed of six different unique numbers, that each of them occurs extremely rarely. The set of ten identical numbers seems psychologically to be "special" among combinations of different numbers, but for probability theory this set is not special. To view an event as special means abolishing the premise of the probabilistic estimate—the postulate of a fair die.

Of course, if the actual event is highly favorable to one of the players, it justifies a suspicion of cheating. The reason for that is our experience, which tells us that cheating is rather probable when a monetary or other award is in the offing. However, the probability of cheating is irrelevant to our discussion, because it is just a peculiar feature of the example of a die; when discussing the question of the spontaneous emergence of life, no analog of cheating is present. Therefore the proper analogy is one in which cheating is excluded so that the postulate of perfect randomness (whose particular cases are the postulates of a fair coin and fair die) can hold. If the possibility of cheating is excluded, only the mathematical probability of any of the over sixty million equally probable events has to be considered. In short, probability theory is a part of science and has been overwhelmingly confirmed to be a valid theory of great power. There is no doubt that the viewpoint of probability theory is correct, even in the face of contradictory common sense. Such a human psychological reaction to an improbable event such as ten identical outcomes in a set is as wrong as the suggestion to a pilot of a spacecraft lagging behind to increase her speed if she wishes to overcome a craft ahead of hers in orbit.

The Case of Multiple Wins in a Lottery

Another example of erroneous attitude to an improbable event, based on psychological reasons, is the case of multiple wins in a lottery.

Consider a simple raffle in which only one hundred tickets are on sale. Assume fraud is excluded. Each of the tickets has the same probability of winning, namely 1/100. Let us assume John Doe is the lucky one. We congratulate him, but nobody is surprised by John's win. Out of the one hundred tickets one must necessarily win, so why shouldn't John be the winner?

Assume now that the raffle has not one hundred but ten thousand tickets sold. In this case the probability of winning was the same for each ticket, namely 1/10,000. Assume Jane Jones won in that lottery. Are we surprised? Of course not. One ticket out of ten thousand had to win, so why shouldn't it be Jane's? The same discussion is applicable to any big raffle where there are hundreds of thousands or even millions of tickets. Regardless of the number of tickets available, one of them, either sold or unsold, must necessarily win.

Now let us return to the small raffle with only one hundred tickets sold. Recall that John Doe won it. Assume now that, encouraged by his win, John decides to play once again. John has already won once; the other ninety-nine players have not yet won at all. What is the probability of winning in the second run? For every one of the one hundred players, including John, it is again the same 1/100. Does John's previous win provide him with any advantages or disadvantages compared to the other ninety-nine players? None whatsoever.

Assume now that John wins again. It is as probable as that any of the other ninety-nine players winning this time, so why shouldn't it be John? However, if John wins the second time in a row, everybody is amazed by his luck. Why the amazement?

Let us calculate the probability of a double win, based on the assumption that no cheating was possible. The probability of winning in the first run was 1/100. The probability of winning in the second run was again 1/100. The events are independent, making the probability of winning twice in a row $1/100 \times 1/100$, which is one in ten thousand. It is exactly the same probability as it was in the raffle with ten thousand tickets played in one run. When Jim won that raffle, we were not surprised at all, despite the probability of his win being only 1/10,000, nor should we have been. So why should we be amazed at John's double win, whose probability was exactly the same?

If a raffle is played only once, and n tickets have been distributed, covering all n possible versions of numbers, of which each one has the same chance to win, then the probability that a *particular* player wins is $p(P) = 1/n$, while the probability that *someone* out of n players (whoever he or she might be) wins is $p(S) = 1$ (i.e., 100 percent).

If, though, the raffle is played k times, and each time x players participate, where $x^k = n$, the probability that a *particular player* wins k times in a row is again the same $1/n$. Indeed, in each game the probability of winning for a particular player now is $1/x$. The games are independent of each other. Hence the probability of winning k times equals the product of probabilities of winning in each game; i.e., it is $(1/x)^k = 1/n$.

However, the probability that *someone* (whoever he or she happens to be) wins k times in a row is now not 1, but not more than x/n, that maximum value corresponding to the situation in which the same x players play in all k games. Indeed, for each particular player the probability of winning k times in a row is $1/n$. Since there are x players, each with the same chance to win k times, the probability of *someone* in that group winning k times in a row is $x \times 1/n$, i.e., x/n. In other words, in a big raffle played only once, somebody necessarily wins ($p = 1$). On the other hand, in a small raffle played k times, while somebody necessarily wins in *each* of k games, it is likely that nobody wins k times in a row, as the probability of such a multiple win is small.

Here is a numerical example. Let the big raffle be such that n = 1,000,000. If all n tickets are distributed, the probability that John Doe wins is 1/1,000,000. However, the probability that somebody (whoever he or she happens to be) wins is 1 (i.e., 100 percent).

If the raffle is small, such that only $x = 100$ tickets are distributed, the probability of any *particular* player winning in a given game is 1/100. If $k = 3$ games are played, the probability that John Doe wins 3 times in a row is $(1/100)^3$, which is again 1/1,000,000, exactly as it was in a one-game raffle with n = 1,000,000 tickets.

However, the probability that *someone* (not a particular player) wins three times in a row, whoever he or she happens to be, is now not 100 percent but not more than x/n = 100/1,000,000—i.e., .0001, which is 10,000 times less than in a one-game raffle with one million tickets. Hence, such a small raffle may be played time after time after time without anybody winning k times in a row. Actually, such a multiple win must be very rare.

When John Doe wins three times in a row, we are amazed not because the probability of that event was one in a million (which is the same as for a single win in a big one-game raffle) but because the probability of *anyone* winning three times sequentially in the three-game small raffle is ten thousand times less than it is in a one-game big raffle. Hence, while in the big raffle played just once the fact that somebody won is 100 percent probable (i.e., it is certain), in the case of a small raffle played three times a triple win is a rare event of low probability (in our example, 1/10,000).

However, if we adhere to the postulate of a fair game, a triple win is still not a special event, despite its low probability—it is as probable as any other combination of three winning tickets, namely in our example 1/1,000,000. To suspect fraud means to abolish the postulate of a fair game. Indeed, if we know that fraud is possible, intuitively, we compare the probability of an honest triple win with the probability of fraud. Our estimate is that the probability of an honest triple win (in our case 1/10,000) is less than the probability of fraud (which in some cases may be quite high).

This discussion relates only to a raffle-type lottery. If the lottery is what sometimes is referred to as the Irish lottery, the situation is slightly different. In this type of a lottery, the players themselves choose sets of numbers for their tickets. For example, in the California state lottery each player has to choose any six numbers between 1 and 50. There are about sixteen million possible combinations of such sets of six numbers. This means that there can be not more than about sixteen million tickets with differing sets of the chosen six numbers. However, nothing prevents two or more players from coincidentally choosing the same set of six numbers. (Such a coincidence is impossible in a raffle, where all tickets distributed among the players each have unique numbers.) If more than one player has chosen the same set of six numbers, this diminishes the probability that *somebody* will win the lottery.

The calculation of the probability of someone (not a particular player) winning in such a lottery, which is necessarily more mathematically complex, is given in the appendix. It shows that in an Irish lottery the probability of someone winning is at least about 37 percent, which is thousands of times larger than the probability of the same player winning more than once in a row.

From these calculations we can conclude that when a particular player wins more than once in consecutive games, we are amazed not because the probability of winning for that particular player is very low, but because the probability of anybody (whoever he or she happens to be) winning consecutively in more than one game is much less than the probability of someone winning only once in an even larger lottery. We intuitively estimate the difference between the two situations. However, the important point is that what impresses us is not the sheer small probability of someone winning against enormous odds. This probability is equally small in the case of winning only once in a big lottery, but in that case we are not amazed. This also illustrates the psychological aspect of probability.

The Role of "Special" Events in Probability

Let us briefly discuss the meaning of the term *special event*. When stating that none of the *n* possible, equally probable events was in any way special, I only meant to say that it was not special from the standpoint of its probability. Any event, while not special in this sense, may be very special in some other sense.

Let us imagine a die, whose six sides bear, instead of numbers 1 through 6, the six letters *A*, *B*, *C*, *D*, *E*, and *F*. Let us imagine further that we toss the die in sets of six trials each. In such a case there are $6^6 = 46,656$ possible, equally probable events. Among those events are the following three: *ABCDEF*, *AAAAAA*, and *FDCABE*. Each of these three events has the same probability, 1/46,656. Hence, from the standpoint of probability none of these three events is special.

However, each of these three events may be special in a different sense. Indeed, for somebody interested in alphabets, the first of the three events may seem to be very special since the six outcomes occur in alphabetical order. Of course, alphabetical order in itself has no intrinsic special meaning—thus a person whose language is, for example, Chinese, would hardly see anything special in that particular order of symbols. The second event, with its six identical outcomes, may seem miraculous to a person inclined to see miracles everywhere and to attach some special significance to coincidences, many of which happen all the time. The third event seems to be not special but rather just one of the large number of possible events. However, imagine a person whose first name is Franklin, middle name is Delano, and whose last name is (no, not Roosevelt!) Cabe. For this person the six trials resulting in the sequence *FDCABE* may look as if his name, F. D. Cabe, was miraculously produced by six throws of dice, despite the probability of such a coincidence being only 1/46,656.

Whatever special significance this or that person may be inclined to attribute to any of the possible, equally probable events, none of them is special from the standpoint of probability. This conclusion is equally valid regardless of the value of the probability, however small it happens to be.

In particular, the extremely small probability of the spontaneous emergence of intelligent life, as calculated (usually not quite correctly) by the opponents of the hypothesis of the spontaneous emergence of life, by no means indicates that the spontaneous emergence of life must be ruled out. (There are many nonprobabilistic arguments both in favor of creationism and against it, which I will not discuss in this chapter.) The spontaneous emergence of life could be an extremely unlikely event, but all other alternatives are extremely unlikely as well. One out of *n* possible events must occur, and there is nothing special in that from the standpoint of probability, even though it may be very special from your or my personal viewpoint.

NOTES

1. For an excellent introductory course on probability theory, combining clarity and a good, rigorous level of difficulty, see Meyer Dwass, *The First Steps in Probability* (New York: McGraw-Hill, 1967).

2. A very well-written course on probability, mathematical statistics, and theory of games is Nikolai B. Tikhomirov, *Teoriya veroyatnostej i matematicheskaya statistika* (Theory of probability and mathematical statistics) (Kalinin, USSR: Kalinin State Institute Press, 1971). In particular, this book contains a concise but quite rigorous explanation of Kolmogorov's axiomatic foundations of probability. Unfortunately, there is no English translation this course well deserves.

3. In many courses on probability theory and mathematical statistics (for example in Dwass, *The First Steps in Probability*, p. 102) the independence of tests is often treated in a manner opposite to that rendered in this chapter; namely, by establishing that individual events are independent if the probability of the combination of the events equals the product of their individual probabilities.

4. Such definitions can be found, for example, in Tikhomirov, *Teoriya veroyatnostej i matematicheskaya statistika*.

5. Dwass, *The First Steps in Probability*, p. 101.

6. The creationist viewpoint using the probability argument has been evinced in multiple publications by early creationists, such as Harry Rimmer, *Modern Science and the Genesis Record* (Grand Rapids, Mich.: Eerdmans, 1937); Henry Morris, *The Bible and Modern Science* (Chicago: Moody Press, 1956), and others, and revived lately in books and articles by William A. Dembski (see chapter 1), Michael J. Behe (chapter 2), Nathan Aviezer (chapter 9), Gerald L. Schroeder (chapter 10), Lee M. Spetner (chapter 11), and many others.

7. Richard Dawkins, *Climbing Mount Improbable* (New York: W. W. Norton, 1996).

8. Doron Witztum, Eliahu Rips, and Yoav Rosenberg, "Equidistant Letter Sequences in the Book of Genesis," *Statistical Science* 9, no. 3 (1994):429–38.

9. For example, Michael Drosnin, *The Bible Code* (New York: Simon and Schuster, 1997); Jeffrey Satinover, *Cracking the Bible Code* (New York: William Morrow, 1997).

10. For a more detailed explanation, see Mark Perakh and Brendan McKay, "Study of Letter Serial Correlation (LSC) in Some Hebrew, Aramaic, Russian, and English Texts" [online], www.nctimes.net/~mark/Texts/ [April 7, 2002].

11. Ibid.

12. Richard J. Larsen and Morris L. Marx, *Introduction to Mathematical Statistics* (New York: Prentice-Hall, 1986), p. 397.

13. For a relatively elementary explanation of the paradoxes of quantum mechanics and relativity, as well as of the seemingly paradoxical effects of orbital motion, see David Halliday, Robert Resnick, and Jearl Walker, *Fundamentals of Physics*, extended ed. (New York: John Wiley and Sons, 1993).

14
THE RISE AND FALL OF THE BIBLE CODE

The subject of this chapter is a story which is both comic and sad. It is a story of how two people—one, Eliyahu Rips, a brilliant mathematician with a record of excellent achievements in modern mathematics, and the other, Doron Witztum, without any scientific credentials but obviously very smart and ingenious—conducted what looked like a serious scientific research and claimed sensationally sounding results. However, when other scientists tried to reproduce the data they obtained with the assistance of Yoav Rosenberg, most of them concluded that the researchers' claims were unsubstantiated.

The history of science knows more than one example of alleged discoveries which turned out to be erroneous. Their "discoverers" reacted to the disproval of their claims in various ways. Many of them promptly admitted their errors; of course, this is the most respectable way to deal with the situation and the best way to preserve a good reputation. The heroes of this story, however, continue to assert the validity of their claims despite all the proofs against them. This is also a story of how an allegedly scientific discovery, despite being shown to be unsubstantiated, was appropriated with enthusiasm by a crowd of people who did not understand the essence of the work of the theory's originators but were delighted by its religious implications and vulgarized it to such an extent that even its original authors had to assert the absurdity of the claims of their epigones.

A BIT OF HISTORY

Statistical Science is a prestigious peer-reviewed scientific journal published by the Institute of Mathematical Statistics. It prints papers on a variety of subjects, dealing with the application of statistical methods to any area of research. Its readership is limited mainly to faculty and researchers interested in the use of mathematical statistics in various fields of science and technology. Unexpectedly, though, one article printed in that journal in 1994—written by the aforementioned Witztum, Rips, and Rosenberg (hereafter referred to as WRR)—found a much broader audience since it was reprinted in full in a best-selling sensational book.[1]

The essence of WRR's paper was as follows. First, these writers showed that in the Hebrew text of Genesis there are large numbers of equidistant letter sequences (ELSs; recall these from previous chapters). This term denotes words formed in a text by letters separated by equal intervals ("skips"). For example, look at the word *DenOteS* in the preceding sentence. The letters *D*, *O*, and *S* are separated by equal intervals of 2 letters (*en* between *D* and *O*, and *te* between *O* and *S*). The three equidistant letters form the acronym *DOS*, which stands for *Disk Operating System* and constitutes an ELS with a *skip* of 3. According to WRR, the ELSs that run in the direction of the text (which in Hebrew is from right to left) and against the text's flow (in Hebrew from left to right) are equally valid. In the latter case the skip is negative. For example, if the word *DenOteS* read from right to left, the letters *S*, *O*, and *D* constitute an ELS spelling *SOD* with a skip of –3. Naturally, every word of the text itself is also an ELS with a skip of 1.

Furthermore, WRR claimed that in the text of Genesis there are multiple *pairs* of ELSs, with the words of the pair related by meaning and situated within the text in close proximity to each other. The authors maintained that the proximity in question far exceeds what could be expected if those pairs of ELSs appeared in the text by chance alone.

To prove their point, WRR conducted a computerized statistical experiment. They first compiled a list of thirty-four famous rabbis who lived between early medieval times and the eighteenth century and whose biographies in *The Encyclopedia of Great Men in Israel*[2] each occupied at least three columns. Later, following advice from a referee, they compiled a second list of thirty-two "less famous rabbis," whose biographies each occupied between 1½ and 3 columns. Then, with the help of an expert in Judaic bibliography, Professor Shlomo Havlin of Bar-Ilan University in

Israel, they matched the rabbis' various appellations with the dates of those rabbis' birth and/or death, according to the Responsa database maintained by Bar-Ilan University. In Hebrew, dates are written using letters of the alphabet (see chapter 6), so WRR searched in the text of Genesis for ELSs both for the appellations and for the dates.

They suggested a formula which supposedly estimates the "distance" between any two ELSs and used it to "measure" distances between ELSs for appellations and dates. Then they shuffled the lists of appellations and dates, creating 999,999 scrambled lists, where the rabbis' names and the pertinent dates were randomly mismatched. Their computer program measured the distance for each pair of appellations/dates, both in the correct and in the mismatched lists. Then the program calculated four aggregate measures of proximity between appellations and dates for each one of the scrambled lists. They denoted the aggregate criteria of proximity as statistics P_1, P_2, P_3, and P_4. The values of these proximity measures were then ranked, assigning rank 1 to the list of appellations/dates which displayed the "closest" proximity (that is, the lowest values for the statistics identified above), rank 2 to the list with the next higher values, etc. Somewhere in the rankings was the original, correct list of appellations/dates.

The results claimed by WRR were nothing less than astonishing. The ranks of the "correct" (unscrambled) list of appellations/dates were very low for each of the four proximity statistics, not exceeding a few thousand out of one million competing scrambled lists, and often much less than that. WRR concluded that the ELSs for the rabbis' appellations and for their correct dates of birth or death are situated in the text of Genesis at *an unusually close proximity*, with the level of confidence 1 in 62,000. WRR stated finally that the proximity of the ELSs for the rabbis' appellations to the ELSs for their correct dates of birth or death in the Book of Genesis was "not due to chance."

The meaning of such a conclusion was quite far-reaching. Since the Book of Genesis was written many centuries before the rabbis in question were born, its author must have known the future. WRR's opinion (shared also by many others who were impressed by the amazing outcome of WRR's experiment) was therefore that their results prove the divine origin of the Book of Genesis. Hence these results were touted as *scientific* proof of the existence of God, and more specifically, of the God of the Torah.

The editor of *Statistical Science*, Robert Kass, published WRR's paper with a comment saying that the paper was a puzzle offered to readers. In subsequent additional comments posted on the Internet, Kass made it clear that the editors did not accept WRR's claims, and hoped that some readers would be willing to invest enough time and effort to unearth the hidden flaws in WRR's methodology.[3]

The publication of WRR's paper had a number of unexpected consequences. One consequence was the appearance of a large number of papers, lectures, Internet postings, etc., whose authors, most of them not capable of comprehending the mathematical apparatus in WRR's article, vulgarized WRR's results by searching for multiple ELSs in the Bible without attempting to verify their claims statistically. Computer programs enabling anyone to search for ELSs in the Bible have been peddled all over the Internet. Some other fast entrepreneurs offer to find, for a fee, the information allegedly "encoded" in the Bible concerning the payer's personal matters. Many other writers and authors of Web postings, referring to WRR's paper as "the proof" of the Bible code, utilize ELSs in the Bible to promote various agendas. One that seems to be most widely spread is that of Christian preachers.

There is an organization named Aish Ha Torah, whose aim is to attract those Jews who have either lost faith or have never been observant back to the fold. This is done by conducting seminars where trained lecturers discuss various facets of Judaism. Even though various topics are presented in these lectures, the subject of the codes in the Torah is strongly emphasized there, as one of the main tools for convincing the doubting participants to re-embrace the faith. The organization maintains a Web site in which it states that its goal is to provide a forum for Witztum and to promote his work.[4]

At least nine books about the Bible code have recently appeared, at least two pseudodocumentaries have been broadcast, and at least one dramatic film has been released, all of them promoting and supporting the code. The largest splash was probably made by the sensational book published by the journalist Michael Drosnin titled *The Bible Code*. This book appeared on the *New York Times* list of bestsellers and was translated into many languages, while its author was featured in *Newsweek* and *Time*, appeared on a number of TV shows, sold movie rights for the book to Warner Brothers, and set out on a worldwide promotion tour, during which he has had some less than polite words for those who dare to doubt the validity of his sensational claims.

Drosnin claims to have discovered in the Bible multiple predictions of future events encoded as clusters of ELSs. The most loudly advertised was the alleged prediction of the assassination of Israeli prime minister Itzhak Rabin (discussed in greater detailed later in this chapter), but Drosnin's book also displays other, similar tables containing ELSs allegedly predicting various events such as the assassination of Anwar Sadat, the terrorist acts by the Palestinians in Jerusalem in our time, an "atomic Holocaust" in the early years of the twenty-first century, and the like.

The book by Drosnin was soon followed by a book titled *Cracking the Bible Code*, authored by psychiatrist Jeffrey Satinover.[5] Strangely, while Drosnin's book was repudiated by Rips and some other adherents of WRR as an invalid vulgarization of WRR's results, Satinover's book met their apparent approval, even though both books contain very similar examples of ELS arrays whose interpretation is not supported by any statistical considerations.

Almost immediately after the publication of WRR's paper, some scientists started expressing skeptical views toward its validity. With time, more and more people involved themselves in the controversy, joining the dispute on both sides.

The dispute in regard to the authenticity of the codes essentially revolves around three questions:

- What is the statistical probability that ELSs occur in the Bible by chance?
- Do similar ELSs occur in texts other than the Bible?
- Are human beings capable, with or without computers, of creating complex arrays of ELSs such as those found in the Bible?

Code proponents usually maintain that the probability sought in the first question is extremely small. To the second and third questions listed above, "code people" usually answer no. All three assertions have been disputed by code opponents.

HOW COMMON ARE ELSs?

The numerous examples of ELSs found in the Bible and demonstrated by Witztum, Drosnin, Satinover, and others have made a strong impression on many. Indeed, it seemed hard, even for me, to imagine that so many coincidences could have happened by chance. Therefore I decided to test first how often various ELSs appear in randomly chosen texts in various languages.

For my first test I chose several short texts in Russian, English, and Hebrew. Following the method used by all Bible code searchers, I removed from the tested texts spaces between words as well as all punctuation marks, converting the texts into continuous strings of letters. As soon as this was done, in all the tested examples, numerous individual ELSs popped into sight from the text. As expected, most of them happened to be three-letter words, but some four-, five-, and six-letter words could be identified on almost every page as well.

Code proponents often demonstrate "arrays" of semantically related ELSs they find in the Hebrew Bible. I could easily find many similar arrays of semantically related words in the form of ELSs in every text I tested. As an example, we will look at a short story I wrote in English many years ago.[6] Figure 14.1 shows the text of this story, in which there are 1,460 letters. This segment was chosen without any preliminary trials, right from the first page.[7] Among the abundant ELSs found in this segment were many three-letter words, fewer four-letter words, still fewer five-letter words, etc. In the figure only a few of these ELSs are marked by straight lines with arrows indicating the direction of reading. (If the skip equals the width of the page, the ELS appears vertically. If the skip differs from the page width, the corresponding ELS appears on a straight line at some angle to the vertical.) In the middle of the page, there is an ELS for the six-letter name *Torvil* (marked 1) situated vertically with a skip of 68 (since the page's width is 68 letters). Parallel to it, in the adjacent vertical, there is an ELS for the word *ice* (marked 2) with the same skip of 68. Across the page, there is an ELS for the name *Dean* (marked 3) with a skip of 70. Close to both *Torvil* and *Dean*, ELSs for the word *win* appear twice, one with a skip of –70 (marked 4) and the other with a skip of –140 (marked 5). Of course, Torvil and Dean were famous champions in figure skating. The story in question was written about ten years before *Torvil* and *Dean* demonstrated their skills on *ice*, *win*ning championships twice. Using the "logic" usually applied by Drosnin and other code proponents, one might suggest that I predicted an event which would occur later, and had "encoded" my prediction in the form of a combination of several ELSs placed close to each other in the text of my story. (I did not.)

Among the ELSs for four-letter words on this page are *land* (not far from the three-letter word *sea*), *lull*, *tilt*, *odor*, etc. I discovered similar occurrences of arrays of ELSs, seemingly related by meaning, on almost any page of any text I tested in Russian, Hebrew, or English.[8] These simple tests have shown that the phenomenon of the ELS is very common, and that many individual ELSs and arrays of ELSs appear in any text. The reason for that is, of course, the fact that any language consists of a vast number of words.

I performed these noncomputerized tests before I came across the Web publications by Brendan McKay and Dave E. Thomas.[9] I was gratified to

```
ASITCOULDBEEXPECTEDTHESTRANGELASTNAMEOFILYABARAKHOKHLOSUPPLIEDANINEX
HAUSTIBLESOURCEOFJOKESBOTHFORTHECAMPSINMATESANDTHEWARDERSITSSIMPLEST
VERSIONALSOEASIERTOPRONOUNCEBARAKHLOWHICHCAMETOMINDATONCEASSOONASTHI
SNAMESOUNDEDDURINGHEADCOUNTSPROVIDEDALOTOFFUNASTHISWORDINTHERUSSIANV
ERNACULARMEANSDEPENDINGONTHECONTEXTEITHERPERSONALBELONGINGSORMOREOFT
ENANYLOWQUALITYSTUFFWHILECONDUCTINGSEARCHESTHEWARDERSWOULDROUTINELYM
OVEAHANDOVERBOTHBARAKHOKHLOSBELONGINGSANDHISHEADANDASKISTHATALLYOURB
ARAKHLOANDCHUCKLEHAPPILYSOMEMORESOPHISTICATEDVERSIONSINCLUDEDSUCHPEA
RLSASKHOKHLOADISTORTEDKHOKHOLTHEDEROGATIVEREFERENCETOUKRAINIANSEVENT
HOUGHHEWASNOTANUKRAINIANORBARABIRAASIMILARREFERENCETOTATARSEVENTHOUG
HHEWASNOTATATARBARAKHOLKAMEANINGAFLEAMARKETOFTENSHORTENEDTOSIMPLYKHO
LKAAHORSENECKWHICHINEVITABLYLEDTOADESIRETOSTRIKEUPONHISNECKINSOMEEVE
NMORECOMPLEXVERSIONSTHECHARACTERSFROMTHEODDNAMEWERECOMBINEDINVARIOUS
PERMUTATIONSWITHWORDSUSUALLYNOTINCLUDEDINLITERARYDICTIONARIESBUTNEVE
RTHELESSQUITECOMMONINTHEEVERYDAYRUSSIANVERNACULARUSUALLYBARAKHOKHLOE
NDUREDTHESEEXERCISESINWITSTOICALLYHEWOULDTHOUGHEAGERLYEXPLAINTHEORIG
INOFHISNAMETOEVERYBODYWILLINGTOLISTENACCORDINGTOTHESTORYTOLDBYBARAKH
OKHLOHEWASBORNINBESSARABIAATTHATTIMESTILLAPARTOFRUMANIAANDLOSTHISPAR
ENTSATAVERYEARLYAGEHEKNEWALMOSTNOTHINGABOUTTHEMEXCEPTFORTHEIRBEINGRU
SSIANTHEIRLASTNAMEBEINGEITHERTIMOKHINORTERYOKHINANDHISGIVENNAMEBEING
```

Fig. 14.1. The first page of a short story by the author. The spaces between the words and punctuation marks have been removed. Certain ELSs are shown by means of arrows running parallel to the ELS and indicating the direction of reading.

find that my conclusion turned out to be in strong agreement with the multiple examples of ELS clusters found by McKay in a number of nonbiblical texts, most notably *Moby Dick*, as well as many examples of similar ELS clusters demonstrated by Thomas.

WAS A SUPERHUMAN MIND NECESSARY TO CREATE THE CODES?

The proponents of the Bible codes claim that the alleged creator of these codes must have possessed superhuman abilities, since neither a human mind nor the best computers available to us are capable of creating such a complex web of ELSs. As evidence to the contrary, I would like to refer the reader to *The Codebreakers* by David Kahn, one of the foremost experts in cryptology.[10] In this book one can find a plethora of information about the ability of men and women to both encode and decode information using methods whose complexity and sophistication make the alleged Bible codes look simply primitive by comparison.

Indeed, let us recall one of the simplest tools used for encoding secret information, believed to have been invented by the renowned Italian mathematician and writer Girolamo Cardano in the sixteenth century. In a sheet of paper, a set of holes is cut, forming the *Cardano grille*. The grille is placed over a sheet of blank paper, and the message to be enciphered is written through the holes, one letter per hole. Then the grille is removed, and the blank spaces between the letters of the secret message are filled with a text that has some innocent contents. The letters of the encoded message thus become part of the overall text, but now are separated by skips. To decode the message, a grille identical to that used for encoding is placed over the text, and the secret message is read through the holes. When the holes are cut at equal distances, a "simple" Cardano grille is obtained. It, of course, produces ELSs exactly like those discovered in the text of the Bible.

When I was a child of about twelve, a few of my friends and I used to send each other secret messages in the classroom right under the nose of our teacher. To encode a message, we used several techniques, including the Cardano grille. (We had no idea that it was invented in Italy in the sixteenth century, though.) Sometimes we used a "simple" Cardano grille, creating sets of ELSs not unlike those found in the Bible. On other occasions we used a grille in which the distance between the holes would either increase or decrease from letter to letter, thus producing encoded messages where the skip changed from character to character. As I will show later in this chapter, similar codes with regularly increasing or decreasing skips can be easily located in any text.

Especially for this book, I decided to try to encode again (as I did as a twelve-year-old) some simple phrase, willing to spend on that task no more than ten to fifteen minutes. First I wrote the following phrase: *Rabin will die*, which, in the parlance of cryptology, would be my *plaintext* (the message I wish to encode). The choice of this particular expression was due to the widely publicized alleged prediction of Rabin's assassination cited in *The Bible Code*. I wrote the letters of my plaintext on a piece of paper, leaving spaces between letters, which would enable me to insert nine other letters between any two consecutive letters of the plaintext. Then I wrote a text between the letters of this expression, which, even if not very sophisticated (and slightly imperfect grammatically) is nevertheless meaningful. The entire exercise took nine minutes. Here is my resulting *ciphertext:* **R**_ivers are d_**A**_mningly ro_**B**_ust in some_ **I**_ndian colo_**N**_ies, moving_ **W**_ater unerr_**I**_ngly, thus a_**L**_leviating_ **L**_ust for the_ **D**_rinks, forc_**I**_ng people r_**E**_concile with the otherwise harsh climate_ (the hidden plaintext is shown in bold capitals).

The intended recipient of this message, knowing that there is a skip of ten, would have no problem decoding the message. If a serious need existed for me to prepare a secret message using a Cardano grille, I would certainly be able to write a much more sophisticated ciphertext, given more time.

The conclusion: a human mind is quite capable of creating arrays of ELSs not unlike those found in the Bible. (An impressive confirmation of this conclusion, demonstrated by Gidon Cohen of York, United Kingdom, was described in chapter 6.)

By refuting the claims about the necessity of a superhuman mind for the creation of the alleged Bible codes, I am not suggesting that the ELSs in the Bible were created by human authors. The explanation best compatible with the factual evidence is that arrays of ELSs appear in the Bible not by design but as sheer coincidences.

CODES RELATED TO JESUS

Among the publications about the Bible code is the book by Grant Jeffrey titled *The Signature of God* (discussed at length in chapter 6).[11] The author discusses six examples of ELSs found in the Hebrew Bible, which in his opinion spell the name *Yeshua*, the Hebrew form for Jesus. The examples of the passages from the Hebrew Bible, as printed in Jeffrey's book, preserve what are called in Hebrew *nekudot*, diacritical marks placed beneath, above, or within letters (which in the Hebrew alphabet are all consonants) to indicate the accompanying vowels. The presence of *nekudot* thus forces a definite reading for each syllable. Accounting for the printed *nekudot*, not a single example in Jeffrey's book does indeed spell *Yeshua*, but rather meaningless combinations of letters spelling nonexistent words such as *Yasvei*, *Yashaua*, and the like. Of course, this observation is only a testimony to Jeffrey's ignorance of Hebrew rather than an argument against the existence of the codes. Despite the errors by Jeffrey, the existence of numerous ELSs in the Hebrew Bible which spell not only the word *Yeshua*, but also combinations of words like *Yeshua shmi* ("my name is Jesus"), *Yeshua yakhol* ("Jesus can"), *dam Yeshua* ("blood of Jesus"), *Yeshua moreh*

("Jesus teacher"), etc., can be indeed easily demonstrated. The question is, though, whether such ELSs are unique to the Bible or occur commonly in all texts.

I decided to search in nonbiblical Hebrew texts for ELSs spelling the name *Yeshua* and its combinations with some other words such as those listed above, all of which have been found by messianic pastor Yacov Rambsel in the Bible and touted as proofs of his beliefs.[12] I randomly pulled from my shelf a few Hebrew books. The first one happened to be a book by a contemporary Israeli writer Dahn Ben-Amotz, the title of which translates as *Screwing Is Not Everything*.[13]

In Hebrew, the name *Yeshua* has four letters (*yud-shin-vav-ayin*). Sometimes Rambsel used, instead of this four-letter form, the shorter, three-letter version (*yud-shin-ayin*). Likewise, I decided to look for occurrences of the three-letter version as well.

I leafed randomly through Ben-Amotz's book, and soon (p. 47) I located ELSs spelling the two words *Yeshua shmi*, one right after the other, both with a skip of only 2. On page 23, within only three lines of text, ELSs for the words *dam* and *Yeshua* occurred one after another, both with a skip of only 3. On page 27, within only three lines of text, the same ELSs appeared, both with a skip of 4. On page 63, within only two lines of text, an ELS for the word *Yeshua* appeared twice, once in the three-letter version with a skip of 3, and once in the four-letter version with a skip of –1. In the same two lines of text the word *moreh* (*mem-resh-hey*), meaning "teacher," appeared three times with skips of 3, 4, and –6. In the same lines the word *mori* (*mem-resh-yud*), meaning "my teacher" appeared with a skip of –5. The characters of the words *moreh* and *mori* appeared interspersed with the characters forming the ELS for *Yeshua*. On page 164, within three paragraphs, an ELS for *Yeshua* appeared four times, three times in the three-letter version (with skips of 2, 4, and –6) and once in the four-letter version, with a skip of 7. On the same page, within the same three paragraphs, the four-letter word *yakhol* (*yud-khaf-vav-lamed*) appeared with a skip of 51. The characters of the ELS for *yakhol* were interspersed with those for *Yeshua*. As mentioned earlier, the phrase *Yeshua yakhol* means "Jesus can" or "Jesus is able," and, when found by Rambsel in the Bible, was interpreted by him as one of the proofs of his claims.

I found many more similar groups of ELSs in Ben-Amotz's book, spelling every single phrase Rambsel found in the Bible, including *Yeshua*. I found also, just as easily, many similar phrases including *Yeshua* in a number of other Hebrew books, such as the textbook *Geography of the*

1. צורות קו החוף של הארץ

שתי עובדות מציינות את צורת חופיה הים־תיכוניים של ארץ־ישראל: אפיו
המיושר של החוף, הנעדר מפרצים גדולים, מן הכרמל ועד לצפון סיני, וצורתו
הקשתית של חוף מיושר זה, המשנה את כיוונו מצפון־צפון־מזרח – דרום־דרום־
מערב בחלק הצפוני של ארץ־ישראל לצפון־מזרח – דרום־מערב בחלקה הדרומי של
הארץ, עד הגיעו בצפון סיני לכיוון מזרח – מערב.
את הסיבה לצורה הקשתית של החוף יש לחפש במבנה הפנימי של אזור זה. כפי
שצוין בפרק על התפתחות הנוף נשענים החלקים הדרומיים של הארץ על הגוש
הקדום הערבי־נובי. קיומו של גוש קדום זה גרם לכך, שבמשך תקופות גיאולוגיות
ארוכות הושקעו משקעי הים בשולי גוש זה בצורת קשת. כיוון החוף תואם איפוא
את מבנם ואת כיוונם של הרי הארץ.

Fig. 14.2. Part of page 91 in *Geography of the Land of Israel* (reproduced with permission). Some ELSs are shown by frames around the appropriate letters.

Land of Israel.[14] Figure 14.2 is a reproduction of part of page 91 in this book.

In the very first line (right under the section heading) there is an ELS with a skip of 6 (i.e., from right to left) which reads *Torah* (*tav-vav-resh-hey*). In the second line from the top, the third letter from the right is *yud*. It starts an ELS with a skip of 4 that reads *Yeshua*. Right after that ELS, on the same line, there is another ELS with a skip of 2 that reads *mori*. One line down, starting from *mem* in the fourth word from the right, there is an ELS with a skip of 2 that reads *moreh*. Finally, in the third line from the bottom, the ninth letter from the left is *yud*. It starts an ELS with a skip of –14 (i.e., from left to right) that reads *yakhol*. Rambsel and Jeffrey touted occurrences of similar ELSs in the Bible as miraculous confirmations of their beliefs.

Is page 91, as shown here, an exception? Figure 14.3 is a reproduction of part of page 140 from the same textbook. There are two groups of ELSs on that page; let us look at the first, at the top of the page. In the third line from the top (not counting the section heading), the second letter from the right is *shin*. It starts an ELS with a skip of 3 that reads *shmo* ("his name"). In the same line, to the left of that ELS, there is another with a skip of –3 that reads *Yeshua*. Two lines down, there is one more ELS with a skip of –4, which reads *shmi* ("my name"). Finally, one more line down, there is an ELS with a skip of –1, which once again reads *shmi*. Of course, if all the above ELSs were found in the text of the Bible, then Rambsel and Jeffrey

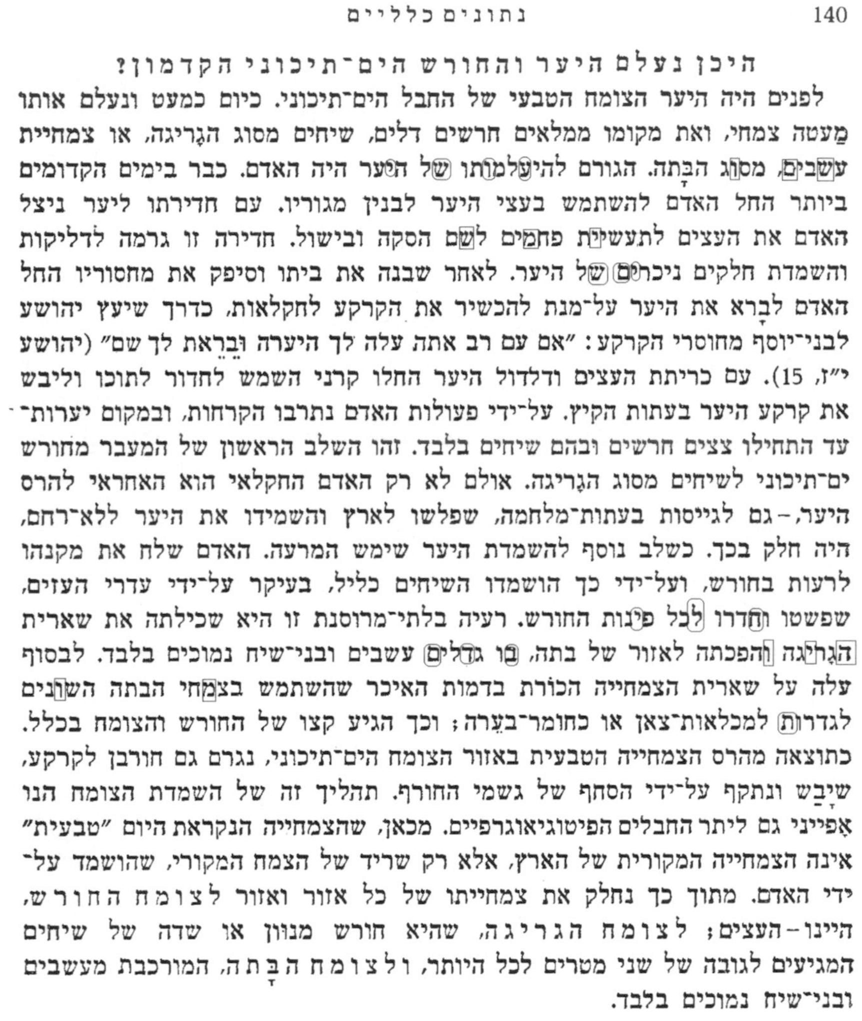

140 נתונים כלליים

היכן נעלם היער והחורש הים־תיכוני הקדמון?

לפנים היה היער הצומח הטבעי של החבל הים־תיכוני. כיום כמעט ונעלם אותו מעטה צמחי, ואת מקומו ממלאים חרשים דלים, שיחים מסוג הגריגה, או צמחיית עשבים, מסוג הבתה. הגורם להיעלמותו של היער היה האדם. כבר בימים הקדומים ביותר החל האדם להשתמש בעצי היער לבנין מגוריו. עם חדירתו ליער ניצל האדם את העצים לתעשיית פחמים לשם הסקה ובישול. חדירה זו גרמה לדליקות והשמדת חלקים ניכרים של היער. לאחר שבנה את ביתו וסיפק את מחסוריו החל האדם לברא את היער על־מנת להכשיר את הקרקע לחקלאות, כדרך שיעץ יהושע לבני־יוסף מחוסרי הקרקע: "אם עם רב אתה עלה לך היערה ובראת לך שם" (יהושע י"ז, 15). עם כריתת העצים ודלדול היער החלו קרני השמש לחדור לתוכו וליבש את קרקע היער בעתות הקיץ. על־ידי פעולות האדם נתרבו הקרחות, ובמקום יערות־עד התחילו צצים חרשים ובהם שיחים בלבד. זהו השלב הראשון של המעבר מחורש ים־תיכוני לשיחים מסוג הגריגה. אולם לא רק האדם החקלאי הוא האחראי להרס היער, – גם לגייסות בעתות־מלחמה, שפלשו לארץ והשמידו את היער ללא־רחם, היה חלק בכך. כשלב נוסף להשמדת היער שימש המרעה. האדם שלח את מקנהו לרעות בחורש, ועל־ידי כך הושמדו השיחים כליל, בעיקר על־ידי עדרי העזים, שפשטו וחדרו לכל פינות החורש. רעיה בלתי־מרוסנת זו היא שכילתה את שארית הגריגה והפכתה לאזור של בתה, בו גדלים עשבים ובני־שיח נמוכים בלבד. לבסוף עלה על שארית הצמחייה הכורת בדמות האיכר שהשתמש בצמחי הבתה השונים לגדרות למכלאות־צאן או כחומר־בערה; וכך הגיע קצו של החורש והצומח בכלל. כתוצאה מהרס הצמחייה הטבעית באזור הצומח הים־תיכוני, נגרם גם חורבן לקרקע, שיבש ונתקף על־ידי הסחף של גשמי החורף. תהליך זה של השמדת הצומח הנו אפייני גם ליתר החבלים הפיטוגיאוגרפיים. מכאן, שהצמחייה הנקראת היום "טבעית" אינה הצמחייה המקורית של הארץ, אלא רק שריד של הצמח המקורי, שהושמד על־ידי האדם. מתוך כך נחלק את צמחייתו של כל אזור ואזור ל צומח החורש, היינו – העצים; ל צומח הגריגה, שהיא חורש מנוון או שדה של שיחים המגיעים לגובה של שני מטרים לכל היותר, ו לצומח הבתה, המורכבת מעשבים ובני־שיח נמוכים בלבד.

Fig. 14.3. Part of page 140 in *Geography of the Land of Israel* (reproduced with permission). Some ELSs are shown by frames around the appropriate letters.

---

would be announcing with delight to their gullible readers a miracle confirming their beliefs.

My conclusion was obvious, namely that a variety of ELSs, including combinations of *Yeshua* with various other words occur quite commonly in nonbiblical texts as well as in the Bible. Therefore their appearance in the Bible does not constitute a proof of anybody's views or beliefs.

THE "AMAZING" PREDICTIONS DISCOVERED BY DROSNIN AND SATINOVER

It was only after dealing with Jeffrey and Rambsel that I learned about the books by Drosnin and Satinover. Could it be that these writers did a better job than Jeffrey and Rambsel and provided some stronger evidence in favor of codes in the Bible? Let us look at these two bestsellers.

In *The Bible Code*, Michael Drosnin claims to have made amazing discoveries in the Hebrew Bible, where many predictions of future events have been allegedly encoded as arrays of ELSs. The most acclaimed of these predictions is probably that of Yitzhak Rabin's assassination. The arrays in question, reproduced as Tables 28 and 29 in Drosnin's book, comprise the name of the late Israeli prime minister (found with a very large skip of 4,772 characters) as well as the phrase *rotzeakh asher irtzakh*, meaning "killer who will kill," the sets of letters denoting the year (in the Hebrew calendar) of Rabin's assassination, and also the name of the killer, *Amir*.

Looking at these tables reveals that of these words, only the name of Yitzhak Rabin is an ELS (with a very large skip). All the rest of the words in the set are just parts of the regular text, not encoded at all. Moreover, the ELS for Rabin's name extends over more than thirty thousand letters of text. Hence, the concept of "proximity" of all the mentioned words to the ELS in question is rather ambiguous. The illusion of proximity has been created by presenting the text of the Bible as a table 4,772 letters wide. Otherwise, all those words allegedly situated close to the ELS in question would actually be found scattered over the length of the text.

Since many people remained skeptical in regard to these claims, Drosnin said in an interview published in *Newsweek* (June 9, 1997) that if somebody found a prediction of a prime minister's assassination in *Moby Dick*, then he, Drosnin, would admit being wrong. I assume that by citing *Moby Dick* Drosnin actually meant any book other than the Bible.

Not to look too far, I turned again to *Screwing Is Not Everything* in order to find some ELSs related to Rabin. Again, I did not use any computer program, thereby limiting myself to relatively short skips and only those ELSs located close to each other. Following Drosnin's practice, I counted only the letters, ignoring the spaces between words, commas, periods, etc.

The ELS for *Rabin* (four Hebrew letters, *resh-beth-yud-nun*) popped up on page 33, in the uppermost paragraph on that page, with a skip of –35. Furthermore, within the two uppermost paragraphs, consisting of only about six hundred characters, the following ELSs appeared, one following the other:

1. An ELS consisting of four Hebrew letters with a skip of only –3: *yud-resh-tzade-khet*, which reads *Irtzakh* (meaning *he will kill*)
2. Three Hebrew characters with a skip of –15: *resh-hey-mem*, which is the Hebrew abbreviation for *rosh ha memshala*, meaning *prime minister* (an abbreviation commonly used in Hebrew, and also used as such by Drosnin).
3. Three Hebrew characters with a skip of –8, *gimel-bet-resh*, which is normally read as *gever* ("man") but is interpreted, for example by Satinover, as *gibor* ("hero" or "mighty man")
4. As mentioned before, the name *Rabin*, with a skip of –35

So, in just two short paragraphs on a randomly chosen page, in a randomly chosen Hebrew book, without employing any rearrangement of the text such as that usually performed by Drosnin, we have a set of four ELSs, all words appearing in one chain in very close proximity: *he will kill, Prime Minister, hero, Rabin.* If we wished to calculate the probability of the occurrence of these four ELSs within these short paragraphs, using methods of calculation routinely employed by code proponents, we would arrive at an astonishing number of about one in several billions. (As was shown in chapter 13, the calculations of probabilities performed by code proponents actually provide meaningless numbers.)

All that seemed to be still missing in these two paragraphs was an ELS for the name of Rabin's killer, Amir. I then recalled my experience with the Cardano grille, when my twelve-year-old friends and I sometimes used a grille with gradually increasing or decreasing skips. Indeed, if the creator of the codes hid information in the form of ELSs, couldn't he also use, for example, DSLSs (decreasing-skip letter sequences), as my friends and I once did? It would be even a better code, wouldn't it?

I looked for any occurrence of such a code. In the same chain of words mentioned above I found the following four Hebrew letters: *ayin-mem-yud-resh*, which reads *Amir*, with the skip between the first and the second letter being –8, between the second and the third letter being –9, and between the third and the fourth letter being –10. What a nice regularity, and what a beautiful code, isn't it? Now all encoded words, all within the same two paragraphs, with small skips, in a chain, read *Amir will kill prime minister hero Rabin.*

Following Drosnin, one might say, "What an amazing discovery!" Obviously, the probability of these words appearing within two short paragraphs, in such close proximity to each other, must be exceedingly small.

Then shall we conclude that in a book published in 1979, a message was deliberately *encoded* predicting the assassination of the *prime minister, the hero, Rabin, by Amir*, some sixteen years before it actually happened?

(This example was actually posted on the Internet in January of 1998.[15] So far, despite his promise, Drosnin has not replied in any form.)

Of course, I found the "prediction" of Rabin's assassination in a Hebrew book; what about *Moby Dick,* specifically mentioned by Drosnin in his interview? Responding to Drosnin's challenge, Australian mathematician and computer expert Brendan McKay applied a computer program searching for ELSs to the text of *Moby Dick.* He discovered there numerous arrays of ELSs predicting assassinations of many prime ministers and other political figures, including Austrian chancellor Engelbert Dollfus, Leon Trotsky, Indira Gandhi, Abraham Lincoln, Martin Luther King Jr., and, of course, Itzhak Rabin.

Cracking the Bible Code also contains similar examples of supposedly amazing finds in the Bible. Although Satinover uses an additional criterion of *nearly minimal skips*, his finds are still almost certainly results of random chance.

For example, in chapter 10 the author presents a number of arrays of ELSs found in the Hebrew Bible by himself as well as by Witztum, Rips, and Moshe Katz. These arrays supposedly contain encoded information about Satinover's ancestor Rabbi Abraham—nicknamed "The Angel"—as well as Emperor Franz Joseph of Austria, diabetes, the AIDS epidemic, and the assassination of Anwar Sadat.

I decided to look for similar arrays in a text other than the Bible. In fact, I chose the very same array as can be found on page 164 of *Cracking the Bible Code*, attributed to Witztum, which deals with the AIDS epidemic. This array contains ELSs which spell the following words: *AIDS* (*aleph-yud-dalet-samekh*), *mavet* (*mem-vav-tet*), meaning "death," *be dam* (*bet-dalet-mem*), meaning "in blood," *the HIV* (*hey-hey-yud-vav*), etc.

Again, I leafed randomly through the book by Ben-Amotz until I came upon page 67. There I found the following ELSs situated within the first two paragraphs, which contained a total of about six hundred characters: *be dam* with a skip of 15, *mavet* with a skip of 10, and *HIV* with a skip of 18. Additionally, in the same paragraphs there was the word *kholi* (*khet-lamed-yud*), meaning "disease."

There seemed to be no ELSs for *AIDS* in that short text. Again, remembering the Cardano grilles I used as a kid, and realizing that the code might not necessarily be limited to ELS, I looked for *AIDS* "encoded" with a regularly increasing or decreasing skip. I found it in the same paragraphs, the skips being –18 between the first and the second letters, –17 between the second and the third, and –16 between the third and fourth. How small is the probability of these occurrences expected to be?

I easily found a similar group of ELSs, again without using a computer, in *Geography of the Land of Israel*. Let us look again at page 140 of that book, as shown in figure 14.3. This time, look at the group of ELSs on the lower part of the page. In the twelfth line from the bottom, the seventh letter from the right is *khet*. It starts an ELS with a skip of 4, which reads *kholi*. One line down, the first letter from the right is *hey*. It starts an ELS with a skip of 3, which reads *HIV*. In the same line, to the left of *HIV*, there is an ELS with a skip of 3, which reads *be dam*. One more line down, in the third word from the left, there is the letter *mem*. It starts an ELS with a skip of 9, which reads *mavet*. Is there any difference between these ELSs and those found in the Bible? None whatsoever.

THE "SIMPLE" CALCULATIONS OF PROBABILITIES

Many proponents of the Bible code have tried to calculate the probabilities of the appearance, by chance, of various ELSs in the text of the Bible. Unfortunately, most of these calculations are characterized by errors and misconceptions and produce unreliable values of probabilities (as discussed in chapter 13). In such calculations the very small probabilities (such as Jeffrey's "one in fifty quadrillions")[16] are usually obtained by multiplying many not very small probabilities. For example, the probability of a particular letter appearing at a certain location in the text is calculated by dividing the total number of occurrences of that letter in the text by the overall text length expressed in the number of all letters in it. Even this first step is actually wrong, though, as it may be applied only to a random conglomerate of letters. Meaningful texts are very far from being random. On the contrary, meaningful texts possess a high degree of order in various forms.[17] Therefore the probability of a particular letter appearing at a certain site in a meaningful text may be very different from the probability calculated for a random text.

However, there is a much more serious error in these calculations. Having calculated (erroneously, as explained above) the probabilities for various letters occurring at various locations in the text, the code proponents then usually *multiply* the probabilities found for each letter in a given ELS and thus arrive at extremely low numbers. Multiplication of individual probabilities is routinely employed in the theory of probability, but the necessary condition to make this multiplication valid is the *independence* of the multiplied probabilities. If the individual probabilities are not independent, probability theory treats them in a different way, as so-called conditional probabilities. To determine whether individual probabilities are independent or not is by no means a trivial task. In particular, the individual probabilities multiplied by code proponents are definitely not independent (see chapter 13). Therefore the extremely small probabilities of the occurrence of that or this ELS, or of ELS arrays—which are so triumphantly claimed by the code proponents like Jeffrey, Rambsel, Drosnin, Satinover, etc.—are meaningless.

Now, let us consider the possibility that the probabilities in question are indeed very small. Does it mean the occurrence of corresponding ELSs must be attributed to a conscious design?

As was shown in chapter 13, probability theory cannot predict whether an event will or will not occur. Even if the calculated probability of an event is very small, it does not mean that the event in question will not happen. All that probability theory can assert is that in a very large number of tests the average number of occurrences of a specific outcome will approach the calculated value of probability for this outcome as the number of tests increases—no more and no less than that.

The chance of finding a particular ELS in the Bible may be small (but still much larger than usually claimed by code proponents), but the chance of finding *some* ELS, for example combining the word *Yeshua* with certain other words, is much larger. The reason for that is that the Hebrew language (and any other language as well) contains so many different expressions and phrasal constructions that the chance of coming across some seemingly meaningful combination of Yeshua with some other words in the Torah is quite large. The fact that many expressions found in the Bible by code proponents were also located in nonbiblical texts may serve as a confirmation that calculations of probabilities, like those often used by the code proponents, are meaningless and cannot be used to justify any conclusions about the reality of the Bible codes.

A GENERAL LOOK AT WRR'S PAPER

When I felt confident that I had sufficiently dealt with such Bible code epigones as Drosnin, Satinover, Jeffrey, etc., I turned to WRR themselves.

All the Bible code proponents and defenders refer to that paper with the utmost esteem, repeating time and time again that WRR's article had "scientifically proven" the genuineness of the Bible code. Among the arguments made in favor of the code's authenticity, WRR's epigones usually emphasize that Witztum, Rips, and Rosenberg are genuine scientists; that their paper was published in a prestigious, refereed magazine; and that no skeptical scientists have yet found any errors in WRR's calculations. However, all these assertions are without merit.

First, of the three authors of the famous article, only Rips is indeed a scientist, a highly qualified mathematician in the field of group theory (but, unlike many of the code opponents, not in mathematical statistics). Rosenberg developed a computer program used by WRR. Many similar programs have been offered (some for free and many more for sale), and none of these programs' creators (such as Kevin Acre, Dave Thomas, Randall Ingermanson, etc., all of whom maintain Web sites) pretend that their programs qualify as a scientific achievement. Witztum, who characterized himself as "the foremost code researcher in the world" in an article published in *Jewish Action* magazine (March 1988), has no scientific credentials but a Master's degree in physics.

WRR's followers have a habit of constantly exaggerating the scientific qualifications of anybody who is or could possibly be in favor of WRR's work. The epithets such as "world-class" mathematician, probabilist, etc., are routinely applied to whoever might, often just in passing and in ambiguous terms, say anything that WRR's defenders choose to construe as support of WRR's conclusions.

(Of course, whether WRR are genuine scientists or amateurs is irrelevant; I have touched on it only as a response to WRR's followers' persistent misuse of that argument.)

Furthermore, if one wishes to use the argument based on the credentials of WRR, then one has to account for the opinion of many real experts in mathematical statistics who signed a letter stating that they had personally analyzed WRR's article and rejected their conclusions.[18] At the time of writing, more than fifty mathematicians had already signed the letter in question. Only experts in mathematical statistics were offered a chance to sign the letter in question. Were it offered also to physicists, linguists, Bible scholars, etc., the number of signatures would certainly be much larger. By contrast, there is no document signed by any number of scientists who would support WRR's conclusions. It can be therefore stated that the scientific community, and more specifically specialists in statistical science, have overwhelmingly rejected WRR's work.

To thoroughly analyze WRR's work and to determine where the authors were at fault took a considerable time. A number of Internet postings and papers have appeared since 1994 in which various facets of WRR's work have been criticized from various viewpoints.[19] Therefore their followers' argument that no scientist could find errors in their article is a blatant distortion of the actual situation. Five years after WRR's article, the same *Statistical Science* published a paper by three mathematicians, Brendan McKay, Dror Bar-Natan, and Gil Kalai and one psychologist, Maya Bar-Hillel (referred to collectively as MBBK), that offered the most comprehensive refutation of WRR's work.[20] Robert Kass, the editor of the journal when WRR's paper was printed, supplied a comment with MBBK's article, stating that the puzzle offered in 1994 had been solved successfully and the supposedly amazing results of WRR's experiment had been given a convincing rational explanation.

While the excellent paper by MBBK provides a devastating criticism of WRR's procedures and conclusions, it does not encompass *all* facets of the controversy. In particular, MBBK actually accept some rules of the game employed by WRR, concentrating mainly on the demolition of the statistical evidence provided by the latter, but neither rejecting their fundamental hypothesis nor analyzing some of the details of their basic approach to statistical procedures. While MBBK's argumentation is more than sufficient to completely destroy the credibility of WRR's conclusions, a look at some other fundamental weaknesses of WRR's work is also of interest.

In the following sections I provide both a brief report on MBBK's formidable analysis of WRR's work and an equally brief consideration of some additional aspects of WRR's basic approach to the problem in question.

The Basic Hypothesis

At the beginning of their paper WRR assert that if one is dealing with any meaningful text, it is reasonable to expect that words which are semantically related—for example, *hammer* and *anvil*—will occur in the text in close proximity to each other. Likewise, they contend, if the ELSs in the Bible constitute a "code" inserted there by a superhuman mind, then the ELSs which are related in regard to content can also be expected to occur in close proximity to each other.

WRR's hypothesis is doubtful even in regard to human-created meaningful texts. It is easy to imagine a text where *hammer* occurs very frequently while *anvil* is completely absent. However, for the sake of argument, let us accept the above hypothesis for human-created texts. WRR's extension of that hypothesis to the occurrences of semantically related ELSs in the Bible is a jump across a logical pit. If the "code" has indeed been created by a superhuman mind, then, by definition, we have no way to know how that mind works. What reason do we have to assume that the alleged superhuman creator of the code wished to place the ELSs related by meaning close to each other?

All ELSs reported by WRR and their followers are just individual words without grammatical connections between them. Despite a number of attempts, so far neither WRR nor any of their followers succeeded in proving that sets of ELSs in the Bible form meaningful phrases, let alone whole sentences. Then what could be the reason to haphazardly place various ELSs in a text "close" to each other? Moreover, it seems more logical (or, better, no more illogical) to guess that the creator of the alleged code would place related ELSs as far from each other as the text would allow. Indeed, as it was shown in one of the preceding sections, the human mind is capable of creating ELSs in texts. This task is relatively easy if the skips are short. If the skip exceeds the length of one page, the creation of an ELS becomes difficult. Hence, if a superhuman creator of codes wished to encode some information using ELSs, it is equally plausible (or, rather, no more implausible) to hypothesize that he would rather opt for placing parts of the code at long distances from each other in order to make clear that his code is not of human origin.

WRR's hypothesis, which is at the core of their entire effort, seems to be arbitrary, as it neither has foundation in factual evidence nor is supported by any argumentation. In other words, even if WRR managed to prove that semantically related pairs of ELSs indeed occur in the Book of Genesis in an unusually close proximity, this in itself cannot serve as a proof of the superhuman origin of the given ELSs.

The proximity of ELSs may have a quite nonmiraculous origin, stemming from the natural properties of meaningful texts. Indeed, research conducted by myself and Brendan McKay revealed a specific pattern of strong organization in all meaningful texts which is absent in randomized texts.[21] Because of the high level of order in meaningful texts, every mathematical function reflecting properties of a text has nearly extreme values for meaningful texts as compared, for example, with most randomized texts. The complex mathematical quantity WRR use as a measure of "distance" between ELSs is no exception. Hence, even if the unusually close proximity of semantically related ELSs in the Book of Genesis were proven, it would not constitute a proof of their supernatural origin. Moreover, as we will see, the unusually close proximity claimed by WRR has by no means been proven in their work.

Unanswered Questions

To verify the results observed for the Book of Genesis, WRR also conducted their experiment on six control texts. Four of those control texts were obtained by permutations of various elements (words, verses, etc.) of the Genesis text. One other control text was the Book of Isaiah, and the final text was the Hebrew translation of Leo Tolstoy's novel *War and Peace*.

It is of interest that among all these texts, the worst result (that is, the highest rank of the "correct" appellations/dates list among all the competing shuffled lists) was observed for the Book of Isaiah. In other words, the experiment which, as WRR maintain, was successful in the Book of Genesis was a complete failure in the Book of Isaiah.

Since many arrays of ELSs found in the Book of Isaiah are often demonstrated by code proponents, it seems puzzling that WRR did not provide any discussion of their failure to get a positive result in the Book of Isaiah.

Moreover, how can the complete absence in WRR's work of any reports on attempts to apply their method to the other four books of the Torah be explained? Perhaps the answer can be inferred from the results obtained by some scientists who filled the void left by WRR. McKay used the computer program supplied by WRR as well as their list of rabbis' appellations/dates, and tried it on the other four books of the Pentateuch. The results were decidedly negative. While the "correct" list of rabbis' appellations/dates used by WRR, in conjunction with their method, produced low ranks among one million shuffled lists in the Book of Genesis, in the other four books of the Pentateuch the ranks were very high—in the hundreds of thousands.[22]

(For the record, Judaic tradition holds that the Torah was given on Mount Sinai to Moses as a string of letters without divisions into words, chapters, or books; the divisions were made only later. Thus the boundaries between the five books of the Torah have no real significance and have been disputed by some prominent figures in Jewish history.)

WRR avoid any discussion of why their experiment produced such good results in Genesis but utterly failed in the other four books of the Torah. The guess that immediately comes to mind is that the list of appellations/dates was somehow optimized in WRR's study toward the positive outcome of their experiment with Genesis. As we will see later, this guess finds some confirmation when a detailed study of the above list is performed. This statement in no way implies that WRR deliberately adjusted their data to produce the desired effect: the history of science knows many examples of perfectly honest and unbiased scientists inadvertently selecting from their data sets those subsets which produced results meeting the experimenters' expectations and ignoring other subsets which might have produced an opposite outcome.

Statistical Methodology Used by WRR

A rebuttal of WRR's work from the standpoint of mathematical statistics and probability theory has been undertaken by A. Michael Hasofer, the prominent expert in mathematical statistics. I will provide here the main points of this rebuttal. This section will necessarily be more technical than most in this book, as it deals with subtle elements of statistical procedure. Readers who are uncomfortable with mathematics can skip it, although they will miss the serious deficiencies in WRR's statistical manipulation of ELSs in the Bible that are demonstrated within.

The part of mathematical statistics relevant to the discussion of WRR's work is hypotheses testing (see chapter 12 for its general role in the scientific endeavor). The proper procedure in testing hypotheses from the standpoint of statistics includes several necessary steps:

1. The first step is setting the null hypothesis. Usually this is a certain statement implying that, *within the framework of a set of conditions*, there is no statistically significant new phenomenon to be discovered. In other words, the null hypothesis reflects the expectations that a test aimed at the discovery of a new phenomenon will fail. In the case of WRR's work, the null hypothesis, in broad terms, is that the proximity of pairs of semantically related ELSs in the Book of Genesis is such as would be expected if the distribution of the ELSs in the text were purely random. When applying a statistical test, the null hypothesis normally is expressed in terms of certain quantities to be measured whose values would indicate whether the null hypothesis must be accepted or rejected. In WRR's work, such quantities are the ranks of the "correct" list of appellations/dates among about one million competing scrambled lists. The lower the ranks in question, the more likely it is that the null hypothesis is

wrong and has to be rejected. As we will see, the legitimacy of WRR's null hypothesis is at best uncertain.

2. The next necessary step is setting the alternative hypothesis. The alternative hypothesis implies that a new statistically significant phenomenon is likely to exist. Hypotheses testing essentially boils down to competition between the null hypothesis and the alternative hypothesis, to determine which is more likely to be true. Unfortunately, contrary to accepted statistical procedure, WRR never expressly formulated any alternative hypothesis, leaving it to the reader to figure out.
3. The following step in a proper statistical test is to define the so-called *critical region* and the *power of test*. The critical region defines the boundaries for the set of quantities to be measured such that if the measured quantities are found within those boundaries, the alternative hypothesis is considered to be more likely than the null hypothesis. The power of test is a quantity related to the size of the critical region. WRR, contrary to proper statistical procedure, defined neither the critical region nor the power of test, thus rendering their conclusions statistically ambiguous.

In particular, trying to interpret what WRR intended as their alternative hypothesis, one comes to the conclusion that the power of test implied in their study must be 1. In plain words, it means that to produce sufficient statistical evidence in favor of their (hidden) alternative hypothesis, the rank of the correct list of appellations/dates must always be 1, which of course almost never happened in their tests.

Besides these deviations from proper statistical procedure, WRR's work suffers from many other irregularities concerning the rules of mathematical statistics. One such problem is the nature of their null hypothesis. As MBBK indicate in their article, the null hypothesis adopted by WRR actually reflects not only the properties of the text of Genesis, but also those of the appellations/dates list.[23] The ranks of the "correct" list are partially determined by the composition of the list and not only by the text of Genesis. Indeed, as MBBK demonstrate, slight variations in the list's composition might result in drastic changes of ranks. I will return to that observation a little later.

Another serious deficiency in WRR's statistical treatment, pointed out by both Hasofer and MBBK is the asymmetry between the scrambled lists. For each "famous rabbi," WRR use several (up to eleven) appellations. They also use several versions of dates of birth or death. A total of 298 combinations of names with dates existed for WRR's list of rabbis. Of those combinations, 135 are *not* found as pairs of ELSs in Genesis; WRR used only the other 163 combinations. Let us assume that rabbi *A* has two dates and three appellations in the "correct" list. (For example, dates of birth and death, and appellations in several forms, one being only the last name, the other a nickname, one more the name plus title, etc.) Each combination of an appellation with a date provides an entry to the list of appellations/dates. Thus rabbi *A* contributes six entries to the "correct" appellations/dates list. However, let us assume that rabbi *B* has only two appellations and one date in that list. Hence, rabbi *B* contributes two entries. These two rabbis contribute eight entries total to the "correct" list. There is, among the mismatched lists, one where the appellations for rabbi *A* are combined with the dates for rabbi *B*, and vice versa. Then rabbi *A* would contribute not six but rather only three entries to the scrambled list, and rabbi *B*, instead of two, would contribute three entries. Hence the two rabbis in question would contribute to the scrambled list a total of six instead of eight entries. Different mismatched lists will have different numbers of entries, and this will affect the values of aggregate proximity criteria. The symmetry of sets of numbers to be compared is a necessary condition for a proper statistical test, but such symmetry is absent in WRR's work.

One more item which met with criticism was WRR's choice of the measure of distance between ELSs, which was defined by WRR using a very complicated procedure. In order to calculate distance, in some cases as many as over six million arithmetic operations are needed. In one of Hasofer's examples, the "distance," according to WRR's formula, between two ELSs that *touched each other* turned out to be much larger than that between two other ELSs that were quite obviously remote from each other. Since the aggregate proximity criteria, and therefore the ranks of the "correct" lists, were calculated by WRR based on their distance, I believe Hasofer's observation alone renders WRR's conclusions meaningless.

Statistical versus General Scientific Hypothesis: Four Statistics Calculated by WRR

So far our discussion of WRR's research has been completely within the framework of statistics. This consideration, based mainly on observations by experts in mathematical statistics, among them Hasofer and MBBK, has shown that the results of WRR's statistical treatment of their experimental data are unreliable and cannot serve as a foundation for their extraordinary claims. Now, however, for the sake of discussion, let us assume that WRR's claims are based on an impeccable statistical procedure and that therefore their null hypothesis has to be rejected and their (not explicitly expressed) alternative hypothesis accepted. From a purely statistical viewpoint, this would be the satisfactory outcome of their experiment. Even with this rather large assumption, there are some considerations which will lead us to problems in WRR's method beyond pure statistical procedure.

In any text on mathematical statistics there are warnings against overestimation of the validity of statistical analysis.[24] Acceptance of the alternative hypothesis completes the statistical analysis, but it is not necessarily the ultimate test of the problem in question. The only firm conclusion established through hypotheses testing is that *under the defined set of conditions*, the alternative hypothesis is more likely than the null hypothesis. There always exists a probability that some hypothesis not formulated in the study is even more likely than the formulated alternative hypothesis.

There is a difference between a statistical hypothesis and a general scientific hypothesis. A statistical hypothesis necessarily deals with *random variables*. Having "proven" the alternative hypothesis does not necessarily mean that it has been proven as the general scientific hypothesis.

Consider the following example. Imagine that a study was conducted in which the frequency of cases of tuberculosis was measured in a certain country, and in the course of that study it had been noticed that the cases of tuberculosis seemed to be less frequent among people who own gold watches. To verify this observation, a systematic statistical research has been initiated. The null hypothesis was that the frequencies of TB and ownership of gold watches are distributed in a random fashion and are independent of each other. The alternative hypothesis was that the cases of TB are negatively correlated with ownership of gold watches, since presumably rich people both possess gold watches more often and, because of better nutrition and living conditions, contract TB less often. For obvious reasons, in the properly conducted statistical study, the alternative hypothesis would overwhelmingly win. From a purely statistical viewpoint, the strong negative correlation between TB and gold watches would be well established. As a general scientific hypothesis, though, this conclusion would be meaningless, and no statistician would recommend curing TB by distributing gold watches.

Unfortunately, despite many similar, well-known examples, some statisticians can develop a mindset that makes them lose perspective of the general scientific hypothesis and view the statistical proof of an alternative hypothesis as the end in itself.

In our example, the meaninglessness of the statistical result becomes obvious if some *material evidence* is added to the study. Namely, attempts to cure tuberculosis by distributing gold watches, the results of which can easily be foreseen, will rebut the statistically valid but nevertheless meaningless results. In some other cases, the material evidence may be not readily available, but the results of a statistical study can still be judged to be meaningless just by analyzing the behavior of the quantities employed to characterize the null and alternative hypotheses. This was, in my opinion, the case with WRR.

To explain what I mean, consider this example. Imagine that a study is being conducted to compare the vitamin C concentration in apples versus that found in oranges. The amount of the vitamin in each individual apple or orange is different; therefore this quantity is a typical random variable and hence the legitimate object of a statistical study. It is impossible to analyze every apple and orange, so a selection is made of, say, one hundred thousand apples and the same number of oranges from various crops and regions. The amount of vitamin C is then measured in each selected apple and orange. It is then necessary to choose a certain aggregate measure characterizing the statistically averaged amount of the vitamin in each of the two types of fruit. Assume that in order to subject the study to self-control, two different aggregate measures are chosen. For example, one measure can be the arithmetic mean of all the measurements, which will be denoted P_1, while the other aggregate measure, denoted P_2, can be, for example, a product of data for all individual examples, excluding the data in the top and bottom 5 percent of the measured vitamin's concentrations. Obviously, P_1 and P_2 will be different numbers, but this fact will not invalidate the results by itself. What if, though, the arithmetic mean, P_1, is found larger for apples, while the product, P_2, is found larger for oranges? In other words, if we rely on P_1, then the rank assigned to apples will be 1, and to oranges, 2. However, if we rely instead on P_2, oranges will be assigned rank 1, and apples 2. Hence, according to $P_{1,}$ we conclude that apples contain on average more vitamin C than oranges, but according to P_2, we conclude that apples contain on average less vitamin C than oranges. The two aggregate criteria produced mutually exclusive results. The only possible conclusion is that at least one of the two aggregate criteria, and perhaps both of them, are unreliable.

This situation is exactly what happened in WRR's research. They suggested two formulas, one for calculating what they denoted statistic P_1 and the other for statistic P_2, both being the supposed cumulative measures of the proximity of pairs of ELSs in a text. Later WRR added two more aggregate measures of proximity, denoted P_3 and P_4, calculated by the same formulas as P_1 and P_2 respectively but applied to a slightly modified list of rabbis' names. The ultimate results reported by WRR were not the values of the four cumulative measures per se, but rather the *ranks* of the correct list among about one million competing scrambled lists. Strangely, the *ranks* of the "correct" list of appellations/dates were different, depending on which of four *P* values was used.

As a specific example, in an experiment reported by Rips to the Israeli Academy of Sciences on March 19, 1996, the rank of the correct list was 14 if statistic P_1 was used, but 2,724 if P_2 was used.[25] If the rank of the correct list is 14, it means that among the 999,999 competing scrambled lists, there are thirteen lists where the proximity of ELSs for rabbis' appellations to those of the corresponding dates is *better* than it is in the correct, unscrambled list. If, though, the rank of the correct list is 2,724, it means that there are not thirteen but 2,723 scrambled lists with a better proximity than in the correct list. Thus at least 2,710 scrambled lists are believed to have a better proximity than the correct list, if we rely on criterion P_1, but the same at least 2,710 scrambled lists are believed to have a *worse* proximity than the correct list if we trust P_2. Which result do we believe?

The only possible conclusion is that either P_1 or P_2—or perhaps both—are unreliable measures of proximity; the same relates to P_3 and P_4.

In WRR's experiments, the ranks of the correct list of appellations/dates were consistently low (even if not equal) for all of four *P* values (mostly between 4 and several thousand). Hence, from a purely statistical viewpoint (and ignoring the deficiencies of WRR's statistical treatment discussed earlier) there are sufficient grounds to reject the null hypothesis and to accept the (undefined) alternative hypothesis as more likely. Most of the critics of WRR's in fact do this, concentrating instead on the weaknesses of WRR's data lists (discussed in the next section). However, even if we were to accept WRR's conclusion as a legitimate result of their statistical study, this conclusion would still remain an unsubstantiated hypothesis under the requirements of a general scientific hypothesis. One of the reasons for this is the contradiction between their four cumulative measures of proximity. (This last consideration is, of course, of purely academic significance, because, as was shown in the preceding sections, the statistical study by WRR is itself fraught with irregularities.

The "Wiggle Room" in WRR's Study

One more question in our critique of WRR remains: namely, how to explain specifically the consistently low (if unequal) ranks of the correct list of appellations/dates obtained for all four *P* values.

Barry Simon, a prominent mathematician and IBM Professor of Mathematics and Theoretical Physics at Caltech, the author of many books and hundreds of scientific papers, and a recipient of many awards, is just one of the many scientists who have unequivocally rebutted WRR's method and conclusions. In order to explain the paradoxical results of their statistical study, Simon suggests the concept of "wiggle room."[26] This term relates to the large number of choices WRR could make when compiling their lists of rabbis' appellations and dates. For example, WRR informed us that the data for the famous rabbis' appellations and pertinent dates were extracted from the Responsa database maintained by Bar-Ilan University. An examination of Responsa revealed that the sixty-six rabbis included in WRR's two experiments had been known by at least twice as many names and appellations as WRR used in their lists. Witztum, Rips, and Rosenberg seem to have made a number of arbitrary choices about which appellations and which dates to include, or not include, in their list.

The natural question is, of course, whether or not the wiggle room available to WRR was sufficient to produce a successful list, and whether other choices available to WRR, if incorporated into their list, would have invalidated their results (i.e., made the ranks of the correct list large enough to deprive them of statistical significance). The answer to this question has been given mainly by MBBK in a number of Internet postings and in their article "Solving the Bible Code Puzzle."

This group has shown how small variations in the composition of a list can drastically change the rank of that list among its scrambled versions. In one such example, Witztum claimed to obtain a rank of 1 for a list of data where he juxtaposed ELSs for a series of words related to the dreadful Nazi extermination camp at Auschwitz. McKay then used a data list which differed from that used by Witztum only in minor details and was at least equally legitimate. For that slightly modified list, the rank was found to be 289,000 instead of 1.[27]

In regard to the "famous rabbis" experiment, McKay and Bar-Natan similarly slightly modified the list of appellations/names. They preserved 80 percent of WRR's list, deleted a few names whose inclusion in WRR's list rested on a shaky foundation, and added a few names whose inclusion had a justification at least as good as all the other names in WRR's list. Using this new list McKay and Bar-Natan found no statistical evidence of unusual proximity of the pertinent ELSs in the Book of Genesis.[28]

Analyzing WRR's work, MBBK found, for example, that if only four of thirty-two rabbis whose inclusion supplies the strongest contribution to WRR's low rank for the correct list of appellations/dates are deleted, the statistical significance of WRR's results changes from 1/62,000 to 1/30. Deletion of only one appellation of 102 changes the significance level by one order of magnitude. Deletion of only five names results in a change of three orders of magnitude.

In another experiment, McKay and Bar-Natan demonstrate how easily the list of names/dates can be "cooked" to achieve any desirable outcome of the experiment. Their slightly "cooked" list produced no significant result in Genesis, but produced what looked like statistically significant results in the Hebrew translation of Tolstoy's *War and Peace*.

In "Solving the Bible Code Puzzle," MBBK show that the "wiggle room" available to WRR was more than sufficient to compile a list of names/dates that would produce low values of rank in any selected text. Of course, the question of whether WRR made their choices deliberately or these choices were the result of an inadvertent optimization is irrelevant. There is no evidence and no reason to suspect WRR of a deliberate manipulation of their data, but the existence of wiggle room in the tradition of scientific discourse is in itself a sufficient reason to subject WRR's claims to serious doubts.

THE QUESTION OF THE TEXT

WRR's claim is, in particular, based on their assertion that the text of the Book of Genesis has been preserved intact, letter by letter, since it was given to Moses on Mount Sinai some three thousand years ago. Indeed, if the ELSs in the Bible constitute a code inserted by God, this code must have been present from the very beginning. Any changes in the text, however small, would necessarily damage the code, which depends on the exact sequence of all letters.

The claim of code proponents in regard to the supposed perfect preservation of the Torah text has been refuted, for example, by Jeffrey Tigay, professor of Hebrew and Semitic languages and literature at the University of Pennsylvania, as well as by Menachem Cohen, professor of biblical studies at Bar-Ilan University in Israel as well as the editor of authoritative editions of the Bible.[29] A similar criticism has been offered also by MBBK. It has been established that the text of the Bible has undergone many changes in the course of its existence.

Code proponents sometimes respond with two different explanations. One is that the codes found in the Bible are just remnants of the original God-designed codes. To that argument, MBBK replied that the deletion even of one letter in a thousand must completely destroy any ELS-based code. Given the extent of changes the text of the Bible must have experienced, there is no chance that any original code could have survived. This claim is supported by experimental data: in one experiment, MBBK deleted only fifty letters from Genesis, which resulted in the complete disappearance of the effect claimed by WRR.

Another explanation by code proponents is that God knew in advance which changes would occur in the text of the Bible and adjusted the sequences of letters in it in such a way as to ensure that the codes would emerge in our time in the altered text. Such explanations are at best appropriate for theological disputes rather than for discussions involving mathematical statistics. Anyone who wishes to *believe* in the codes is of course entitled to that belief or to any other. This chapter, though, is not about beliefs but about the alleged "scientific" or, more specifically, statistical proofs of the codes' authenticity.

CONCLUSION

The arguments discussed in this chapter show that neither WRR nor their supporters and other code proponents have so far succeeded in scientifically proving that ELSs in the Bible are anything but accidental sequences of letters which naturally occur in any sufficiently long text. The argumentation put forwad by WRR has been debunked from various viewpoints. Specifically, it was shown that the statistical procedures employed by these researchers were in several ways contrary to the established rules of mathematical statistics; it was demonstrated that the "distance" between ELSs had been defined by WRR in an unnatural manner, depriving that quantity of a meaningful interpretation; it was indicated that the aggregate criteria of ELS proximity behave in a haphazard way, thus destroying the credibility of WRR's statistical conclusions from the standpoint of the general scientific approach; it was demonstrated that WRR had considerable "wiggle room," i.e., sufficient freedom to make arbitrary choices to fit their data to the desired outcome; and finally, it was indicated that the results of WRR's experiment, despite considerable effort, could not be reproduced by unbiased scientists.[30]

As for WRR's epigones of all persuasions, such as Drosnin, Satinover, Jeffrey, etc., their work simply does not meet even the minimal scientific requirements to be considered seriously. (It is therefore not surprising that in 1997 WRR and Drosnin were awarded the "IgNobel Prize" at a ceremony at Harvard. The IgNobel Prize was awarded annually by a committee which included real Nobel Prize winners, given for "the discoveries which cannot and must not be reproduced.")

The publication of "Solving the Bible Code Puzzle" deprived code proponents of their argument that the publication of WRR's paper in 1994 in *Statistical Science* meant the approval of their claims by the scientific community. However, rather than admitting, at least with their silence, the faults of their study, WRR instead organized the International Torah Code Society. They conduct annual meetings of that society, whose status in the scientific community, in my view, is not much different from that of the Flat Earth Society. To give a talk at these meetings, the potential presenters must claim adherence to the *Halakha*, i.e., to the tenets of the Orthodox Judaism.

On the other hand, the Internet is full of Web sites where endless examples of alleged codes in the Bible are demonstrated, mostly from a Christian standpoint, many of them blatantly absurd, often revealing their authors' ignorance both of mathematics and of Hebrew. Of course, Rips, Witztum, and the International Torah Code Society as a whole disdainfully dismiss all these publications as nonsense while stubbornly adhering to their own "discoveries." This gives the entire story a distinctively comic tone.

NOTES

1. Doron Witztum, Eliyahu Rips, and Yoav Rosenberg, "Equidistant Letter Sequences in the Book of Genesis," *Statistical Science* 9, no. 2 (1994):429–438; reprinted in Michael Drosnin, *The Bible Code* (New York: Simon and Schuster, 1997).

2. M. Margaliot, ed., *The Encyclopedia of Great Men in Israel: A Bibliographical Dictionary of Jewish Sages and Scholars from the Ninth to the End of the Eighteenth Century* (Tel Aviv: Joshua Chachik, 1961).

3. Robert E. Kass, "Statement on Bible Codes" [online], lib.stat.cmu.edu/~kass/biblecodes/ [July 31, 2003].

4. Aish Ha Torah, "Torah Codes" [online], www.torahcodes.co.il [October 12, 2001].

5. Jeffrey Satinover, *Cracking the Bible Code* (New York: William Morrow, 1997).

6. A Russian-language version of this story can be found in Mark Perakh, "Don't Be Afraid, Barakhokhlo," *Kontinent*, no. 102 (2000):13–40.

7. Mark Perakh, "Four for the Price of One," Short Stories by Mark Perakh [online], www.nctimes.net/~mark/Engstories/barakh.htm [July 24, 2003].

8. Many examples can be seen at Mark Perakh, "Some Bible-code Related Experiments and Discussions," B-codes Page [online], www.nctimes.net/~mark/fcodes/elsyesh.htm [June 29, 2003].

9. Brendan McKay and friends, "Scientific Refutation of the Bible Codes," In Search of Mathematical Miracles [online], cs.anu.edu.au/~bdm/dilugim/torah.html [July 31, 2003]; David E. Thomas, "Hidden Messages and the Bible Code," Committee for the Scientific Investigation of Claims of the Paranormal [online], www.csicop.org/si/9711/bible-code.html [July 31, 2003].

10. David Kahn, *The Codebreakers* (London: Weidenfeld and Nicolson, 1967).

11. Grant Jeffrey, *The Signature of God: Astonishing Biblical Discoveries* (Toronto: Frontier Research, 1996).

12. Yacov A. Rambsel, *Yeshua: The Hebrew Factor* (San Antonio, Tex.: Messianic Ministries, 1996).

13. Dahn Ben-Amotz, *Ziunim ze lo hakol* (Screwing is not everything) (Tel Aviv: Metziut, 1979).

14. Menashe Har El and Dov Nir, *Geography of the Land of Israel* (Tel Aviv: Am Oved, 1975).

15. Mark Perakh, "Rabin," B-Codes Page [online], www.nctimes.net/~mark/fcodes/rabin.htm [July 31, 2003].

16. Jeffrey, *The Signature of God*, p. 206.

17. See, for example, Mark Perakh and Brendan McKay, "Study of Letter Serial Correlation (LSC) in Some Hebrew, Aramaic, Russian, and English Texts" [online], www.nctimes.net/~mark/Texts/ [July 31, 2003].

18. Barry Simon, "Mathematicians' Statement on the Bible Codes" [online], www.math.caltech.edu/code/petition.html [July 31, 2003].

19. Including Perakh, B-Codes Page; McKay and friends, "Scientific Refutation of the Bible Codes"; Simon, "Mathematicians' Statement on the Bible Codes"; Barry Simon, "The Case against the Codes," *Jewish Action* (March 1998); Maya Bar-Hillel, Dror Bar-Natan, and Brendan McKay, "The Torah Code: Puzzle and Solution," *Chance* 11, no. 2 (1998):13–16.

20. Brendan McKay et al., "Solving the Bible Code Puzzle," *Statistical Science* 14, no. 2 (1999):150–73.

21. Perakh and McKay, "Study of Letter Serial Correlation (LSC) in Some Hebrew, Aramaic, Russian, and English Texts."

22. A. Michael Hasofer, "A Statistical Critique of the Witztum et al. Paper," B-Codes Page [online], www.nctimes.net/~mark/fcodes/hasofer.htm [July 31, 2003].

23. McKay et al., "Solving the Bible Code Puzzle," p. 158.

24. See, for example, Michael G. Kendall and Andrew Stuart, *The Advanced Theory of Statistics*, vol. 2 (London: Griffin, 1978), p. 169.

25. Doron Witztum, Eliyahu Rips, and Yoav Rosenberg, "Hidden Codes in Equidistant Letter Sequences in the Book of Genesis: The Statistical Significance of the Phenomenon," B-Codes Page [online], www.nctimes.net/~mark/fcodes/WRR3.htm [April 5, 2002]. (In fact, Rips alone made the presentation, but the accompanying paper was authored by all three members of WRR.)

26. Simon, "The Case against the Codes"; Barry Simon, "Dispute on Codes" (in Russian), *Vremya Iskat*, no. 2 (1999):89–107.

27. Brendan McKay, "An Objective Experiment of Doron Witztum," In Search of Mathematical Miracles [online], cs.anu.edu.au/~bdm/dilugim/witztum/camps.html [August 1, 2003].

28. Brendan McKay and friends, "Scientific Refutation of the Bible Codes."

29. Jeffrey Tigay, "The Bible 'Codes': A Textual Perspective" [online], www.sas.upenn.edu/~jtigay/codetext.html [July 31, 2003]; Menachem Cohen, "The Religious and the Scientific Aspects of the Debate on the Codes Hidden in the Torah as Equidistant Letter Sequences" [online], www.nctimes.net/~mark/fcodes/cohen.htm [July 31, 2003].

30. An experiment which claimed to confirm the presence of a code in Genesis was conducted by Harold Gans, who is director of research at Aish Ha Torah and who thus makes his living by promoting these "codes" (see, for example, Aish.com [online], www.aish.com/issues/biblecodes/s-gans.htm [August 1, 2003]). Simon attempted to reproduce Gans's experiment but slightly changed the procedure in order to eliminate any possible ambiguity caused by "wiggle room" (see Barry Simon, "Barry Simon's Cities Experiment—an Overview," Barry Simon on Torah Codes [online], www.wopr.com/biblecodes/Cities_Overview.htm [August 1, 2003]). Simon included for consideration the names of all the cities where the "famous rabbis" were born, died, worked, or studied, without any changes in spelling or any addition of prefixes, thus substantially reducing the possible wiggle room. The result? No trace of a code found in the Book of Genesis. Simon concluded that Gans's results are in fact most likely due to wiggle room. Recently one more experiment was conducted by a committee comprising both code proponents—professors Eliyahu Rips, Robert (Yisrael) Aumann, and Harry (Hillel) Furstenberg—and an opponent—Professor Dror Bar-Natan—as well as Isaak Lapides, who has not expressed an opinion on the codes but was included in the committee according to a request by Rips. The committee followed a protocol mutually agreed upon (see Dror Bar-Natan, "The Gans Report," Bible Codes [online], www.math.toronto.edu/~drorbn/Codes/Gans/index.html [August 1, 2003]). Bar-Natan spells out the conclusion as "A Complete Failure of the 'Codes.'"

AFTERWORD

As this book covers a range of writers and their beliefs on related but various topics, I would like to summarize briefly the preceding reviews of the books that either promote the intelligent design theory or assert the compatibility of the Bible with scientific data. Among the authors reviewed are professional Christian preachers and Jewish rabbis, as well as scientists—a strange collection. At one extreme we see among the authors of the books reviewed some writers who, while possessing scientific degrees from prestigious institutions (such as Hugh Ross, discussed in chapter 5, or Gerald Schroeder in chapter 10), propagandize notions which they believe meet scientific criteria but which actually only show their dilettante level of understanding of the topics under discussion. Some other writers (such as Grant Jeffrey in chapter 6) do not pretend to be scientists themselves, but still offer quasi-scientific arguments, often distorting scientific data in favor of their beliefs. At the other extreme we see some scientists (such as Nathan Aviezer in chapter 9, Cyril Domb in chapter 8, or Michael Behe in chapter 2), sometimes prominent in their fields of research, who try either to make excursions into the dispute about the veracity of the biblical story, or to offer arguments in favor of intelligent design, wherein they somehow forsake the rigor of a scientific discourse and resort to argumentation of the type to which they would not normally acquiesce in their professional fields.

There is a wide range of levels of discourse in these publications as well. In some of them we see egregiously incorrect statements (e.g., Schroeder, Ross, Jeffrey). The authors of others (Aviezer, Spetner, Behe) avoid gross misinterpretations of fundamental facts of science, but offer conclusions based not on the evidence but rather on the writers' strong desire to keep their religious faith, even though it contradicts the requirements of evidence-based veracity.

Such variation within the group of writers discussed here testifies to the variety of motivations behind their efforts: some of them make their living by exploiting the popularity of religious beliefs, whereas others satisfy their personal desire to find proof supporting whatever they wish to believe.

Science and religion are both human endeavors providing outlets, one to the curiosity which seems to be an inseparable part of human nature, and the other to a hope that human life is something more than a purely biological existence without reason and purpose and that the tragedy of unavoidable death can somehow be rationalized.

The records of both science and religion are mixed. Science and its offspring—technological innovation—have opened the door to an enormous improvement of living standards and of the well-being of billions of men and women, conquering many diseases, immensely enhancing food supplies, increasing manifold the time available for pleasure, and freeing many people from the drudgery of monotonous, boring, and tiring jobs. On the other hand, science has also opened the door to a heretofore unprecedented ability to conduct mass killings and, along with the increase in human ability to master the forces of nature, has done very little to make people behave more reasonably. Religion has arguably served to improve the moral fiber of society by instilling in people ethical notions such as those succinctly presented in the Ten Commandments. On the other hand, religion is to blame for explosions of hatred, senseless and merciless mass murders of those who dare to believe differently, and suppressions of the progress of scientific exploration.

The fact that some scientists invest considerable effort in trying to reconcile science with the Bible testifies to the lack of bravery necessary to face the possibility that there is no supernatural purpose in our life and that death means the complete disappearance of an individual conscience.

The fear of a possible absence of any supernatural meaning to our life is understandable. How immensely better it is to believe that our life means something above and beyond mere biological existence and that death will not exterminate that whole world that each person has inside his or her soul. However, the desire to believe in a supernatural meaning is far from an actual proof of religion's claims. In fact, this desire itself creates a powerful incentive to ignore any evidence negating religious beliefs and to adhere to faith against reason.

On the other hand, atheists may be suspected of another kind of fear.

We all know that each of us, in the course of our lifetime, has committed small (and sometimes not so small) misdeeds, meeting the definition of sin according to many religious beliefs. Refusing to admit the existence of God, whom everyone is supposedly destined to face and to possibly compensate in some way for earlier misdeeds, atheists thus satisfy their desire to avoid that responsibility toward the alleged Supreme Being.

I believe these two questions—the existence or nonexistence of God, and the veracity of particular claims of this or that religion—are fundamentally distinct. I suspect that the question of the existence of God cannot be solved by rational arguments. Predicting the future is an unrewarding job, and therefore I would not dare to insist that no rational proof of God's existence or nonexistence will ever be found, though I suspect that, unfortunately, this may be the case. Therefore, believing either of the two alternatives remains a matter of personal choice, with agnosticism being a middle way.

By contrast, the question of the veracity of this or that particular religion, or of some specific image of God, seems to be within the human capacity for rational judgment. Having read many books and articles propagandizing various religious systems, I have formed the opinion that all of these systems have a human origin. I have not seen a good reason to accept the claims of any religion, including Judaism, Christianity, Islam, etc., as plausible. All of them contradict each other, are controversial within themselves, and often contradict established facts of science and/or history.

Therefore, when I read the following statement by William Dembski, made in his introduction to the collection *Mere Creation* that "as Christians we know naturalism is false,"[1] I view it as a display of arrogance and contempt for readers who may adhere to different beliefs (among which I include atheism). If Dembski "knows," why does he bother to look for proof of his alleged knowledge? If he already knows, then obviously his mind is closed to any arguments that may be directed against his so-called knowledge. His statement means an end to the discussion, and renders all of the rest of his discourse suspicious in regard to his adherence to facts, and hence to truth.

In view of my assertion that I consider both religious faith and atheism irrational (while at the same time realizing that one of these two attitudes must be correct) the reader may wish to ask why the chapters in this book are critical only of those books which are on the side of creationism. The reason is simple: while there is a multitude of books promoting creationism, there are much fewer books taking the opposite slant—and the latter rarely resort to such irrational arguments as those used by the adherents of the supernatural origin of the universe and of life. Atheism is not very popular, especially in the United States, where every presidential candidate bends over backwards to show his heartfelt religiosity. Hence, books promoting the creationist viewpoint are more likely to find readers and even become bestsellers. On the other hand, trying to offer a view denying creationism can sometimes cause problems for authors.

Therefore, while viewing both religious faith and atheistic views as irrational, I did not see a need and have had no opportunity to pounce on any false arguments proposed by the adherents of naturalism. Since the creationists rarely miss an opportunity to "disprove" even those arguments of so-called evolutionists that seem to be uncontroversial and based on indisputable facts, there is little chance that some false arguments by anticreationists have been missed by believers in the supernatural Creation.

I do not expect that this book will sway those readers who have already formed their own opinions in regard to the discussed questions; I only hope that those readers who are searching for a reasonable worldview, but who instead find themselves in a shadow world of uncertainty, will find in the preceding chapters some food for thought that may help them to make their own conclusions.

NOTE

1. William A. Dembski, introduction to *Mere Creation: Science, Faith, and Intelligent Design*, ed. William A. Dembski (Downers Grove, Ill.: InterVarsity Press, 1998).

APPENDIX

CALCULATION OF PROBABILITY OF WINNING A LOTTERY BY A NONSPECIFIED PLAYER

Consider a lottery wherein each player chooses a set of n numbers out of M available numbers. (For example, in the California Lottery, $n = 6$ and $M = 50$; i.e., each player chooses a set of any six numbers between 1 and 50). Let N be the number of possible combinations of n numbers chosen by players (of which one combination, chosen randomly, will be the winning set). Let T be the number of tickets sold ($T \leq N$).

Now calculate $p(L)$, the probability that exactly L players select the winning combination.

The number of choices of L tickets out of T is given by the binomial distribution

$$\text{bin}(T,L) = \frac{T!}{L!(T-L)!}.$$

For those L players to win, they must all select the winning combination. The probability of that is $(1/N)^L$. All the other $T - L$ players must select a nonwinning combination, the probability of that being $(1 - 1/N)^{(T-L)}$. Multiplying those three quantities yields the formula

$$p(L) = \text{bin}(T,L)\ (1/N)^L\ (1 - 1/N)^{(T-L)}.$$

The formula can be simplified, preserving a good deal of its accuracy. Since N and T are usually very large, and L is very small, we can use the following approximations:

$$\frac{T!}{(T\text{-}L)!} \approx T^L$$

$$(1 - 1/N)^{(T\text{-}L)} \approx \exp(-T/N).$$

Now the formula becomes

$$p(L) \approx (T/N)^L \frac{\exp(-T/N)}{L!.}$$

This approximate (but very accurate) formula is the Poisson distribution with the mean of T/N. In the case when $T = N$ (i.e., when all available tickets are sold) we have a simpler formula:

$$p(L) \approx \frac{\exp(-1)}{L!.}$$

(A complication in practice may be that when one person buys more than one ticket, he makes sure that all the combinations he chooses are different. However, the approximate formula will still be very accurate unless someone is buying a large fraction of all tickets, which is unlikely.)

The probability that only one player wins in this type of a lottery is now less than 100 percent; in fact, it is (assuming that $L = 1$) $p(1) = 1/E = .368$, i.e., about 37 percent, which is still thousands of times more likely than the probability of *the same player* winning consecutively in more than one drawing. (Since some players buy more than one ticket, the probability in question is actually larger than 37 percent.)

BIBLIOGRAPHY

Abell, George O. *Realm of the Universe*. Philadelphia: Saunders, 1984.

Aviezer, Nathan. *In the Beginning: Biblical Creation and Science*. Hoboken, N.J.: KTAV, 1990.

———. "The Anthropic Principle." *Jewish Action* 19 (spring 1999): 9–15.

Bar-Hillel, Maya, Dror Bar-Natan, and Brendan McKay. "The Torah Codes: Puzzle and Solution." *Chance* 11, no. 2 (1998):13–19.

Bar-Natan, Dror. "The Gans Report." Bible Codes [online]. www.math.toronto.edu/~drorbn/Codes/Gans/index.html [August 1, 2003].

Barrow, John D., and Frank J. Tipler. *The Anthropic Cosmological Principle*. New York: Oxford University Press, 1986.

Behe, Michael J. *Darwin's Black Box: The Biochemical Challenge to Evolution.* New York: Simon and Schuster, 1996.

Berlinski, David. "Gödel's Question." In *Mere Creation: Science, Faith, and Intelligent Design*, edited by William A. Dembski, pp. 402–26. Downers Grove, Ill.: InterVarsity Press, 1998.

Blahut, Richard E. *Principles and Practice of Information Theory*. New York: Addison-Wesley, 1990.

Borel, Emil. *Probability and Life*. New York: Dover, 1962.

Broglie, Louis V. de. "Ondes et quanta" (Waves and quanta). *Comptes rendus* 177 (1923):507–16.

Bugge, T. H., et al. "Loss of Fibrinogen Rescues Mice from the Pleiotropic Effect of Plasminogen Deficiency." *Cell* 87 (1996):709–19.

Carmell, Aryeh and Cyril Domb, eds. *Challenge: Torah Views on Science and Its Problems*, New York: Feldheim, 1978. Originally published 1976.

Chaitin, Gregory J. "Randomness and Mathematical Proof." *Scientific American* 232 (1975):47–51.

———. "Randomness and Mathematical Proof." in *From Complexity To Life: On the Emergence of Life and Meaning*, edited by Niels Henrik Gregersen, pp. 19–33. New York: Oxford University Press, 2003.

Chiprout, Eli. "A Critique of *The Design Inference*." Talk Reason [online]. www.talkreason.org/articles/Chiprout.cfm [August 14, 2003].

Cohen, Menachem. "The Religious and the Scientific Aspects of the Debate on the Codes Hidden in the Torah as Equidistant Letter Sequences." B-Codes Page [online]. www.nctimes.net/~mark/fcodes/cohen.htm [April 5, 2002].

Cooper, George R. "Information Theory." In *Van Nostrand's Scientific Encyclopedia*, edited by Douglas M. Considine, pp. 1354–60. New York: Van Nostrand Reinhold, 1976.

Cotterell, Arthur, ed. *The Encyclopedia of Ancient Civilizations*. London: Rainbird, 1980.

Cromer, Alan. *Uncommon Sense: The Heretical Nature of Science*. New York: Oxford University Press, 1993.

Culberg, Jan. *Mathematics from the Birth of Numbers*. New York: W. W. Norton, 1997.

Dawkins, Richard. *The Blind Watchmaker: Why the Evidence of Evolution Reveals a Universe without Design*. New York: W. W. Norton, 1996. Originally published 1986.

———. *Climbing Mount Improbable*. New York: W. W. Norton, 1996.

Dembski, William A. "Randomness by Design." Design Inference Website [online]. www.designinference.com/documents/2002.09.rndmnsbydes.pdf [June 3, 2003].

———. *The Design Inference: Eliminating Chance through Small Probabilities*. Cambridge: Cambridge University Press, 1998.

———. *Intelligent Design: The Bridge Between Science and Theology*. Downers Grove, Ill.: InterVarsity Press, 1999.

———. *No Free Lunch: Why Specified Complexity Cannot Be Purchased without Intelligence*. Lanham, Md.: Rowman and Littlefield, 2002.

———. "The Design Revolution: Answering the Toughest Questions about Intelligent Design." Access Research Network [online]. www.arn.org/docs2/news/designrev061003.htm [June 14, 2003].

———, ed. *Mere Creation: Science, Faith, and Intelligent Design*. Downers Grove, Ill.: InterVarsity Press, 1998.

Dembski, William A., and James M. Kushiner, eds. *Signs of Intelligence: Understanding Intelligent Design*. Grand Rapids, Mich.: Brazos Press, 2001.

Dembski, William A., Michael J. Behe, and Stephen C. Meyer, eds. *Science and Evidence for Design in the Universe*. San Francisco: Ignatius Press, 2000.

Doolittle, Russell F. "A Delicate Balance." *Boston Review* 22, no. 1 (1997):28–29.

Drosnin, Michael. *The Bible Code*. New York: Simon and Schuster, 1997.

Dwass, Meyer. *The First Steps in Probability*. New York: McGraw-Hill, 1967.

Edis, Taner. "Darwin in Mind: 'Intelligent Design' Meets Artificial Intelligence." Committee for the Scientific Investigation of Claims of the Paranormal [online]. www.csicop.org/si/2001-03/intelligent-design.html [June 23, 2002].

Eells, Ellery. Review of *The Design Inference*, by William Dembski [online]. philosophy.wisc.edu/eells/papers/direv.pdf [August 6, 2003]. Also published in *Philosophical Books* 40, no. 4 (1999).

Einstein, Albert. "Über einen die Erzeugung und Verwandlung des Lichtes betreffenden heuristischen Gesichtspunkt" (On a heuristic point of view regarding the generation and transformation of light). *Annalen der Physik* 17 (1905):132–41.

———. Letter to Guy H. Raner, September 28, 1945. *Skeptic* 5, no. 2 (1997):64.

———. "Autobiographical Notes." In *Albert Einstein: Philosopher-Scientist*, pp. 3–5. New York: Harper and Bros., 1949.

Elsberry, Wesley R. "Review of W. A. Dembski's *The Design Inference*." Talk Reason [online]. www.talkreason.org/articles/inference.cfm [August 14, 2003].

Espinola, Thomas. *Introduction To Thermophysics*. Dubuque, Iowa: Wm. C. Brown, 1994.

Even-Shoshan, Avraham. *Hamilon haivri hamerukaz* (The abridged Hebrew dictionary). Jerusalem: Kriyat Sefer, 1974.

Falk, Raphael. Review of *In the Beginning*, by Nathan Aviezer (in Hebrew). *Alpai'im* 9 (spring 1994):133–42.

Feit, Carl. "Not By Chance! The Fall of Neo-Darwinian Theory by Dr. Lee M. Spetner." *Jewish Action* (spring 1999):87–89.

Feynman, Richard. *The Character of Physical Law*. New York: Random House, 1994. Originally published 1965.

Fitelson, Branden, Christopher Stephens, and Elliott Sober. "How Not to Detect Design—Critical Notice: William A. Dembski, *The Design Inference*." *Philosophy of Science* 66 (September 1999):472–88.

Gray, Robert M. *Entropy and Information Theory*. Berlin: Springer Verlag, 1991.

Gruber, Boris. *Theory of Crystal Defects*. New York: Academic Press, 1964.

Halliday, David, Robert Resnick, and Jearl Walker. *Fundamentals of Physics*. Extended ed. New York: John Wiley and Sons, 1993.

Har El, Menashe, and Dov Nir. *Geography of the Land of Israel*. Tel Aviv: Am Oved, 1975.

Hasofer, A. Michael. "A Statistical Critique of the Witztum et al. Paper." B-Codes Page [online]. www.nctimes.net/~mark/fcodes/hasofer.htm [April 5, 2003].

Hawking, Stephen J. *A Brief History of Time*. New York: Bantam Books, 1996.

Heeren, Fred. *Show Me God: What the Message from Space Is Telling Us about God*. Wheeling, Ill.: Day Star Publications, 2000.

Hertzog, Zeev. "The Bible: No Findings on the Locations" (in Hebrew). *Haaretz* (Tel Aviv), October 29, 1999. Russian translation published in *Vremya Iskat*, no. 3 (2000):115–22.

Howson, Colin, and Peter Urbach. *Scientific Reasoning: The Bayesian Approach*. La Salle, Ill.: Open Court, 1993.

Hoyle, Fred. *The Intelligent Universe*. New York: Holt, Rinehart, and Winston, 1983.

Ikeda, Michael, and Bill Jefferys. "The Anthropic Principle Does Not Support Supernaturalism." Talk Reason [online]. www.talkreason.org/articles/super.cfm [August 14, 2003].

Jeffrey, Grant R. *The Signature of God: Astonishing Biblical Discoveries*. Toronto: Frontier Research, 1996.

Johnson, Phillip E. *Darwin on Trial*. Downers Grove, Ill.: InterVarsity Press, 1991.

———. *Reason in the Balance: The Case against Naturalism in Science, Law, and Education*. Downers Grove, Ill.: InterVarsity Press, 1995.

———. *Defeating Darwinism by Opening Minds*. Downers Grove, Ill.: InterVarsity Press, 1997.

———. *The Wedge of Truth: Splitting the Foundations of Naturalism*. Downers Grove, Ill.: InterVarsity Press, 2000.

Kahn, David. *The Codebreakers*. London: Weidenfeld and Nicolson, 1967.

Kass, Robert E. "Statement on Bible Codes" [online]. lib.stat.cmu.edu/~kass/biblecodes/ [July 31, 2003].

Kauffman, Stuart. *Investigations*. New York: Oxford University Press, 2001.

———. "The Emergence of Autonomous Agents." In *From Complexity To Life*, edited by Niels Henrik Gregersen, pp. 47–71. New York: Oxford University Press, 2003.

Kendall, Michael G., and Andrew Stuart. *The Advanced Theory of Statistics*. Vol. 2. London: Griffin, 1978.

Kitcher, Philip. "Born Again Creationism." In *Intelligent Design Creationism and Its Critics: Philosophical, Theological, and Scientific Perspectives*, edited by Robert T. Pennock, pp. 257–88. Cambridge: MIT Press, 2001.

Kittel, Charles. *Introduction to Solid State Physics*. New York: John Wiley and Sons, 1976. Originally published 1953.

Kolmogorov, Andrei N. "Three Approaches to the Quantitative Definition of Information" (in Russian). *Problemy Peredachi Informatsii* 1, no. 1 (1965):3–11. English translation published in *Problems of Information Transmission* 1 (1965):1–7 and *International Journal of Computer Mathematics* 2 (1968):157–68.

Korthof, Gert. "On the Origin of Information by Means of Intelligent Design: A Review of William Dembski's *Intelligent Design*." Was Darwin Wrong? [online]. home.planet.nl/~gkorthof/kortho44.htm [August 1, 2003].

———. "Could It Work? Does Evolution Work by Accumulation of Random Mutations as Neo-Darwinism Claims or by 'Adaptive' Mutations as Spetner Claims?" Was Darwin Wrong? [online]. home.planet.nl/~gkorthof/kortho36.htm [June 23, 2003].

Kuhn, Thomas S. *The Structure of Scientific Revolutions*. Chicago: University of Chicago Press, 1970.

Landau, Lev D., and Evgeniy M. Lifshits. *Statisticheskaya fizika* (Statistical physics). Moscow: Gosfizmatizdat, 1971.

Larsen, Richard J., and Morris L. Marx. *Introduction to Mathematical Statistics and Its Applications*. New York: Prentice Hall, 1986.

Margaliot, M. *The Encyclopedia of Great Men in Israel: A Bibliographical Dictionary of Jewish Sages and Scholars from the Ninth to the End of the Eighteenth Century*. Tel Aviv: Joshua Chachik, 1961.

McDonald, John H. "A Reducibly Complex Mousetrap" [online]. udel.edu/~mcdonald/oldmousetrap.html [June 23, 2002].

———. "A Reducibly Complex Mousetrap" [online]. udel.edu/~mcdonald/mousetrap.html [August 14, 2003]. Revised version.

McGowan, John F., III. "Jigsaw Model of the Origin of Life" [online]. www.jmcgowan.com/JigsawPreprint.pdf [August 6, 2003]. Also published in *Instruments, Methods, and Missions for Astrobiology*, vol. 4, Proceedings of the SPIE [Society of Photo-Optical Instrumentation Engineers] 4495 (2001).

McKay, Brendan. "Mathematical Miracles" [online]. cs.anu.edu.au/~bdm/dilugim [March 25, 2003].

———. "Encoding Messages Is EASY." Mathematical Miracles [online]. cs.anu.edu.au/~bdm/dilugim/longels.html [June 23, 2003].

———. "An Objective Experiment of Doron Witztum." Mathematical Miracles [online]. cs.anu.edu.au/~bdm/dilugim/witztum/camps.html [August 1, 2003].

McKay, Brendan, et al. "Solving the Bible Code Puzzle." *Statistical Science* 14, no. 2 (1999):150–65.

McKay, Brendan, and friends. "Scientific Refutation of the Bible Codes." Mathematical Miracles [online]. cs.anu.edu.au/~bdm/dilugim/torah.html [July 31, 2003].

Miller, Kenneth R. "Life's Grand Designs." *Technology Review* 97, no. 2 (1994):24–32.

———. *Finding Darwin's God: A Scientist's Search for Common Ground between God and Evolution*. New York: Cliff Street Books, 1999.

Millikan, Robert Andrews. "A Direct Photoelectric Determination of Planck's '*h*.'" *Physical Review* 7 (1916):355–71.

Morris, Henry. *The Bible and Modern Science*. Chicago: Moody Press, 1956.

Musgrave, Ian. "Spetner and Biological Information." Talk Reason [online]. www.talkreason.org/articles/spetner_v2.cfm [August 14, 2003].

Nalimov, Vasiliy V. *Teoriya eksperimenta* (Theory of experiment). Moscow: Nauka, 1971.

Orr, H. Allen. "Darwin v. Intelligent Design (Again): The Latest Attack on Evolution Is Cleverly Argued, Biologically Informed—and Wrong." *Boston Review* [online]. www.bostonreview.net/BR21.6/orr.nclk [August 6, 2003]. Also published in *Boston Review* 21, no. 6 (1996–97).

———. Review of *No Free Lunch*, by William Dembski. *Boston Review* [online]. bostonreview.mit.edu/BR27.3/orr.html [December 22, 2002]. Also published in *Boston Review* 27, no. 3 (summer 2002).

Paley, William. *Natural Theology: Or, Evidences of the Existence and Attributes of Deity Collected from the Appearances of Nature.* 1802. Reprint, Boston: Gould and Lincoln, 1852.

Panin, Ivan. *The Writings of Ivan Panin.* Agincourt, Ont.: Book Society of Canada, 1972.

Pearcey, Nancy R. "You Guys Lost." In *Mere Creation: Science, Faith, and Intelligent Design*, edited by William A. Dembski, pp. 73–92 . Downers Grove, Ill.: InterVarsity Press, 1998.

Pennock, Robert T. *Tower of Babel: The Evidence against the New Creationism.* Cambridge: MIT Press, 1999.

———, ed. *Intelligent Design Creationism and Its Critics: Philosophical, Theological, and Scientific Perspectives.* Cambridge: MIT Press, 2001.

Perakh, Mark. "Anisotropy of Spontaneous Macrostress in Ferromagnetic Films Induced by Magnetization." Parts 1 and 2. *Journal of the Electrochemical Society* 122, no. 9 (1975):1260–62, 1263–67.

———. "Calculation of Spontaneous Macrostress in Deposits from Deformation of Substrates and Restoring (or Restraining) Factors." *Surface Technology* 8 (1979):265–309.

———. "Slot-Type Field-Shaping Cell: Theory, Experiment and Application." *Surface and Coatings Technology* 31 (1987):409–26.

———. "B-Codes Page" [online]. www.nctimes.net/~mark/fcodes/ [August 14, 2003].

———. "Some Bible Code–Related Experiments and Discussions." B-Codes Page [online]. www.nctimes.net/~mark/fcodes/elsyesh.htm [June 29, 2003].

———. "Rabin." B-Codes Page [online]. www.nctimes.net/~mark/fcodes/rabin.htm [July 31, 2003].

———. "Don't Be Afraid, Barakhokhlo" (in Russian). *Kontinent*, no. 102 (2000):13–40.

———. "Four for the Price of One." Short Stories by Mark Perakh [online]. www.nctimes.net/~mark/Engstories/barakh.htm [July 24, 2003].

———. "The Rise and Fall of the Bible Code" (in Russian). *Kontinent*, no. 103 (2000):240–70.

———. "The Anthropic Principles—Reasonable and Unreasonable." Talk Reason [online]. www.talkreason.org/articles/anthropic.cfm [August 14, 2003].

———. "A Presentation without Arguments: Dembski Disappoints." *Skeptical Inquirer* 26, no. 6 (2002):31–34.

———. "There Is a Free Lunch after All: William Dembski's Wrong Answers to Irrelevant Questions." In *Why Intelligent Design Fails: A Scientific Critique of the New Creationism*, edited by Matt Young and Taner Edis. Piscataway, N.J.: Rutgers University Press, forthcoming.

Perakh, Mark, Aaron Peled, and Zeev Feit. "Photodeposition of Amorphous Selenium Films by the Selor Process." Part 1. *Thin Solid Films* 50 (1978):273–82.

Perakh, Mark, and Aaron Peled. "Photodeposition of Amorphous Selenium Films by the Selor Process." Parts 2 and 3. *Thin Solid Films* 50 (1978):283–92, 293–302.

———. "Light-Temperature Interference Governing the Inverse/Combined Photoadsorption and Photodeposition of a-Selenium Films." *Surface Science* 80 (1979):430–38.

Perakh, Mark, and Brendan McKay. "Study of Letter Serial Correlation (LSC) in Some Hebrew, Aramaic, Russian, and English Texts" [online]. www.nctimes.net/~mark/Texts [August 14, 2003].

Perakh, Mark, and Matt Young. "Is Intelligent Design Science?" In *Why Intelligent Design Fails: A Scientific Critique of the New Creationism*, edited by Matt Young and Taner Edis. Piscataway, N.J.: Rutgers University Press, forthcoming.

Pigliucci, Massimo. "Design Yes, Intelligent No: A Critique of Intelligent Design Theory and Neo-Creationism." *Skeptical Inquirer* 25, no. 5 (2001):34–39.

Popereka, M. Ya. (a.k.a. Mark Perakh). "Investigation of Critical Speeds of Rapidly Rotating Shafts" (in Russian). *Doklady Akademii Nauk Tajikistana* (Reports of the Academy of Sciences of Tajikistan), no. 10 (1954):81–85.

———. "On the Origin of Internal Stresses in Electrolytically Deposited Metals" (in Russian). *Fizika metallov i metallovedeniye* 20 (1965):753–62.

———. *Vnutrenniye napryazheniya elektrolitichski osazhdayemykh metallov* (Internal stress in electrolytically deposited films). Novosibirsk, USSR: Zapadno-Sibirskoye Knizhnoye Izdatelstvo, 1966. English translation published for the National Bureau of Standards and the National Science Foundation by the Indian National Scientific Documentation Centre, New Delhi, 1970.

———. "A Theory of Internal Stresses in Electrolytically Deposited Metals" (in Russian). In *Trudy Tret'ego Mezdunarodnogo Kongressa po Korrozii Metallov* (Proceedings of the Third International Congress on Metallic Corrosion), vol. 3, pp. 350–56. Moscow: Mir, 1966.

———. "Methods of Quantitative Investigation of Adsorption on Solid Electrodes." In *Twenty-fifth Meeting of the International Society of Electrochemistry: Extended Abstracts*, pp. 305–307. Brighton, U.K.: University of Southampton Press, 1974.

Popereka M. Ya. (a.k.a. Mark Perakh), and V. N. Lebedeva. "Potentiostatic Chronofaradometry: Method For a Quantitative Study of Adsorption on Solid Electrodes" (in Russian). Part 1. In *Adsorptsionnye plyonki*, edited by A. I. Baranov, M. Ya. Popereka, and V. M. Rudyak, pp. 3–25. Fizika plyonok 2. Kalinin, USSR: Kalinin University Press, 1972.

Popereka, M. Ya. (a.k.a. Mark Perakh), and V. M. Romanov. "Potentiostatic Chronofaradometry: Method For a Quantitative Study of Adsorption on Solid Electrodes" (in Russian). Part 2. In *Adsorptsionnye plyonki*, edited by A. I. Baranov, M. Ya. Popereka, and V. M. Rudyak, pp. 26–30. Fizika plyonok 2. Kalinin, USSR: Kalinin University Press, 1972.

Popovsky, Mark. *Upravlyaemaya nauka* (Manipulated science). London: Overseas Publications Interchange, 1978.

Popper, Karl R. *Conjectures and Refutations: The Growth of Scientific Knowledge.* London: Routledge and Kegan Paul, 1963.

Rambsel, Yacov A. *Yeshua: The Hebrew Factor.* San Antonio, Tex.: Messianic Ministries, 1996.

Ratzsch, Del. *The Battle of Beginnings: Why Neither Side Is Winning the Creation-Evolution Debate.* Downers Grove, Ill.: InterVarsity Press, 1996.

———. *Nature, Design, and Science: The Status of Design in the Natural World.* New York: State University of New York Press, 2001.

Rimmer, Harry. *Modern Science and the Genesis Record.* Grand Rapids, Mich.: Eerdmans, 1937.

Ross, Hugh. *The Fingerprint of God: Recent Scientific Discoveries Reveal the Unmistakable Identity of the Creator.* Orange, Calif.: Promise, 1989.

———. *Creation and Time: A Biblical and Scientific Perspective on the Creation-Date Controversy.* Colorado Springs, Colo.: Navpress, 1994.

———. *The Genesis Question: Scientific Advances and the Accuracy of Genesis.* Colorado Springs, Colo.: Navpress, 1998.

———. "About Our Founder." Reasons to Believe [online]. www.reasons.org/about/staff/ross.shtml?main#Curriculum [June 21, 2003].

Satinover, Jeffrey. *Cracking the Bible Code.* New York: William Morrow, 1997.

Schaff, Philip. *The Person of Christ.* N.p.: American Trust Society, 1913.

Schanks, Niall, and Karl H. Joplin. "Redundant Complexity: A Critical Analysis of Intelligent Design in Biochemistry." *Philosophy of Science* 66, no. 2 (1999):268–77.

Schlesinger, G. N. *Tradition* 23 (spring 1988):1–8.

Schneider, Thomas D. "Effect of Ties on the Evolution of Information by the EV Program." Molecular Information Theory and the Theory of Molecular Machines [online]. www.lecb.ncifcrf.gov/~toms/paper/ev/dembski/claimtest.html [June 23, 2003].

———. "Evolution of Biological Information." *Nucleic Acids Research* 28, no. 14 (2000):2794–2803.

Schroeder, Gerald L. *Genesis and the Big Bang: The Discovery of Harmony between Modern Science and the Bible.* New York: Bantam Books, 1990.

———. *The Science of God: The Convergence of Scientific and Biblical Wisdom.* New York: Free Press, 1997.

———. *The Hidden Face of God: How Science Reveals the Ultimate Truth.* New York: Free Press, 2001.

———. "The Age of the Universe" [online]. www.geraldschroeder.com/age.html [May 12, 2002].

Serway, Raymond. *Physics for Scientists and Engineers.* Philadelphia: Saunders, 1990.

Shallit, Jeffrey. Review of *No Free Lunch*, by William Dembski [online]. www.math.uwaterloo.ca/~shallit/nflr3.txt [August 7, 2003].

Shannon, Claude E. "A Mathematical Theory of Communication." Parts 1 and 2. *Bell System Technology Journal* (July 1948):379–90; (October 1948):623–37.

Sheratt, Andrew, ed. *The Cambridge Encyclopedia of Archaeology*. New York: Crown, 1980.

Shumaker, David, ed. "The Hebrew-English Dictionary." In *Seven Language Dictionary*, pp. 235–90. New York: Avenel Books, 1978.

Simon, Barry. "The Case against the Codes." *Jewish Action* (March 1998).

———. "Dispute on Codes" (in Russian). *Vremya Iskat*, no. 2 (1999):89–98.

———. "Mathematicians' Statement on the Bible Codes" [online]. www.math.caltech.edu/code/petition.html [July 31, 2003].

———. "Barry Simon's Cities Experiment—an Overview." Barry Simon on Torah Codes [online]. www.wopr.com/biblecodes/Cities_Overview.htm [August 1, 2003].

Spetner, Lee M. *Not by Chance: Shattering the Modern Theory of Evolution*. Brooklyn, N.Y.: Judaica Press, 1998. Originally published 1997.

Spitzer, Brian. "The Truth, the Whole Truth, and Nothing but the Truth? Why Phillip Johnson's *Darwin on Trial* and the 'Intelligent Design' Movement are neither Science—nor Christian." Talk Reason [online]. www.talkreason.org/articles/honesty.cfm [August 6, 2003].

Stenger, Victor J. *Not by Design: The Origin of the Universe*. Amherst, N.Y.: Prometheus Books, 1988.

———. *The Unconscious Quantum: Metaphysics in Modern Physics and Cosmology*. Amherst, N.Y.: Prometheus Books, 1995.

———. "Intelligent Design: The New Stealth Creationism." Talk Reason [online]. www.talkreason.org/articles/Stealth.pdf [June 12, 2003].

Thomas, Dave E. "Hidden Messages and the Bible Code." Committee for the Scientific Investigation of Claims of the Paranormal [online]. www.csicop.org/si/9711/bible-code.html [March 10, 2002].

Tigay, Jeffrey. "The Bible 'Codes': A Textual Perspective" [online]. www.sas.upenn.edu/~jtigay/codetext.html [June 15, 2003].

Tikhomirov, Nikolai B. *Teoriya veroyatnostej i matematicheskaya statistika* (Theory of probability and mathematical statistics). Kalinin, USSR: Kalinin State Institute Press, 1971.

Velikovsky, Immanuel. *Worlds in Collision*. New York: Macmillan, 1950.

Vitányi, Paul. "Meaningful Information." Front for the Mathematics ArXiv [online]. arxiv.org/PS_cache/cs/pdf/0111/0111053.pdf [August 12, 2003]. Also published in *Proceedings of the Thirteenth International Symposium on Algorithms and Computation (ISAAC): Lecture Notes in Computer Science* (Berlin: Springer Verlag, 2002).

Wein, Richard. "What Is Wrong with the Design Inference?" [online]. website.lineone.net/~rwein/skeptic/whatswrong.htm [June 23, 2002].

———. "Not a Free Lunch but a Box of Chocolates: A Critique of William Dembski's Book *No Free Lunch*." Talk Reason [online]. www.talkreason.org/articles/choc_nfl.cfm [August 15, 2003].

Weinberg, Steven. *The First Three Minutes*. New York: Basic Books, 1977.

Will, Clifford M. "The Renaissance of General Relativity." In *Physics for Scientists and Engineers*, by Raymond Serway, pp. 1136–45. Philadelphia: Saunders, 1990.

Witztum, Doron, Eliyahu Rips, and Yoav Rosenberg. "Equidistant Letter Sequences in the Book of Genesis." *Statistical Science* 9, no. 2 (1994):429–38. Reprinted in Michael Drosnin, *The Bible Code* (New York: Simon and Schuster, 1997).

———. "Report to the Israeli Academy of Science." B-Codes Page [online]. www.nctimes.net/~mark/fcodes/WRR3.htm [April 5, 2003].

Wolpert, David. "Dembski's Treatment of the NFL Theorems Is Written in Jello." Talk Reason [online]. www.talkreason.org/articles/jello.cfm [August 14, 2003].

Young, Matt. "Intelligent Design Is Neither" [online]. www.mines.edu/~mmyoung/DesnConf.pdf [June 23, 2002].

———. "How to Evolve Specified Complexity by Natural Means." *Pacific Coast Theological Society Journal* [online]. www.pcts.org/journal/young2002a.html [August 6, 2003].

Young, Matt, and Taner Edis, eds. *Why Intelligent Design Fails: A Scientific Critique of the New Creationism*. Piscataway, N.J.: Rutgers University Press, forthcoming.

Zakharenkov, A. "A Poem" (in Russian). In *Strofy Veka*, edited by E. Evtushenko, p. 985. Moscow: Polyfact, 1997.

INDEX